Research Advances in
Alcohol and Drug Problems

Volume 8

RESEARCH ADVANCES IN
ALCOHOL AND DRUG PROBLEMS

Series Editors

Reginald G. Smart
Howard D. Cappell
Frederick B. Glaser
Yedy Israel
Harold Kalant
Robert E. Popham
Wolfgang Schmidt
Edward M. Sellers

A Continuation Order Plan is available for this series. A continuation order will bring delivery of each new volume immediately upon publication. Volumes are billed only upon actual shipment. For further information please contact the publisher.

Research Advances in
Alcohol and
Drug Problems

Volume 8

Edited by

Reginald G. Smart, Howard D. Cappell,

Frederick B. Glaser, Yedy Israel, Harold Kalant,

Robert E. Popham, Wolfgang Schmidt,

and Edward M. Sellers

Addiction Research Foundation and
University of Toronto
Toronto, Ontario, Canada

PLENUM PRESS · NEW YORK AND LONDON

The Library of Congress cataloged the first volume of this title as follows:

Research advances in alcohol & drug problems. v. 1–

   New York [etc.] J. Wiley, 1974–
     v. 24 cm. annual.
     "A Wiley biomedical health publication."
     ISSN 0093-9714

   1. Alcoholism — Periodicals. 2. Narcotic habit — Periodicals.
RC565.R37   616.8′6′005   73-18088

ISBN-13: 978-1-4612-9687-4   e-ISBN-13: 978-1-4613-2719-6
DOI: 10.1007/978-1-4613-2719-6

© 1984 Plenum Press, New York
Softcover reprint of the hardcover 1st edition 1984

A Division of Plenum Publishing Corporation
233 Spring Street, New York, N.Y. 10013

ADVISORY PANEL

Contributors

TIMO ALANKO, Finnish Foundation for Alcohol Studies, Helsinki, Finland

MARY JANE ASHLEY, Department of Preventive Medicine and Biostatistics, University of Toronto, Toronto, Ontario, Canada

BARRY S. BROWN, National Institute on Drug Abuse, Rockville, Maryland

RICHARD A. DEITRICH, Department of Pharmacology and Alcohol Research Center, University of Colorado School of Medicine, Denver, Colorado

ROBERTA G. FERRENCE, Addiction Research Foundation, Toronto, Ontario, Canada and Queen's University, Kingston,Ontario,Canada

STEPHEN ISRAELSTAM, Addiction Research Foundation, Toronto, Ontario, Canada

LYNN T. KOZLOWSKI, Addiction Research Foundation, Toronto, Ontario, Canada

RICHARD A. MEISCH, Departments of Psychiatry and Pharmacology, University of Minnesota, Minneapolis, Minnesota

ROBERT E. POPHAM, Addiction Research Foundation, Toronto, Ontario, Canada

WOLFGANG SCHMIDT, Addiction Research Foundation, Toronto, Ontario, Canada

HARVEY A. SKINNER, Addiction Research Foundation, Toronto, Ontario, Canada

KAREN SPUHLER, Department of Pharmacology and Alcohol Research Center, University of Colorado School of Medicine, Denver, Colorado

LARS TERENIUS, Department of Pharmacology, Uppsala University, Uppsala, Sweden

GEORGE E. VAILLANT, Dartmouth Medical School, Hanover, New Hampshire

PAUL C. WHITEHEAD, University of Western Ontario, Addiction Research Foundation, London, Ontario, Canada

Preface

This volume is the eighth in the *Research Advances* Series and the fifth published by Plenum Press. The purpose of the series is to review new work in rapidly changing fields. We do not expect reviews to cover the whole field of work on alcoholism and addiction. Nor do we expect that they will be like annual reviews covering all work in a delimited field. Our reviews are designed to explore only the most exciting parts of the total field and to focus on conclusions that can be made about them. The series publishes one volume each year.

Volume 8 is an omnibus rather than a theme volume in that a wide range of topics is covered, including research on alcohol, opiates, and tobacco. As usual, the greater emphasis is on alcohol research, reflecting the importance of the problem and the volume of work to be reviewed.

With Volume 8 come some changes in the Editorial Board. It will be the last volume in the series for Robert E. Popham who has resigned from the Board. He has been with the series since its inception and has contributed a great deal to its development. The members of the Board are grateful for his help. We are adding two new members: Dr. Howard Cappell, whose field is experimental psychology, and Dr. Edward M. Sellers, in clinical medicine and pharmacology.

The Editors

Toronto

Contents

Opiate Tolerance and Dependence
Roles of Receptors and Endorphins

LARS TERENIUS

1.   INTRODUCTION

Opiates are not the only drugs that produce tolerance or dependence. However, tolerance to and dependence on these drugs develop very rapidly and reach a high magnitude. The clinical significance of opiate tolerance and dependence is well-known. Conditions for the use of these drugs are rigorously controlled and they can be used only on very strict clinical indications. Modern societies and less developed ones seem equally vulnerable to illegal abuse of these drugs. There is consequently a great need to define the basic conditions and mechanisms underlying opiate tolerance and dependence development.

Definitions and Scope

The focus of this review is biochemical mechanisms. It should be realized that, for instance, "dependence" in an isolated piece of intestine in an organ bath, as demonstrated by production of a contraction when naloxone is added, is a phenomenon which may have only some characteristics in common with the whole spectrum of dependence phenomena in humans. However, recent research has more and more emphasized the similarities between the physiology of synapses in the intestinal nerve plexuses and in neuronal populations in brain. These similarities extend to the neuronal circuitry as well as to the interaction between various neuronal systems. The similarities also hold across species. So far, virtually every system with opiate receptors has been found to show the phenomena of tolerance and dependence. Dependence is defined as a reaction

LARS TERENIUS ● Department of Pharmacology, Uppsala University, Uppsala, Sweden.

to abrupt withdrawal of opiate, or to antagonism by naloxone. Although this view is not proved, the standpoint taken here is that tolerance and dependence mechanisms at a biochemical level in various systems are similar. For obvious reasons, the review is limited to the direct targets for opiates, such as the opiate receptors (binding sites, transducing and effector mechanisms), the endorphinergic neurons, and the regulation of activity over the endorphinergic synapses.

Opiate tolerance and dependence have been investigated for many years. Before the discovery of the endogenous opiates (Hughes et al., 1975; Terenius and Wahlström, 1975), these phenomena were never attributed to a disturbance of an endogenous system but thought to be due to a highly artificial situation whereby foreign drugs caused changes in the normal balance between the various functions of the brain, in recent decades thought to be expressed in changes in neurotransmitter systems. The changes induced by the opiates might have been considered harmful just because the drugs were considered so artificial. Ever since the endogenous opiates, the endorphins, were discovered this is an untenable view. However, dependence to endogenous opiates does not seem to occur normally, whereas, for instance, one is definitely "dependent" on acetylcholine or insulin. Interference with the synthesis and/or release of these latter agents will lead to fatal consequences. This is because these systems have a vital function and are tonically active in carrying out this function. The functions of endorphins are definitely not vital: in fact, it is hard to establish any tonic activity in endorphin systems at all. The narcotic antagonist naloxone, even if it is not an ideal antagonist (e.g., it differs in affinity between different opiate subreceptors, see below), has been instrumental in the studies of the importance of endorphins. Naloxone will produce very mild effects on sensory thresholds, mood, and behavior, and fairly sophisticated experimental design is necessary to establish any effect at all. However, both acute and chronic administration of naloxone produce significant effects as will be discussed at some length below.

On the other hand, the dynamic range of the receptor–effector mechanisms in opioid systems must be considerable, since an opiate drug can cause powerful analgesia. It can also be shown that strong somatic stimulation can increase the release of opioid peptides into a spinal perfusate at least 30-fold (Yaksh et al., 1983). Thus, endorphin systems are intrinsically powerful, although they seem functionally dormant unless challenged. If an opiate is administered, it will perturb a system that normally operates at a very different and much lower dynamic range.

It may be worthwhile to give some thoughts to the time-scale for development of opiate tolerance and dependence. Declining receptor function occurring in fractions of a second (desensitization) or in minutes (tachyphylaxis) will not be considered here. In general, development of tolerance and dependence require from a few hours to a few days. In the former case it is customary to use the term acute tolerance (or dependence). On an even longer time-scale, chronic opiate administration will cause more or less permanent changes in higher CNS function—carryover effects—which may last for months or years. Short exposure at critical points during ontogeny may also lead to apparently permanent alterations. Such changes that occur over longer time intervals are reminiscent of memory storage and may be mechanistically related to it. Such long-term phenomena will also not be discussed here.

2. RECEPTORS

The development of tolerance to and dependence on opiates is of dramatic character, suggesting changes in receptor numbers or affinities. Such changes have been fervently sought using biochemical techniques during the last decade.

Receptor Subtypes

The opiate receptor was once postulated to be complementary in structure to morphine (the "lock-and-key concept") and to be a pharmacologic "artifact." During the Second World War and the decade following it, a large number of synthetic agents, more or less closely related to morphine, were found to possess morphine-like activity. To date, several thousand compounds, belonging to several apparently different chemical families have been classified as opiates. The effects of most of them would be qualitatively indistinguishable from those of morphine, but some compounds, particularly those generically related to partial agonists such as nalorphine or pentazocine, show anomalous pharmacologic profiles. This fact led Martin and co-workers (1976) to postulate several distinct opiate receptors which were labeled μ-receptor (homonym for m in morphine), κ-receptor (for k in ketocyclazocine, a prototype for partial agonists), and σ-receptor (for S in SKF 10,047, the prototype of opiates with hallucinogenic activity). A summary of the different characteristics of these receptors is given in Table 1. The differences between the receptors extend to differences in the consequences of protracted receptor activation. μ-Agonists are always capable

Table 1. Opiate Receptor Classification in Spinal Dogs[a]

| | Receptor | | |
Effect	μ	κ	σ
Pupillary size	Miosis	Miosis	Mydriasis
Respiratory rate	Stimulation, then depression	No change	Stimulation
Heart rate	Bradycardia	No change	Tachycardia
Body temperature	Hypothermia	No change	No change
Affect	Indifference	Sedation	Delirium
Nociceptive flexor reflexes	Decrease	Decrease	Modest decrease
Naloxone antagonism	Strong	Moderate	Weak or absent

[a] Modified from Iwamoto and Martin, 1981.

of evoking marked degrees of tolerance and dependence. Tolerance to κ does occur but several of these agents do not cause marked dependence in the monkey; there is no marked withdrawal reaction on injection of naloxone. Furthermore, the kind of dependence on κ-agonists is different from that induced by μ-agonists and there is little cross-tolerance between them (Woods et al., 1982). Cross-tolerance experiments in the guinea-pig ileum also point to distinct μ- and κ-receptors, since tolerance development is receptor-selective (Schulz et al., 1981). Tolerance to and dependence on SKF 10,047 does occur, although not as severely as with μ-agonists.

The discovery of endorphins has awakened an interest in opiate receptor subtypes, and several hundred publications are entirely devoted to the distinctive properties of these subtypes. For authoritative reviews on this subject, see Iwamoto and Martin (1981) and Wüster et al. (1981). To account for observations from direct binding studies using radioactive ligands and brain membranes, or from work with isolated tissues, notably the guinea-pig ileum and the mouse vas deferens, it became necessary to introduce as an operational receptor, the δ-receptor, with high preference for the opioid pentapeptides, the enkephalins. Later work showed the rat vas deferens to be particularly sensitive to β-endorphin, and much less so to morphine or enkephalin. Following this observation another receptor, named ε was proposed, which was thought to be relatively specific for β-endorphin. There are several reasons why this area of research must be considered here. First, we have the important observations that drugs classified as κ-agonists, while being strong analgesics, may not cause strong dependence and do not substitute for morphine in a morphine-dependent monkey (Woods et al., 1982). Several drugs of this kind have in fact been introduced in clinical use and found to be useful even if their partial antagonistic character renders them of rather low analgesic potency. Second, dependent on receptor subtype, there may be different types of effector mechanisms. Thus Pert and co-workers have introduced the potentially useful concept of type I and II receptors, the former being coupled to adenylate cyclase, the latter to some other effector mechanism (Bowen et al., 1981). Unfortunately, there is at present no consensus on how to translate from the receptor subtype terminology to the type I/II dichotomy.

Table 2. Receptor Preferences of Some
Opioid Peptides

Peptide	Receptor preference
Enkephalin	δ
Met-enkephalin Arg^{6}Phe^{7a}	κ
Dynorphin (1–17), (1–8)	κ
β-Endorphin	μ, κ, ε (nonselective)

[a] And other enkephalins extended at the C-terminus with one to three amino acids.

Third, there is increasing evidence that the opioid peptides themselves show differences in receptor profiles (Table 2) and therefore may interact differently in tolerance and dependence mechanisms.

The full implications of receptor subtypes for tolerance and dependence development are not yet understood. It is clear, however, that future drug therapies will be strongly governed by the potential selectivity of such receptors with regard to tolerance and dependence. Some preliminary data indicate that δ-receptor agonists do have low dependence-producing liability (Frederickson et al., 1981).

Receptor Numbers

In the early 1970s it became possible to perform direct measurements of opiate receptor numbers using biochemical techniques (Pert and Snyder, 1973; Simon et al., 1973; Terenius, 1973). The standard technique uses crude subcellular fractions of brain tissue, or more rarely, purified membrane preparations thereof. Chronic opiate administration appears not to affect the number of binding sites, irrespective of whether tritium-labeled dihydromorphine or naloxone is used as receptor probe (Klee and Nirenberg, 1974; Cox and Padhya, 1977). Contrarily, acute treatment with opiate seemed to cause a small increase in the number of available sites tested in vivo. A methodologically different approach was taken by Takemori and collaborators, who studied receptor binding in sliced brain tissue (striatum) in vitro. In these preparations, the receptor affinity for tritium-labeled naltrexone (an opiate antagonist) was greater in tissue from morphine-tolerant and -dependent mice than from controls (Oishi and Takemori, 1982). In contrast, the affinity for tritiated dihydromorphine (an opiate agonist) was not altered. The maximum amount of naltrexone bound was slightly reduced in tissue from tolerant–dependent mice, but this was attributed to morphine still present in the brains of chronically morphine-treated animals. This morphine occupied some of the receptor sites and thus made them unavailable to the tritiated marker compounds. The failure to find any clear alteration of receptor binding of opiate agonists is somewhat surprising. It is more or less a rule that chronic exposure to other neurotransmitters or neuromodulators leads to a reduction in numbers of the respective receptors. One possibility that should be considered is that of conversion between opiate receptor subtypes in the stage of opiate tolerance; such conversion has been reported (Bowen et al., 1981). Conceivably the lack of change in receptor numbers after chronic morphine exposure (Oishi and Takemori, 1982) might be due to replacement of lost μ receptors by conversion of other types into μ receptors.

While chronic agonist exposure does not lead to observable changes in receptor numbers, chronic treatment with the antagonist naloxone results in an increase (Lahti and Collins, 1978) and enhances sensitivity to morphine (Tang and Collins, 1978). This is analogous to what has been found for other neurotransmitter–modulator systems but is unexpected, in view of the lack of effect of chronic agonist exposure (see above). This seems completely incomprehensible. One may again turn to the fact that under physiological conditions only a very few receptors become activated. The total potential of the system is never approached. Regulation of receptor numbers may therefore exist within the dynamic range where the system operates under physiological conditions. Under pathologic conditions, or with heavy, chronic opiate use, a much larger fraction of the receptors is activated. These receptors may be frozen in an inactive form but not physically downregulated. Thus, massive opiate treatment may cause a qualitative change in the receptor properties, which might explain the decreasing slopes of the dose–response curves in long-term morphinized rats (Mucha et al., 1978). A measure of the efficacy of receptor function is the fraction of receptors that must be occupied in order to produce a maximum effect. If not all receptors are required, there is a receptor reserve or "spare" receptors. Chavkin and Goldstein (1982) used an irreversible selective narcotic antagonist, β -chloronaltrexamine, to inactivate receptors gradually in the guinea-pig ileum *in vitro*. They found that 80–90% of the receptors were spare. Interestingly, in a 6-day morphine-tolerant preparation, the receptor reserve was less. These results suggest that efficacy of receptor coupling is under stimulus control.

As the primary biochemical target, the opiate receptor appears comparatively rigid. The absence of apparent downregulation as a consequence of chronic opiate exposure is unexpected. It is possible that more refined techniques will reveal minor changes in receptor numbers or interconversion between subtypes.

Receptor Microenvironment

Loh and co-workers (1978) showed that a pure biochemical, cerebroside sulfate, showed many of the properties expected of an opiate receptor. Using an ingenious two-phase system, they could show that agonists and antagonists affected the cerebroside differently. Even if the receptor is not cerebroside itself, membrane lipids may well be important for receptor function. Age-related changes in morphine potency in rats have also been attributed to changes in membrane lipid content. A recent publication suggests that changes in membrane fluidity will affect opiate-receptor binding (Heron et al., 1981). The receptor microenvironment therefore seems very sensitive to changes in lipid content. The euphoric effects of ethyl alcohol have been partly ascribed to effects on membrane lipid fluidity and it is noteworthy that ethanol and its higher homologues at concentrations in the promille range, affect ligand binding to opiate receptors, particularly of the δ-subtype (Hiller et al., 1981).

Receptor Transducing and Effector Mechanisms

If a suitable ligand transforms a receptor into an active conformation, the signal is translated and amplified to produce a macroscopically identifiable change in membrane potential. Essentially two transducing pathways are known, one

A.

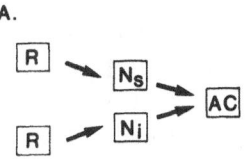

Figure 1. General mechanisms for intracell-
ular signal transduction via membrane receptors. N_s, GTP binding protein stimulating
adenylate cyclase (AC); N_i, GTP binding protein inhibiting AC; PI, phosphatidylinositol.

B.

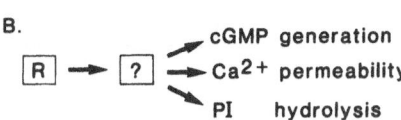

through the formation of cAMP and the other through a complex series of biochemical reactions involving cGMP formation, increased Ca^{2+} conductance, and hydrolysis of phosphatidylinositol (Fig. 1). Different receptors may couple to different transducing mechanisms, with type I receptors coupling to adenylate cyclase and type II receptors coupling to other mechanisms. An effect of opiates on cAMP generation was discovered quite early (Klee and Nirenberg, 1974; Traber et al., 1975) and this pathway of signal transduction has been studied extensively. There are also elaborate data on the effects of opiates on Ca^{2+} levels, Ca^{2+} uptake, and, conversely, on the effect of Ca^{2+} depletion on opiate action. Although the significance of these studies on calcium relative to the pathways of signal transduction is not known, these studies will also be considered here.

Adenylate Cyclase, cAMP, and Opiates. The action of opiates on adenylate cyclase has been preferentially studied in a neuroblastoma x glioma hybrid cell system. This system has considerable experimental advantages over brain tissue, because it permits many fundamental mechanisms to be studied in isolation. These cells have opiate receptors which probably can be classified as δ, being moderately sensitive to the morphine-type alkaloids but fully sensitive to enkephalins. Adenylate cyclase is readily and rapidly inhibited by opioids and this effect is reversible upon removal of the drug. Long-term (12 hr or longer) exposure of the cells to an opioid leads to a gradual increase in cyclase activity which is long-lived. After chronic exposure, abrupt withdrawal of opiate or addition of naloxone to the medium will cause an excessive production of cAMP. This gradual loss of inhibition during chronic opiate exposure and excessive activity on removal of opiate seems to be a fitting mechanism for tolerance and dependence at a cellular level (Sharma et al., 1975). A graphic representation is given in Fig. 2.

This model has several attractive features. In its original form it was thought to represent an example of the enzyme expansion theory: a compensatory increase in the activity of a drug-inhibited enzyme at some key step in the receptor–effector mechanisms of opiate action. The mode of regulation of adenylate cyclase has remained obscure until recently. Work by Klee and co-workers (Sharma et al.,

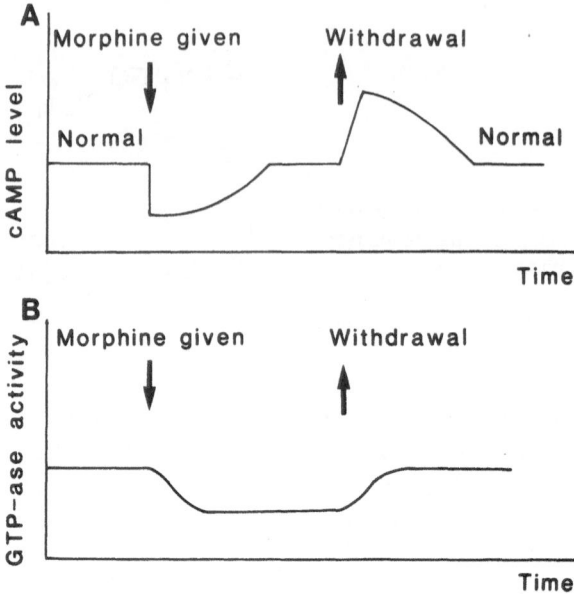

Figure 2. (A) Cyclic AMP regulating receptors and opiate tolerance. (B) GTP-ase activity and opiate tolerance.

1977) indicated that the increase in cyclase activity was blocked by cyclohexi-mide. This observation suggests that synthesis of new protein is involved in the compensatory increase. However, if the system is stimulated with sodium fluoride or with a stable analogue of guanosine triphosphate (directly at the N-protein-cyclase level, bypassing the receptor) there is practically no difference between the opiate-naïve and -tolerant preparations. This finding and others suggest that the mechanism for cyclase activation is not a proliferation of enzyme molecules but rather a compensatory conversion of the enzyme into an active form. It should be noted that dual regulation of cyclase may be a general phenomenon extending to α-adrenergic and muscarinic receptors.

Knowledge about the regulation of adenylate cyclase activity in general has increased recently. Regulatory roles have been proposed for guanosine triphos-phate (GTP) which activates adenylate cyclase, and for a GTP-ase which, by degrading GTP, mediates inhibition of the cyclase. The GTP mechanisms for modulation of adenylate cyclase appear to be general (Cassel and Selinger, 1976). The neuroblastoma x glioma hybrid cell system has been instrumental in studying the importance of GTP for opiate action. This nucleotide will inhibit the binding of opiate agonists to the receptors, and opiate inhibition of adenylate cyclase is not observed in the presence of GTP analogue (Blume et al., 1980). Thus, the presence of GTP leads to more efficient coupling of the stimulatory N-protein and consequently to increased cyclic AMP formation. Klee and co-workers have now reported that opiates may inhibit adenylate cyclase by stimulating hydrolysis of GTP (Koski and Klee, 1981). Opiates and opioid peptides seem to stimulate a specific GTP-ase in a stereospecific manner and in analogy with the structural requirements for interacting with adenylate cyclase activity. It is therefore postulated that the opiate receptor in this cellular system regulates the GTP concentration. Extended exposure of the cells to opiates leads to a slow compensatory inhibition of GTP-ase activity (Fig. 2).

The advantages of studying tolerance and dependence mechanisms in this cellular system have been emphasized, but the disadvantages are that the phenomena studied are only mediated via a special subclass of receptors (δ) and that, after all, the cells may show properties different from those expressed in intact brain tissue. However, several groups following the pioneering work of Collier and Roy (1974) have reported inhibitory effects of opiates on adenylate cyclase in brain homogenates. Collier and colleagues have also studied the behavioral consequences of interference with cAMP levels. The enzyme phosphodiesterase that degrades cAMP has a control function with respect to cell functions that depend on the cAMP level. Caffeine and other methylxanthine derivatives inhibit this enzyme and, when given to experimental animals in high dosages, cause behavioral excitation (Butt et al., 1979). The pattern of behavior elicited has strong similarities to naloxone-precipitated opiate withdrawal reaction. Collier and colleagues have coined the word "quasi-withdrawal reaction" to account for this phenomenon. The authors postulate that the behavioral reaction following opiate withdrawal may be of similar nature. More recently the same group has reported on the existence of endogenous substances, perhaps peptides, that inhibit phosphodiesterase activity (Collier et al., 1982).

Interaction between Opiates and Ca^{2+}. Calcium has multiple functions in nerve tissue as a membrane stabilizer, in the regulation of various enzymes, in impulse propagation, and in transmitter release. Any neurotropic agent will therefore interact with Ca^{2+}, directly or indirectly. However, a number of observations link Ca^{2+} more directly to opiate action and pain perception (Chapman and Way, 1980). Behavioral studies indicate that Ca^{2+} depletion by the chelating agent EGTA potentiates morphine analgesia, while direct administration of Ca^{2+} has a dampening effect on it. In the morphine-tolerant rat, administration of Ca^{2+} will reduce withdrawal reactions precipitated by naloxone. It has also been observed that acute morphine treatment decreases synaptosomal Ca^{2+} content, while its chronic administration leads to Ca^{2+} levels above control. Naloxone treatment will quickly reverse this elevation of Ca^{2+}.

These observations have been followed by extensive studies on the influence of opiates on binding and fluxes of radioactive Ca^{2+} in brain tissue or synaptosomal preparations. Several groups have reported that opiates inhibit Ca^{2+} uptake and reduce steady-state levels (Guerrero-Munoz et al., 1979). Depolarization induced by a high K^+ concentration increases Ca^{2+} uptake into synaptosomes; this K^+-stimulated uptake of Ca^{2+} is inhibited by morphine (End et al., 1981). In contrast, chronically morphine-treated rats show a progressive increase in Ca^{2+} uptake.

Although these findings point to a key role of calcium ion in opiate action, the nature of this role is uncertain. It has been suggested that the opiate-induced reduction in Ca^{2+} flux (or binding) may occur at sites in the synapse that are involved in transmitter release. Since Ca^{2+} is known to be essential for this release (Henderson et al., 1978), such a mechanism could explain the known inhibitory effect of opiates on the release of transmitters such as acetylcholine and dopamine. However, several alternative explanations are also possible, and it is perhaps more likely that the opiate-sensitive Ca^{2+} binding reflects more general changes in receptor or membrane properties. For instance, such a mechanism seems more likely to account for the slow homeostatic increase in Ca^{2+} uptake during chronic opiate treatment. Even if some structural specificity has been demonstrated for this effect, it cannot be specifically related to any special receptor subgroup. It is therefore also quite possible that Ca^{2+} has more to do with more nonspecific effects of opiates such as general CNS depression.

3. EFFECTS OF CHRONIC OPIATE TREATMENT ON ENDORPHIN LEVELS

Much of the recent work on opiate tolerance and dependence has been directed to the possible participation of endorphins. A common approach has been to analyze brain or other tissue levels during chronic opiate exposure. This approach has inherent methodologic problems since changes in neuronal dynamics may not appear as changes in tissue levels. These levels will be a function of the biosynthesis, processing, and metabolic degradation, and it is quite possible that the active pool of a particular peptide may be small in comparison with total peptide levels. It may therefore be assumed that levels in brain perfusates or cerebrospinal fluid (CSF) may give relevant information as to the actual release of endorphins from brain or spinal cord, while plasma levels would more accurately reflect the dynamic activity in endorphin-producing tissues outside the blood–brain barrier.

Chronic Opiate Treatment and Tissue Endorphin Levels

Studies of brain tissue levels of endorphins during chronic opiate treatment or in naloxone-precipitated withdrawal have shown inconclusive results. Several reports indicate no effect of opiates on either Met-enkephalin or β-endorphin content in different areas of rat brain. Treatment for more than a month may cause some reduction in striatal met-enkephalin levels (Przewlocki et al., 1979). These studies would therefore point to an absence of significant negative feedback mechanisms during chronic opiate treatment. However, enkephalin content is decreased in the striatum of rats undergoing spontaneous withdrawal for 24 hr (Bergström and Terenius, 1979).

Studies on human addicts have been conducted by several groups, and the results have been quite varied. Plasma or CSF has been analyzed. Clement-Jones and colleagues (1979) have reported that β-endorphin levels in plasma were generally elevated in mild withdrawal (19 of 22 samples from 12 patients) and β-endorphin levels in CSF were elevated in 5 of the 6 subjects studied. Comparative studies of Met-enkephalin were negative. Acupuncture given to these patients in withdrawal raised the CSF concentration of Met-enkephalin but not of β-endorphin (Clement-Jones et al., 1979). A study on plasma levels of β-endorphin-like material (including the precursor β-lipotropin and the 31-K prohormone) revealed lower levels in addicts than in control samples (Ho et al., 1980). Finally, our group has observed a broader range of receptor-active endorphins in the CSF of heroin addicts than of controls. Samples from these subjects were taken after withdrawal for 3 weeks and again after 3 weeks of methadone maintenance therapy. At both times, the levels of endorphinlike material in the CSF of the addicts tended to be either higher or lower than in normals. The patients were released from hospital while receiving methadone, and on subsequent follow-up studies the patients with abnormal endorphin levels during withdrawal seemed to show a better clinical response (social readjustment) to methadone maintenance therapy than the patients with one value within the normal range (Holmstrand et al., 1981). In another series of 27 opioid-dependent or postdependent subjects, CSF levels of receptor-active opioid peptides were elevated, suggesting a compensatory release in response to desensitized receptor function (O'Brien et al., 1983). The reason for this variation of results in different studies is not yet clear. However, it should be noted that these investigations have involved different methods for the measurement of endorphins, with different degrees of specificity or cross-reactivity to substances other than β-endorphin or Met-enkephalin.

Gold et al. (1981) used an indirect approach to study endorphin release in seven methadone addicts after a 16-day detoxification period. An intravenous dose of naloxone in these subjects failed to raise the plasma ACTH level significantly, while in healthy volunteers it more than tripled the level. Since ACTH and β-endorphin are released concurrently from the anterior pituitary, Gold et al. interpreted their findings as an indication that the pool of β-endorphin available for secretion was also low in the addicts, even 16 days after their last methadone. However, it is far from certain that the production and release of β-endorphin from the anterior pituitary into the plasma are controlled by the same factors as production and release from the hypothalamus into other parts of the brain.

Chronic Opiate Treatment and the Biosynthesis and Posttranslational Processing of Endorphin Precursors

During the last few years, there has been remarkable progress in the understanding of the biosynthesis of endorphins. They constitute a whole family of peptides deriving from three distinct protein precursors, each of which is formed from about 250 amino acids. The three different systems, proopiocortin (or proopiomelancortin, POMC), proenkephalin, and proenkephalin B are precursors of β-endorphin, enkephalin, and dynorphin, respectively (Fig. 3). Each of these systems has a unique distribution throughout the body. In the CNS, there is no overlap between the systems.

The β-endorphin precursor, POMC (Nakanishi et al., 1979), has the β-endorphin sequence at the C-terminus. It also contains other sequences with biological activity: ACTH/α-MSH, β-MSH, and γ-MSH. β-Endorphin can be processed further, for instance, by acetylation of the N-terminus (Zakarian and Smyth, 1979) leading to loss of opioid activity. The full sequence of the enkephalin precursor in bovine adrenal gland was very recently described (Gubler et al., 1982). The precursor contains no less than six Met-enkephalin sequences and one Leu-enkephalin sequence. The exact posttranslational processing of this precursor is not known, although it is clear that in addition to the enkephalins, a fairly large number of longer peptide sequences with opioid activity may be formed. The significance of these putative opioid peptides is so far unknown. Human CSF is rich in peptides which presumably contain an enkephalin sequence with C-terminal elongation (Nyberg et al., 1983). Such peptides may be significant because of their resistance to metabolic degradation. The metabolic degradation of the enkephalins has been studied extensively with an aminopeptidase eliminating the N-terminal tyrosine, and an endopeptidase, "enkephalinase," which causes fragmentation between Gly^3 and Phe^4 (Schwartz et al., 1981). To this latter enzyme has been attributed a key role in the termination of enkephalin action (see further below). The dynorphin precursor, pro-enkephalin B (Kakidani et al., 1982) contains three separate sequences which may be processed to opioid peptides including dynorphin. Since dynorphin contains the Leu-enkephalin sequence, it was at first thought to be a possible precursor of Leu-enkephalin *in vivo*. However, there is probably little conversion to Leu-enkephalin. The posttranslational processing in this system is not well known at present, but it may well be significant since the larger peptides have mainly κ-receptor selectivity, and the shorter ones an increasingly δ-receptor profile.

It has already been pointed out that the level of an endorphin peptide in tissue is equal to the net balance of its biosynthesis, release, and degradation. For several hypophyseal hormones, the rate of biosynthesis is related to the rate of release. To approach the rate of biosynthesis and posttranslational processing, pulse–chase precursor incorporation experiments are performed. The tissue is incubated with a radioactive amino acid to probe protein synthesis and with

PROOPIOMELANOCORTIN

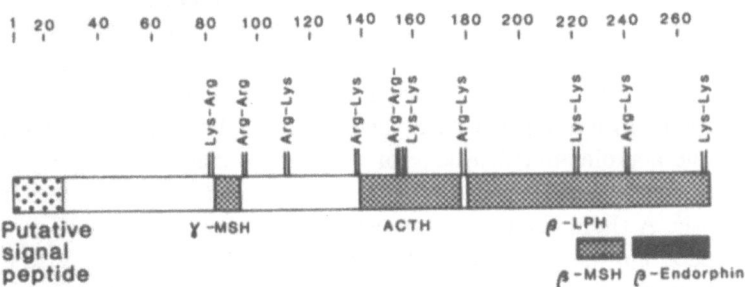

PROENKEPHALIN A

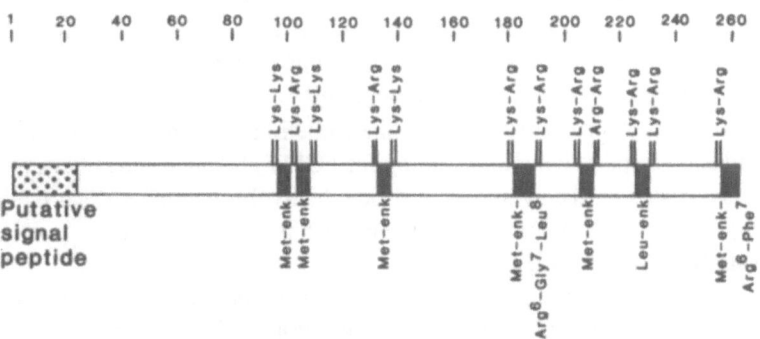

PROENKEPHALIN B

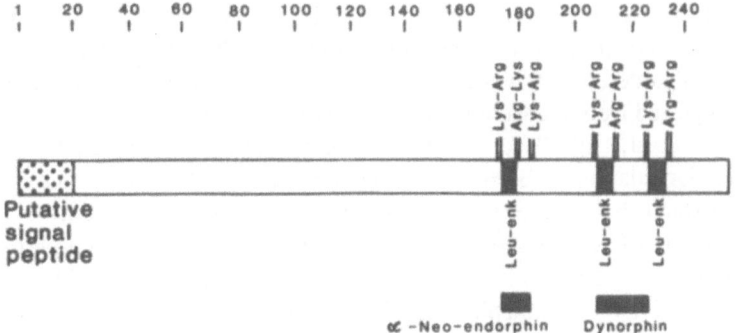

Figure 3. Endorphin precursors. Numbers refer to amino acid residue number. Lys, lysine; Arg, arginine; MSH, melanocyte-stimulating hormone; Gly, glycine; Leu, leucine; Met, methionine; Phe, phenylalanine; enk, enkephalin.

nucleotides to probe nucleic acid synthesis. Work done so far on opiate tolerance has been focused on the neurointermediate lobe of the rat pituitary. Although, an early study (Przewlocki et al., 1979) reported a decrease in β-endorphin-like material after chronic morphine treatment, the same group noted no change after chronic administration of etorphine or levorphanol (Wüster et al., 1980). Studies by Höllt et al. (1981) indicated that chronic morphine treatment reduced the rate of incorporation of tritium-labeled phenylalanine into POMC, β-LPH, and β-endorphin in the neurointermediate lobe of rat pituitary gland. This effect appeared to be selective, since it was more pronounced than the effect on general protein synthesis. A slight delay of posttranslational processing of POMC to β-LPH and β-endorphin was also indicated. Very similar results were reported by a Canadian group (Gianoulakis et al., 1981, 1982). Chronic opiate treatment for 3–21 days caused a marked decrease in incorporation of [^{3}H]phenylalanine into β-endorphin-like material. Reduction of incorporation into β-LPH and POMC was less marked. The effect of chronic ethanol treatment was opposite to that of morphine, giving increased incorporation into β-LPH and β-endorphin. The effect on POMC levels was less pronounced, emphasizing that the posttranslational processing may be particularly susceptible.

A more direct approach to primary biosynthetic regulation is to probe the rate of synthesis of specific m-RNA coding for POMC, by pulse–chase analysis. One study suggests that the synthesis may be selectively suppressed, that is, the effect on specific m-RNA is more pronounced than the general effect on m-RNA biosynthesis (Höllt et al., 1981).

A complicating factor in these studies is that chronic morphine treatment is known to have a general effect on neuronal metabolism. The morphine dosages used in these experiments were also considerable and the animals became strongly tolerant. Even if the rate of specific mRNA and protein synthesis appears more affected than corresponding bulk synthesis, the effect might still be indirect. It is therefore probably too early to draw a definite conclusion about negative feedback mechanisms for β-endorphin biosynthesis. On the other hand, it seems fairly clear that there may be an effect of chronic morphine treatment on the posttranslational processing. Whether this is secondary to changes in β-endorphin release, as suggested from measurements of β-endorphin-like peptides in plasma of addicts (Ho et al., 1980), or whether there are differences in enzymatic maturation or degradation of β-endorphin, remains to be tested. It will also be important to extend this kind of study from pituitary glands to the brain.

Augmentation of enkephalin degradation in the striatum of the morphine-tolerant rat has been described (Schwartz et al., 1981). The effect is said to be due to altered activity of "enkephalinase," either by an increased number of enzyme molecules (enzyme expansion) or due to enzyme activation. In view of the probably complex regulation of enkephalin biosynthesis, a detailed analysis using pulse–chase analysis for the enkephalin system would be necessary to establish the relative importance of increased enkephalinase activity.

4.   HOMEOSTATIC REGULATION OF ENDORPHIN SYSTEMS

Several neuropeptides are of a size and a flexibility that they may easily be able to interact with different kinds of receptors. Similarly to an antigenic determinant, which in peptides or protein is a sequence of about four to five amino acids, the minimum active sequence of an endorphin for receptor interaction is five amino acids (e.g., enkephalin). As already pointed out, C-terminus elongation may alter receptor selectivity; with increasing chain lengths, N-terminal dynorphin fragments acquire an increasing degree of κ-receptor selectivity. In molecules of the size of ACTH or β-endorphin (with more than 30 amino acids), there are definitely possibilities for several, perhaps separate, message sequences. Schwyzer (1980) recently summarized evidence for multiple messages within the ACTH sequence and de Wied and colleagues (1978) have demonstrated nonopioid messages in des-Tyr analogues of γ-endorphin, which are N-terminal fragments of β-endorphin devoid of opioid activity. Another kind of multiple message within a peptide system is given by POMC, the β-endorphin/ACTH precursor, in which several distinct molecular sequences have different messages. Thus, a particular peptide system has the potential for automodulation and for generating different kinds of signals.

Another possible mode of modulation is via the interaction of neuropeptides stemming from different peptide precursors and normally from different neurons. Thus, several laboratories have reported that the administration of certain peptides modifies the development of tolerance to opiates. There are also reports on substances with antiopioid character.

Endorphin Systems with Multiple Message Sequences

It should be pointed out that very little is known about which peptides are actually released from any of the endorphin precursors into the synapse area. It might be that peptides extracted from the brain do not represent the major products of release. Identification of these products is therefore of prime importance in the neuropeptide field. An interesting example is given by dynorphin. This peptide is known from the work of Lee, Loh, and collaborators to interact in a complex fashion with morphine analgesia. In a morphine-naïve mouse, dynorphin antagonizes morphine analgesia at dosages at which dynorphin itself has no appreciable effects; in the morphine-tolerant animal it potentiates morphine action (Tulunay et al., 1981). Spontaneous withdrawal symptoms in the morphine-dependent mouse were also blocked by dynorphin. These latter results were confirmed in the monkey (Aceto et al., 1982) and in humans (Wen and Ho, 1982), in whom intravenous doses of dynorphin could substitute for morphine. These dynorphin doses were by themselves not morphinelike.

These observations are not easy to interpret. They cannot be explained by the κ-agonist activities of the compound. The recent observation of antiopiate

activity of des-Tyr-dynorphin (Walker et al., 1982), which lacks affinity for μ-
or κ-type opiate receptors, suggests the existence of modulatory receptors for
dynorphin, the characteristics of which it will be important to delineate.

The POMC system may be even more complex. This precursor can generate
ACTH and several different MSH molecules in addition to β-endorphin. Interactions between β-endorphin and ACTH have been observed by several laboratories. Receptor binding and behavioral studies have suggested ACTH to be
a partial agonist to opiate receptors (Wiegant et al., 1977). Jacquet (1979, 1982)
postulated the existence of an excitatory receptor (for ACTH?) besides the μ-
receptor in the periaqueductal grey in rat brain. Opiates are presumed to interact
with both receptors. During tolerance development, the analgesia receptors are
downregulated, whereas the ACTH receptor is not, and therefore becomes un-
masked. Injection of naloxone to an opiate-dependent individual will unmask
the excitatory receptor and lead to behavioral excitation.

Another example of possible dual regulation stems from studies on the
fragmentation of β-endorphin in an isolated preparation of myenteric plexus from
the morphine-dependent guinea-pig ileum (Opmeer et al., 1980, 1982). Here
there is an increase in the formation of γ-endorphin-related fragments and β-
endorphin. The former show nonopioid neurotropic activity (de Wied et al.,
1978), the significance of which in morphine-dependence remains unclear.

Neuropeptide Modulators

Several neuropeptides have been reported to influence the development of
opiate tolerance, but there is no agreement about the direction or magnitude of
this influence. The tripeptide prolyl-leucyl-glycinamide was reported to facilitate
the development of morphine tolerance and dependence (van Ree and de Wied,
1976; Contreras and Takemori, 1981). However, the same peptide was reported
by other investigators to have no effect on morphine tolerance (Mucha and Kalant,
1979) or actually to inhibit it (Walter et al., 1979). Morphine tolerance was also
reported to be inhibited by a dipeptide, cyclo(prolyl-leucine), and other analogues
of melanocyte inhibitory factor (Walter et al., 1978, 1979). Since the reasons
for these discrepancies are not clear, the physiological significance of these
peptide effects, if any, remains unknown.

Another approach has been taken by Wahlström and Terenius (1980) and
Han and collaborators (1980). They reasoned that the rapid development of
tolerance to opiates might be due not only to reduction in opiate receptor sensitivity, but also to the induced synthesis of morphine antagonist substances.
Brain tissue from opiate-tolerant rats, CSF from opiate addicts, or human brain
tissue obtained at autopsy of terminal cancer patients given chronic morphine,
were subjected to chromatographic separations. Antiopioid activity was assessed
on the electrically stimulated guinea-pig ileum by the ability to block the depressant actions of opiates. Both groups found evidence for antiopioid activity
which showed increased levels in the tolerant stage. Recent data indicate that

the material studied by Wahlström and Terenius (1981) has the characteristics of a physiological antagonist. The functional significance will probably become clear when the material has been structurally identified.

5. OUTLOOK

This review has emphasized a commonality in biochemical mechanisms associated with the development of tolerance to, and dependence on, opiates. Every cell with a set of opiate receptors has been considered a target for biochemical changes. Biochemical consequences of chronic opiate administration would therefore be expected to be very generalized, for instance, to affect most brain areas in view of the widespread distribution of opiate receptors and therefore be easily observable. It is therefore almost a paradox that available experimental data give very little indication of changes; gross numbers of receptors or overall peptide levels are largely unaffected. However, it has been pointed out that receptors and peptides may occur in such surplus that changes become unobservable. Experiments designed to measure surplus receptors (Chavkin and Goldstein, 1982) do indicate significant changes. In the same vein, experiments need to be designed in which the dynamic activity in the peptidergic synapse is measured. A complicating factor in studies of the central nervous system is the complex circuitry of opioid systems. Studies so far have not been addressing particular pathways or systems. In whole-animal studies, it may well be that certain areas of the CNS show stronger changes than others. A recent study of cerebral glucose utilization in morphine addiction and in acute withdrawal in the rat (Wooten et al., 1982) is an instructive example. During addiction, there is increased utilization in the hippocampal formation, the subiculum, and entorhinal cortex. This returns to normal on withdrawal, which, on the other hand, induces increased utilization in the amygdala and other limbic areas.

Of all possible experimental variables, the drug factor has been most extensively studied. The lack of cross-tolerance between drugs of slightly different receptor profiles shows that several receptors, perhaps with different kinds of receptor-coupling mechanisms, are involved. Even if a strong nonaddictive analgesic does not yet exist, and may in fact be unachievable as long as opioid mechanisms are recruited, the wealth of information from this area at the whole-animal level has not yet been extensively used in biochemical studies. Such work done on a comparative basis might prove to be a fertile area of research. Other neglected areas of research include the importance of the age factor for tolerance and dependence development. It may also be rewarding to examine the relevance of genetic factors. A recent study suggests, in fact, that of two closely related species of monkeys, rhesus and cynomolgus, the former is highly susceptible to the development of tolerance and dependence to hydromorphone, whereas the latter is almost resistant (Ternes et al., 1983). If it proves possible to link such differences to biochemical indices, this would facilitate research on

the etiology of opiate dependence, which at present is mainly based on epidemiologic evidence (Cloninger and Reich, 1983).

ACKNOWLEDGMENTS

Supported by the National Institute on Drug Abuse (Washington, D.C.)
and the Swedish Medical Research Council.

REFERENCES

Aceto, M. D., Dewey, W. L., Chang, J.-K., and Lee, N. M., 1982, Dynorphin (1–13): Effects in
    nontolerant and morphine-dependent rhesus monkeys, *Eur. J. Pharmacol.* **83**:139–142.
Bergström, L., and Terenius, L., 1979, Enkephalin levels decrease in rat striatum during morphine
    abstinence, *Eur. J. Pharmacol.* **60**:349–352.
Blume, A. J., Lichtshtein, D., and Boone, G., 1980, Receptor-mediated inhibitions of NG108-15
    adenylate cyclase: Essential role of Na$^+$ and GTP, in: *Receptors for Neurotransmitters and
    Peptide Hormones,* (G. Pepeu, M. J. Kuhar, and S. J. Enna, eds.), Raven Press, New York.
Bowen, W. D., Gentleman, S., Herkenham, M., and Pert, C. B., 1981, Interconversion and forms
    of the opiate receptor in rat striatal patches, *Proc. Natl. Acad. Sci. USA* **78**:4818–4822.
Butt, N. M., Collier, H. O. J., Cuthbert, N. J., Francis, D. L., and Saeed, S. A., 1979, Mechanism
    of quasi-morphine withdrawal behaviour induced by methylxanthines, *Eur. J. Pharmacol.*
    53:375–378.
Cassel, D., and Selinger, Z., 1976, Catecholamine-stimulated GTP-ase activity in turkey erythrocyte
    membranes, *Biochem. Biophys. Acta* **452**:538–551.
Chapman, D. B., and Way, E. L., 1980, Metal ion interactions with opiates, *Annu. Rev. Pharmacol.
    Toxicol.* **20**:553–579.
Chavkin, C., and Goldstein, A., 1982, Reduction in opiate receptor reserve in morphine tolerant
    guinea pig ileal, *Life Sci.* **31**:1687–1690.
Clement-Jones, V., McLoughlin, L., Lowry, P. J., Besser, G. M., Rees, L. H., and Wen, H. L.,
    1979, Acupuncture in heroin addicts: Changes in met-enkephalin and β-endorphin in blood and
    cerebrospinal fluid, *Lancet* **2**:380–383.
Cloninger, C. R., and Reich, T., 1983, Genetic heterogeneity in alcoholism and sociopathy, in:
    Genetics of Neurological Psychiatric Disorders, (S. S. Kety, L. P. Rowland, R. L. Sidman,
    and S. W. Matthysse, eds.), Raven Press, New York.
Collier, H. O. J., and Roy, A. C., 1974, Morphine-like drugs inhibit the stimulation by E pros-
    taglandins of cyclic AMP formation by rat brain homogenate, *Nature* **248**:24–27.
Collier, H. O. J., Butt, N. M., and Saeed, S. A., 1982, Endogenous peptides that inhibit brain
    cyclic AMP phosphodiesterase, *J. Neurochem.* **38**:275–277.
Contreras, P. C., and Takemori, A. E., 1981, Facilitation of morphine-induced tolerance and physical
    dependence by prolyl-leucyl-glycinamide, *Eur. J. Pharmacol.* **71**:259–268.
Cox, B. M., and Padhya, R., 1977, Opiate binding and effect in ileum preparations from normal
    and morphine pretreated guinea-pigs, *Br. J. Pharmacol.* **61**:271–278.
End, D. W., Carchman, R. A., and Dewey, W. L., 1980, Interactions of narcotics with synaptosomal
    calcium transport, *Biochem. Pharmacol.* **30**:674–676.
Frederickson, R. C. A., Smithwick, E. L., Shuman, R., and Bemis, K. G., 1981, Metkephamid,
    a systemically active analog of methionine enkephalin with potent opioid δ-receptor activity,
    Science **211**:603–605.

Gianoulakis, C., Woo, N., Drouin, J.-N., Seidah, N. G., Kalant, H., and Chrétien, M., 1981, Biosynthesis of β-endorphin by the neurointermediate lobes from rats treated with morphine or alcohol, *Life Sci.* **29:**1973–1982.

Gianoulakis, C., Drouin, J.-N., Seidah, N. G., Kalant, H., and Chrétien, M., 1982, Effect of chronic morphine treatment on β-endorphin biosynthesis by the rat neurointermediate lobe, *Eur. J. Pharmacol.* **72:**313–321.

Gold, M. S., Pottash, A. L. C., Extein, I., Martin, D. A., Finn, L. B., Sweeney, D. R., and Kleber, H. D., 1981, Evidence for an endorphin dysfunction in methadone addicts: Lack of ACTH response to naloxone, *Drug Alcohol Depend.* **8:**257–262.

Gubler, U., Seeburg, P., Hoffman, B. J., Gage, L. P., and Udenfriend, S., 1982, Molecular cloning establishes proenkephalin as precursor of enkephalin-containing peptides, *Nature* **295:**206–208.

Guerrero-Munoz, F., Lourdes Guerrero, M. de, Way, E. L., and Li, C. H., 1979, Effect of β-endorphin on calcium uptake in the brain, *Science* **206:**89–91.

Han, J. S., Tang, J., Huang, B. S., Liang, X. N., and Zhang, W. Q., 1980, Acupuncture tolerance in rats: Anti-opiate substrates implicated, *Chin. Med. J.* **92:**625–627.

Henderson, G., Hughes, J., and Kosterlitz, H. W., 1978, In vitro release of Leu- and Met-enkephalin from the corpus striatum, *Nature* **271:**677–679.

Heron, D., Israeli, M., Hershkowitz, M., Samuel, D., and Shinitzky, M., 1981, Lipid-induced modulation of opiate receptors in mouse brain membranes, *Eur. J. Pharmacol.* **72:**361–364.

Hiller, J. M., Angel, L. M., and Simon, E. J., 1981, Multiple opiate receptors: Alcohol selectively inhibits binding to Delta receptors, *Science* **214:**468–469.

Ho, W. K. K., Wen, H. L., and Ling, N., 1980, Beta endorphin-like immunoactivity in the plasma of heroin addicts and normal subjects, *Neuropharmacology* **19:**117–120.

Holmstrand, J., Gunne, L.-M., Wahlström, A., and Terenius, L., 1981, CSF-endorphins in heroin addicts during methadone maintenance and during withdrawal, *Pharmacopsychiatria* **14:**126–128.

Höllt, V., Haarmann, I., and Herz, A., 1981, Long-term treatment of rats with morphine reduces the activity of messenger ribonucleic acid coding for the β-endorphin/ACTH precursor in the intermediate pituitary, *J. Neurochem.* **37:**619–626.

Hughes, J., Smith, T. W., Kosterlitz, H. W., Fothergill, L. A., Morgan, B. A., and Morris, H. R., 1975, Identification of two related pentapeptides from the brain with potent opiate agonist activity, *Nature* **258:**577–579.

Iwamoto, E. T., and Martin, W. R., 1981, Multiple opioid receptors, *Med. Res. Rev.* **4:**411–440.

Jacquet, Y. F., 1979, β-Endorphin and ACTH-opiate peptides with coordinated roles in the regulation of behaviour?, *Trends Neurosci.* **2:**140–143.

Jacquet, Y. F., 1982, Opposite temporal changes after a single central administration of β-endorphin: Tolerance and sensitization, *Life Sci.* **30:**2215–2219.

Kakidani, H., Furukani, Y., Takahashi, H., Noda, H., Morimoto, Y., Hirose, T., Asai, M., Inayama, S., Nakanishi, S., and Numa, S., 1982, Cloning and sequence analysis of cDNA for porcine α-neo-endorphin/dynorphin precursor, *Nature* **298:**245–249.

Klee, W. A., and Nirenberg, M., 1974, A neuroblastoma x glioma hybrid cell line with morphine receptors, *Proc. Natl. Acad. Sci. USA* **71:**3474–3477.

Koski, G., and Klee, W. A., 1981, Opiates inhibit adenylate cyclase by stimulating GTP hydrolysis, *Proc. Natl. Acad. Sci. USA* **78:**4185–4189.

Lahti, R. A., and Collins, R. J., 1978, Chronic naloxone results in prolonged increases in opiate binding sites in brain, *Eur. J. Pharmacol.* **51:**185–186.

Loh, H. H., Law, P. Y., Ostwald, T., Cho, T. M., and Way, E. L., 1978, Possible involvement of cerebroside sulfate in opiate receptor binding, *Fed. Proc.* **37:**147–152.

Martin, W. R., Eades, C. G., Thompson, J. A., Huppler, R. E., and Gilbert, P. E., 1976, The effects of morphine- and nalorphine-like drugs in the non-dependent and morphine-dependent chronic spinal dog, *J. Pharmacol. Exp. Ther.* **197:**517–532.

Mucha, R. F., and Kalant, H., 1979, Failure of prolyl-leucyl-glycinamide to alter analgesia measured by the Takemori test in morphine-pretreated rats, *J. Pharm. Pharmacol.* **31:**572–573.

Mucha, R. F., Niesink, R., and Kalant, H., 1978, Tolerance to morphine analgesia and immobility measured in rats by changes in log-dose–response curves, Life Sci. 23:357–364.

Nakanishi, S., Inoue, A., Kita, T., Nakamura, M., Chang, C. Y. A., Cohen, S. N., and Numa, S., 1979, Nucleotide sequence of cloned cDNA for bovine corticotropin-β -lipotropin precursor, Nature 278:423–427.

Nyberg, F., Wahlström, A., Sjölund, B., and Terenius, L., 1983, Characterization of electrophoretically separable endorphins in human CSF, Brain Res. 259:267–274.

O'Brien, C. P., Terenius, L., Wahlström, A., McLellan, A. T., and Krivoy, W., 1983, Endorphin levels in opioid dependent human subjects: A longitudinal study, Ann. N.Y. Acad. Sci. 398:377–383.

Oishi, R., and Takemori, A. E., 1982, Stereospecific accumulation of dihydromorphine and naltrexone by corpus striatal slices of morphine-dependent mice, Neuropharmacology 21:57–61.

Opmeer, F. A., Loeber, J. G., and van Ree, J. M., 1980, Altered levels of β -endorphin fragments after chronic morphine treatment of guinea-pig-ileum in vitro and in vivo, Life Sci. 27:2392–2400.

Opmeer, F. A., Peter, J., Burbach, H., Wiegant, V. M., and van Ree, J. M., 1982, β-endorphin proteolysis by guinea-pig ileum myenteric plexus membranes: Increased β-endorphin turnover after chronic exposure to morphine, Life Sci. 31:323–328.

Pert, C. B., and Snyder, S. H., 1973, Opiate receptor: Demonstration in nervous tissue, Science 179:1011–1014.

Przewlocki, R., Höllt, V., Duka, T., Kleber, G., Gramsch, C., Haarmann, I., and Herz, A., 1979, Long-term morphine treatment decreases endorphin levels in rat brain and pituitary, Brain Res. 174:357–361.

Schulz, R., Wüster, M., Rubini, P., and Herz, A., 1981, Functional opiate receptors in the guinea-pig ileum: Their differentiation by means of selective tolerance development, J. Pharmacol. Exp. Ther. 219:547–550.

Schwartz, J.-C., Malfroy, B., and de la Baume, S., 1981, Biological inactivation of enkephalins and the role of enkephalin-dipeptidyl-carboxypeptidase ("enkephalinase") as neuropeptidase, Life Sci. 29:1715–1740.

Schwyzer, R., 1980, Organization and transduction of peptide information, Trends Pharmacol. Sci. 1:327–331.

Sharma, S. K., Klee, W. A., and Nirenberg, M., 1975, Dual regulation of adenylate cyclase accounts for narcotic dependence and tolerance, Proc. Natl. Acad. Sci. USA 72:3092–3096.

Sharma, S. K., Klee, W. A., and Nirenberg, M., 1977, Opiate-dependent modulation of adenylate cyclase, Proc. Natl. Acad. Sci. USA 74:3365–3369.

Simon, E. J., Hiller, J. M., and Edelman, I., 1973, Stereospecific binding of the potent narcotic analgesic ^{3}H-etorphine to rat-brain homogenate, Proc. Natl. Acad. Sci. 70:1947–1949.

Tang, A. H., and Collins, R. J., 1978, Enhanced analgesic effects of morphine after chronic administration of naloxone in the rat, Eur. J. Pharmacol. 47:473–474.

Terenius, L., 1973, Stereospecific interaction between narcotic analgesics and a synaptic plasma membrane fraction of rat cerebral cortex, Acta Pharmacol. Toxicol. 32:317–320.

Ternes, J. W., Ehrman, R., and O'Brien, C. P., 1983, Cynomolgus monkeys do not develop tolerance to opioids, Behav. Neurosci. 97:327–330.

Traber, J., Fischer, K., Latzin, S., and Hamprecht, B., 1975, Morphine antagonizes action of prostaglandin in neuroblastoma and neuroblastoma x glioma hybrid cells, Nature 253:120–122.

Tulunay, F. C., Jen, M.-F., Chang, J.-K., Loh, H. H., and Lee, N. M., 1981, Possible regulatory role of dynorphin on morphine- and β-endorphin-induced analgesia, J. Pharmacol. Exp. Ther. 219:296–298.

van Ree, J. M., and de Wied, D., 1976, Prolyl-Leucyl-Glycinamide (PLG) facilitates morphine dependence, Life Sci. 19:1331–1340.

Wahlström, A., and Terenius, L., 1980, Factor in human CSF with apparent morphine-antagonistic properties, Acta Physiol. Scand. 110:427–429.

Wahlström, A., and Terenius, L., 1981, Search for morphine-antagonistic agents in human CSF and brain, in: *Advances in Endogenous and Exogenous Opioids* (H. Takagi, ed.), pp. 138–140, Kodanska, Tokyo, and Elsevier, Amsterdam.

Walker, J. M., Tucker, D. E., Coy, D. H., Walker, B. B., and Akil, H., 1982, Des-tyrosine-dynorphin antagonizes morphine analgesia, *Eur. J. Pharmacol.* **85**:121–122.

Walter, R., Ritzmann, R. F., Bhargava, H. N., Rainbow, T. C., Flexner, L. B., and Krivoy, W. A., 1978, Inhibition by Z-Pro-D-Leu of development of tolerance to and physical dependence on morphine in mice, *Proc. Natl. Acad. Sci. USA* **75**:4573–4576.

Walter, R., Ritzmann, R. F., Bhargava, H. N., and Flexner, L. B., 1979, Prolyl-leucyl-glycinamide, cyclo (leucylglycine), and derivatives block development of physical dependence on morphine in mice, *Proc. Natl. Acad. Sci. USA* **76**:518–520.

Wen, H. L., and Ho, W. K. K., 1982, Suppression of withdrawal symptoms by dynorphin in heroin addicts, *Eur. J. Pharmacol.* **82**:183–186.

de Wied, D., Kovacs, G. L., Bohus, B., van Ree, J. M., and Greven, H. M., 1978, Neuroleptic activity of the neuropeptide β-LPH 62–67 (des-Tyr-γ-endorphin: DTγE), *Eur. J. Pharmacol.* **82**:183–186.

Wiegant, V. M., Gispen, W. H., Terenius, L., and de Wied, D., 1977, ACTH-like peptides and morphine: Interaction at the level of the CNS, *Psychoendocrinology* **2**:63–69.

Woods, J. H., Young, A. M., and Herling, S., 1982, Classification of narcotics on the basis of their reinforcing, discriminative, and antagonistic effects in rhesus-monkeys, *Fed. Proc.* **41**:221–227.

Wooten, G. F., DiStefano, P., and Collins, R. C., 1982, Regional cerebral glucose utilization during morphine withdrawal in the rat, *Proc. Natl. Acad. Sci. USA* **79**:3360–3364.

Wüster, M., Schulz, R., and Herz, A., 1980, Inquiry into endorphinergic feedback mechanisms during the development of opiate tolerance/dependence, *Brain Res.* **189**:403–411.

Wüster, M., Schulz, R., and Herz, A., 1981, Multiple opiate receptors in peripheral tissue preparations, *Biochem. Pharmacol.* **30**:1883–1887.

Yaksh, T. L., Terenius, L., Nyberg, F., Jhamandas, K., and Wang, J.-Y., 1983, Studies on the release by somatic stimulation from rat and cat spinal cord of active materials which displace dihydromorphine in an opiate-binding assay, *Brain Res.* **264**:119–128.

Zakarian, S., and Smyth, D., 1979, Distribution of active and inactive forms of endorphins in rat pituitary and brain, *Proc. Natl. Acad. Sci. USA* **76**:5972–5976.

2

Alcohol Self-Administration by Experimental Animals

RICHARD A. MEISCH

1. INTRODUCTION

During the last several years a number of reviews of animal studies of alcohol intake have been published (Cicero, 1979, 1980; Meisch, 1977, 1981; Mello, 1976; Myers, 1978; Pohorecky, 1981), and several books on the subject of such studies, usually proceedings of scientific meetings, have also appeared (Eriksson et al., 1980; McClearn et al., 1981).

In this review, animal studies of alcohol intake are analyzed from a specific perspective. Alcohol intake is viewed as operant behavior, that is, behavior controlled by its consequences. Consequences that result in increases in behavior are termed reinforcers. For example, food-deprived rats can be trained to press a lever if lever presses result in delivery of a food pellet. Under these conditions, the food pellet functions as a positive reinforcer, and the behavior of pressing the lever is an instance of operant behavior.

Drugs from several pharmacologic classes can function as positive reinforcers for animals. These drug classes include narcotic analgesics, general depressants, psychomotor stimulants, and dissociative anesthetics (for reviews see Johanson and Schuster, 1981; Pickens et al., 1978; Spealman and Goldberg, 1978). The results of these studies of drug self-administration provide a context within which results of alcohol self-administration studies can be evaluated. In this review it is assumed that alcohol is a drug and that alcohol-taking behavior is a specific instance of the more general phenomenon of drug-taking behavior.

RICHARD A. MEISCH ● Departments of Psychiatry and Pharmacology, University of Minnesota, Minneapolis, Minnesota.

2. METHODS USED TO STUDY ALCOHOL INTAKE

Two-Bottle Water–Alcohol Preference Paradigm

Most studies of alcohol drinking have used a technique introduced by Richter (Richter and Campbell, 1940). Two drinking bottles are attached to a rat's cage. One bottle contains an alcohol solution and the other contains water. The water can be viewed as the vehicle, and the alcohol solution can be thought of as the drug plus vehicle. Thus, rats have a choice of either the vehicle alone or the vehicle plus alcohol. Under these conditions rats prefer alcohol concentrations from 1% [weight per volume (w/v)] to about 6% (Richter and Campbell, 1940). Many subsequent studies have shown that with this two-bottle choice method, rats do not show signs of intoxication and do not develop physiological (physical) dependence. Absence of physiological dependence under these conditions is not surprising since rats consume alcohol at a rate lower than their rate of alcohol metabolism (cf. Goldstein, 1978).

In addition to resulting in low amounts of alcohol consumed, the two-bottle choice method has a number of limitations. Position and tube preferences have been described (Gillespie and Lucas, 1958; Goodrick, 1972; Korn, 1960). Leakage sometimes occurs even when spouts have stainless-steel ball tips (Amit et al., 1973), and if a drop forms at the end of the tube, evaporation of alcohol can occur quickly (Sohler et al., 1969). The volume consumed is usually measured only once every 24 hr, and this is too infrequent an interval since different patterns of intake within a 24-hr interval can result in different effects. For example, the effects of a given volume of alcohol solution would vary depending on whether the volume was consumed gradually over a 24-hr interval or abruptly within a 20-min period. This problem of recording the time course of drinking could be diminished by the use of drinkometer devices to record the pattern. However, this has only occasionally been done (Meisch, 1977).

The most serious criticism is that with the two-bottle preference technique alcohol intake is probably determined by palatability and not by the interoceptive stimuli that occur once alcohol is absorbed. Thus, the variables that control intake in the two-bottle choice are different from those that control intake when alcohol is serving as a more effective reinforcer. Cicero (1979) has concluded that the studies conducted with this two-bottle preference paradigm are ". . . essentially worthless and contribute little except confusion to an already overburdened literature." I agree.

Liquid Diets

Animals will ingest large amounts of alcohol when alcohol is a component of a liquid diet. These diets have been used with mice (Freund, 1969), rats (Walker and Freund, 1971), rhesus monkeys (Pieper and Skeen, 1972), baboons

(Lieber and DeCarli, 1974), and chimpanzees (Pieper et al., 1972). Liquid diets are an effective way to get animals to dose themselves with alcohol. Intake of alcohol is determined by physiological need for intake of the vehicle, namely the liquid diet. When animals have been given the option of drinking a liquid diet without alcohol or a liquid diet with alcohol, they have chosen the diet lacking alcohol (Freund, 1969; Hunter et al., 1974). Thus, liquid diets with alcohol are usually not appropriate for analyzing determinants of alcohol intake, and in fairness it should be noted they were devised to achieve chronic high intake and not to reveal the determinants of alcohol drinking.

Operant Conditioning Techniques

A number of techniques involve operant conditioning procedures. These techniques have in common that access to alcohol is made contingent upon an arbitrary response such as a lever press. These procedures have been used to study intake of alcohol via the oral, intragastric, and intravenous routes. The use of the term "operant" to designate only these techniques is arbitrary in that the behavior of drinking liquid from a bottle is also an instance of operant behavior. However, for descriptive purposes procedures that employ an additional response such as a lever press will be described as operant procedures.

Intravenous Route. Procedures have been developed that permit an animal to self-inject drugs intravenously (Deneau et al., 1969; Thompson and Schuster, 1964; Weeks, 1962). A chronic indwelling venous catheter is inserted into the internal or external jugular vein and threaded down to the level of the right atrium. The distal end is run subcutaneously to a stab wound between the scapulae where it exits, and the external portion is protected so that the animal cannot dislodge it. The distal end is connected to an infusion pump that the animal can activate by pressing a lever. Rhesus and Bonnett monkeys reliably learn to self-inject large amounts of alcohol (Deneau et al., 1969; Carney et al., 1976; DeNoble and Begleiter, 1978; Karoly et al., 1978; Winger and Woods, 1973). The amounts taken ranged as high as 8 g/kg per day, and signs of intoxication were noted.

Rats will also intravenously self-administer alcohol (Numan, 1979; Oei and Singer, 1979; Smith and Davis, 1974; Smith et al., 1975, 1976). However, the amounts (g/kg) self-administered are usually much lower than the amounts monkeys self-administer, and there have been no reports in rats of the sustained intoxication that monkeys display.

Intragastric Route. Procedures similar to the ones used in intravenous studies have been developed for intragastric drug self-administration studies. The principal difference is that the catheter is inserted into the stomach instead of into a vein (Smith et al., 1975; Yanagita and Takahashi, 1973). Both rats and rhesus monkeys will intragastrically self-administer alcohol (Altshuler and Talley, 1977; Altshuler et al., 1975; Davis et al., 1976; Marfaing-Jallat et al., 1974;

Smith et al., 1976; Werner et al., 1977; Yanagita and Takahashi, 1973). Again, in the monkey studies the dosages used have been higher and signs of intoxication have been reported more often than in rats.

Oral Route.   A problem in the use of the oral route is that in order for a drug to come to function as a reinforcer animals must self-administer the drug, and they are unlikely to self-administer a drug unless it already functions as a reinforcer. When the intravenous route is used, animals can be enticed to press the lever by several simple strategies such as taping a raisin to the lever (Deneau et al., 1969). If a highly reinforcing drug such as cocaine is used, learning occurs rapidly. Subsequently, other drugs such as alcohol can be substituted for cocaine (Winger and Woods, 1973).

With oral alcohol intake more extensive procedures are necessary to establish alcohol as a reinforcer. Two difficulties must be circumvented (Mello and Mendelson, 1971a,b). One is that the taste of alcohol in concentrations above 6% (w/v) is aversive for most animals. A second problem is that there is a substantial delay between drinking an alcohol solution and the onset of the effects that follow absorption. This delay makes learning more difficult. One method for solving these problems involves the use of schedule-induced polydipsia. This describes the excessive drinking that occurs when food-deprived animals intermittently receive small pellets of food (Falk, 1961, 1969). When pellets are delivered at a rate of one/min, rats will drink up to one-half their body weight in water within a 3-hr session (Falk, 1961). Schedule-induced polydipsia has been used for two reasons in alcohol studies: to establish alcohol as a reinforcer and to maintain chronic high intake. The latter topic will be discussed in a subsequent section.

By using schedule-induced polydipsia to engender high liquid intake it is possible to substitute low alcohol concentrations, such as 2% (w/v), for water. Across subsequent sessions the concentration can be increased to 8% (w/v) or higher. Once a stable pattern of drinking is established, the intermittent food schedule can be eliminated, and alcohol-maintained responding persists at levels substantially above those maintained by water (Freed et al., 1970; Freed and Lester, 1970; Meisch and Thompson, 1971, 1974b; Meisch et al., 1975). This procedure works with both monkeys and rats, and, importantly, it can also be used to establish other drugs as reinforcers (Carroll and Meisch, 1978, 1980; Meisch and Stark, 1977).

Alcohol and other drugs can be established as reinforcers without using schedule-induced polydipsia. If rats or rhesus monkeys are deprived of food and reduced in body weight, they will drink water soon after receiving their daily food ration. This drinking has been termed food-induced drinking. As in the polydipsia studies, once the pattern of water drinking is stable it is possible to substitute a low concentration of alcohol for water. The alcohol concentration is then gradually increased across sessions. Once drinking is stable at a concentration such as 8% (w/v), the time of feeding is shifted to after the experimental session. Alcohol intake persists at higher levels than water intake (Meisch, 1975; Meisch and Henningfield, 1977). These results indicate that schedule-induced polydipsia is not necessary for establishing alcohol as a reinforcer.

In some instances alcohol-reinforced responding can be developed simply by providing food-deprived rats or rhesus monkeys with the opportunity to drink low alcohol concentrations. Alcohol concentrations are increased across sessions. Responding gradually increases despite the absence of food to induce drinking. Once a concentration such as 8% is reached and behavior is stable, the response requirement for alcohol can be increased from one response per delivery to two, four, or eight responses. When water is substituted for the alcohol, responding decreases, and when alcohol is reintroduced, responding again increases. Thus, this simple exposure technique also can establish alcohol as a reinforcer. The common features of these three procedures consist of using food-deprived animals and starting at low concentrations which are then gradually increased. Once alcohol is functioning as a reinforcer, the pattern and quantity of intake are similar regardless of the acquisition procedure (Meisch, 1975).

3. FACTORS AFFECTING ALCOHOL INTAKE

Genetic Variables

Inbred lines of mice differ in their alcohol intake (McClearn and Rodgers, 1959, 1961). This is a well-documented finding (for a review see McClearn, 1981). Both rats and mice have been selectively bred for high and low alcohol intake (Anderson et al., 1979; Eriksson and Rusi, 1980; Li et al., 1979; Mardones et al., 1953). With two lines of rats selectively bred for low alcohol intake, the NP and the ANA lines, it is possible to increase their alcohol consumption markedly. Alcohol intake of the NP rats was increased by reducing the rats to 80% of their free-feeding weights and by also adding sodium chloride (1.0 g%) and saccharin (0.125 g%) to a 10% [volume per volume (v/v)] alcohol solution. These manipulations increased intake from a baseline of 1.0 g/kg per day to 12.0 g/kg per day (Waller et al., 1982). Alcohol intake of the ANA rats was increased to levels shown by the AA alcohol-preferring strain by presenting the alcohol in a punch or wine rather than in water (York, 1981). Thus the low intake usually shown by these strains is probably not due to some physiological or biochemical limitation. A major weakness of the genetic studies is their reliance on the two-bottle water–alcohol preference test. Further progress in this area may depend on better methods to initiate and maintain alcohol-reinforced behavior. Two review articles and a monograph on genetics and animal studies of alcohol drinking have been published (Crabbe and Belknap, 1980; McClearn, 1981; McClearn et al., 1981).

History of Alcohol-Reinforced Behavior

Alcohol intake may increase over time as a function of experience. In two studies alcohol was established as a reinforcer for food-deprived rats (Meisch and Thompson, 1973, 1974a). When the rats were subsequently food-satiated, there was a pronounced decrease in their alcohol intake. However, after the rats had remained food-satiated for several months, they were retested with alcohol at 8% (w/v). On retest their intake was markedly increased but still below the values obtained when they were food-deprived.

In one study monkeys intravenously self-injected a range of alcohol dosages (32–560 mg/kg per injection) under a variable interval 2-min reinforcement schedule (Carney et al., 1976). When the effects of a range of dosages were determined a second time, the monkeys injected substantially greater amounts of alcohol at all dosages. The investigators concluded that the greater intake was due to the development of tolerance. However, if the greater intake was solely due to tolerance, then one might expect less rather than more responding at the lower dosages, since with tolerance the lower dosages would be less effective as reinforcers. Increases in drug-reinforced behavior as a function of experience have been noted with other drugs (Spealman and Goldberg, 1978).

Current Circumstances

Food Deprivation.   Alcohol intake is frequently studied under conditions of food deprivation. All of the studies that employ schedule-induced polydipsia use food-deprived animals. Additionally, as mentioned above, in operant studies alcohol is established as a reinforcer when the subjects are food-deprived. Thus, the role of food deprivation needs to be examined.

The general finding is clear: Food deprivation increases alcohol intake. This is well-known and documented (Meisch, 1977). The correct interpretation of this finding is not so clear. In the past the usual interpretation was that the increases in alcohol intake were due to the caloric property of alcohol; this has often been termed the caloric hypothesis (Freed and Lester, 1970). A number of findings now suggest that this interpretation is not correct. These findings derive from studies conducted with alcohol and studies conducted with other drugs.

One finding contrary to the caloric hypothesis is that when food-deprived rats are limited to an alcohol solution as their sole source of calories, they will choose a saccharin solution over 5% (v/v) alcohol with resulting weight loss and death (Samson and Falk, 1974a). In another experiment rats were deprived to 80% of their free-feeding weight, and alcohol was established as a reinforcer (Beardsley et al., 1978). Subsequently during daily 3-hr sessions the rats were given concurrent access to 8% (w/v) alcohol and water; access to a 0.1-ml dipper cup of each liquid was contingent on a lever press. Under these conditions the rats responded almost exclusively to obtain alcohol. In the next phase their daily ration of food was placed in the operant chamber. Responding for both alcohol and water increased, and alcohol-maintained responding continued to exceed water-maintained responding. This increase in alcohol responding is inconsistent with a caloric hypothesis, for such a hypothesis would lead one to predict either a decrease or no change in alcohol intake. A similarly designed study was conducted with rhesus monkeys (Henningfield and Meisch, 1981). However, in contrast to the experiment with rats, the rhesus monkeys were given their daily ration of food 15 min prior to the session. Once again, this procedure resulted in an increase in alcohol intake.

In two studies alcohol-reinforced behavior was studied under conditions first of food deprivation and then of food satiation (Meisch and Thompson, 1973, 1974a). In both studies alcohol-reinforced behavior declined to low levels immediately after food satiation. However, over several months responding increased but not to the level obtained when the rats were food-deprived. Such a recovery in responding is not consistent with the caloric hypothesis. In one of these studies food-satiated rats responded under fixed-ratio schedules to obtain 32% (w/v) alcohol (Meisch and Thompson, 1974a). The rates of responding maintained by 32% alcohol exceeded rates maintained by water. In a third study alcohol was established as a reinforcer for rats under conditions of food deprivation (Beardsley et al., 1978). Later, the rats were food-satiated; they received unlimited access to food both in their home cages and in the operant conditioning chambers. Each chamber contained two levers and two liquid delivery systems. Alcohol-maintained responding exceeded water responding even when the alcohol concentration was 32% (w/v). Again these results do not support a caloric interpretation.

The results of studies with other drugs also aid in interpreting the role of food deprivation. As with alcohol, food deprivation increases consumption of drugs that function as reinforcers. Most of these studies have been conducted only in the last few years. In brief, food deprivation increases behavior maintained by drug reinforcers from several pharmacologic classes: narcotic analgesics (Carroll and Meisch, 1978, 1979; Meisch and Kliner, 1979), dissociative anesthetics (Carroll, 1982; Carroll and Meisch, 1980), psychomotor stimulants (Carroll et al., 1981; de la Garza et al., 1981; Papasava et al., 1981), and general depressants (Kliner and Meisch, 1982). These increases occur with both rats and rhesus monkeys and with both the oral and intravenous routes (Carroll et al., 1979). The increases in drug intake produced by food deprivation are large, and are not secondary to an increased intake of the water or saline vehicle; they are not due to random increases in activity; and they are not accounted for by changes in the amount of body fat. Together these findings suggest that the increases in alcohol intake may have nothing to do with its caloric value. Nevertheless, the increases due to food deprivation can no longer be attributed simply to alcohol's caloric value.

Drug Effects on Alcohol Intake.   Many studies have examined the effects
of drugs on alcohol intake. Drugs that alter neurotransmitter function have received much attention (for a review see Kiianmaa, 1980). Unfortunately in most
studies the two-bottle choice was used, and little can be concluded.

One series of studies is noteworthy because of the control procedures used
(Davis et al., 1978, 1979). In these studies 25 mg/kg of alcohol was infused
intragastrically after each lever press. Rats served as subjects. The effects of
alpha-methyl-*p*-tyrosine, U-14,624, FLA 57, and haloperidol were studied. Al-
pha-methyl-*p*-tyrosine reduces brain levels of norepinephrine and dopamine.
Haloperidol is a dopamine-receptor blocker, and U-14,624 and FLA 57 are
inhibitors of dopamine-beta-hydroxylase. During an initial 10-hr session each
lever press resulted in an intragastric infusion of saline; and rates of lever pressing
were low. During the next 10-hr session each lever press produced an intragastric
infusion of alcohol, and rates of lever pressing increased. Other changes such
as impairment in locomotor behavior were not reported. In the third session
saline was substituted for alcohol, and responding was extinguished. In the fourth
session, termed a reacquisition session, lever presses again resulted in alcohol
infusions. Prior to this session rats were pretreated with one of the drugs being
studied. The reacquisition of alcohol-reinforced lever pressing was blocked by
pretreatment with alpha-methyl-*p*-tyrosine, U-14,624, and FLA 57 whereas reacquisition was not blocked by pretreatment with saline or haloperidol.

One interpretation of these findings is that the drugs were nonspecific sup-
pressors of lever pressing. Two experiments were conducted to evaluate this
possibility (Davis et al., 1978, 1979). In one experiment sweet milk, instead of
alcohol, was used as an intragastric reinforcer. The drugs that had blocked the
reacquisition of alcohol-reinforced responding did not block the reacquisition of
milk-reinforced responding. These results are contrary to the idea that the drug
effects on alcohol-maintained responding were due to a general suppression of
operant behavior.

In the second experiment the drug FLA 57 was studied to determine if it
would block the establishment of a buzzer as a conditioned reinforcer. The
response lever was removed, and two groups of rats were given noncontingent
intragastric infusions of alcohol (25 mg/kg per infusion). Each infusion was
paired with a buzzer. Four 10-hr sessions were conducted, and prior to each
session the experimental rats received an intraperitoneal injection of FLA 57 (50
mg/kg), and the control rats received an intraperitoneal injection of saline. After
the fourth session of buzzer-alcohol pairings, 4 days were allowed to elapse, so
that the rats could recover from the effects of FLA 57. The response lever was
then reinserted, and during a test session each lever press produced a simultaneous
activation of the buzzer and infusion of saline. Rats that in the past had received
FLA 57 responded at low rates. In contrast, rats that had previously received
saline responded at significantly higher rates. The investigators concluded that
FLA 57 prevented the establishment of the buzzer as a conditioned reinforcer
(Davis et al., 1979). They noted that on the test day the differences between the

two groups could not be attributed to nonspecific depressant actions of FLA 57 since 4 days had elapsed since the last injection of this drug. Their more general conclusion was that ". . . a cerebral noradrenergic system plays an important role in the reinforcing effect of ethanol without an involvement of dopaminergic systems" (Davis et al., 1978). Their conclusion is strengthened by the similar results obtained with two different drugs, FLA 57 and U-14,624, that share a mechanism of action. Even stronger evidence would be orderly dose–response data, but such data are lacking. However, these studies contain control procedures of the type necessary in any experiments seeking to attribute changes in alcohol consumption to changes in neurotransmitter action.

Liquid Options.   Alcohol intake can be altered by the availability of other liquids. If, in addition to water and an alcohol solution, a palatable liquid such as a sucrose or dextrose solution is available, then alcohol intake by rats and mice decreases (Lester and Greenberg, 1952; Mardones et al., 1955). Rats that were physiologically dependent on alcohol preferred a 5% (v/v) alcohol to dextrose solution until the dextrose concentration was increased to 3%. In contrast, rats that were not physiologically dependent upon alcohol preferred the 5% (v/v) alcohol only until the dextrose concentration reached 1.4% (Samson and Falk, 1974a).

Lever pressing reinforced by 5% (v/v) alcohol was studied as a function of the concentration of a concurrently available sucrose solution (Samson et al., 1982). Each liquid was available under a fixed-ratio 8 schedule of reinforcement. When water rather than sucrose was the alternative liquid, the rats responded predominantly for the alcohol solution. After this pattern of concurrent water–alcohol responding had become stable, the water was replaced by various sucrose solutions. At sucrose concentrations between 1.00% and 1.25%, responding was distributed about equally between the two liquids. Moreover, the responding for sucrose was not accompanied by a decrease in responding for alcohol. However, at higher sucrose concentrations alcohol-maintained responding decreased while sucrose-maintained responding increased.

Solution Composition.   Intake of alcohol solutions can be increased by the addition of sweetening agents such as sucrose and saccharin (Eriksson, 1969; Gilbert, 1974). Of special interest are the effects of adding these agents to an alcohol solution under conditions where the baseline intake of alcohol is already high. In two studies saccharin was added to alcohol solutions under conditions of schedule-induced drinking. In one of these studies 0.2% sodium saccharin was added to 5% (w/v) alcohol (Gilbert, 1978a) and in another study 0.25% sodium saccharin was added to 5% (v/v) alcohol (Samson and Falk, 1974b). The addition of saccharin significantly increased alcohol drinking. In one experiment rats increased their alcohol consumption from 13.1 to 15.1 g/kg per day and when alcohol was withdrawn, severe tonic–clonic seizures occurred (Samson and Falk, 1974b). In the other study alcohol was available for 50 min each day. The addition of saccharin increased the rate of alcohol drinking; intake was as high as 3 g/kg per hr (Gilbert, 1978a). These findings indicate that even when alcohol intake is high, it is not limited by physical incapacitation, since increases could still be obtained by the addition of saccharin.

Physiological Dependence.   Physiological (physical) dependence on alcohol has been demonstrated with mice (Freund, 1969; Goldstein and Pal, 1971), rats (Begleiter, 1975; Branchey et al., 1971; Falk et al., 1972), dogs (Essig and Lam, 1968); rhesus monkeys (Deneau et al., 1969; Ellis and Pick, 1970; Pieper and Skeen, 1972), and chimpanzees (Pieper et al., 1972). The results of these and other studies with animals make it possible to evaluate the role of physiological dependence in determining alcohol intake. The results show that physiological dependence is not a sufficient condition for maintaining alcohol intake. For example, when four physiologically dependent mice were given a choice of drinking a liquid diet containing sucrose in place of alcohol, all four mice exclusively selected the solution containing sucrose; an abstinence syndrome resulted (Freund, 1969). Similar results have been obtained with rats (Hunter et al., 1974). In one study with rhesus monkeys physiological dependence was established by the experimenter administering alcohol through a nasogastric tube. After the monkeys were physiologically dependent, the administration of alcohol was stopped, and the monkeys were given access to an alcohol solution. It was thought that the monkeys might avoid withdrawal by drinking alcohol. However, none of the monkeys drank alcohol, and withdrawal reactions occurred (Myers et al., 1972).

Physiological dependence is also not necessary for generating high alcohol intake (cf. Mello, 1973). In studies of intravenous alcohol self-administration, rhesus monkeys reliably self-injected alcohol even when access was limited to 3 hr or 1 hr per day (Carney et al., 1976; Karoly et al., 1978; Winger and Woods, 1973; Woods et al., 1971). Large amounts were taken, but physiological dependence was not seen. This is not surprising since between sessions the alcohol that was taken within sessions could be completely metabolized. With the oral route, once alcohol has been established as a reinforcer, rats and rhesus monkeys consistently drink volumes of alcohol solutions that substantially exceed water intake (Beardsley et al., 1978; Freed et al., 1970; Henningfield and Meisch, 1978; Meisch and Beardsley, 1975; Meisch and Thompson, 1971, 1973, 1974a,b). With other drugs, substantial drug-seeking behavior and drug intake can also occur in the absence of physiological dependence (Spealman and Goldberg, 1978; Young et al., 1981).

There have been no animal studies in which alcohol was first established as a reinforcer and then the effects of physiological dependence on alcohol intake determined (Meisch, 1984). Thus, it is not known if alcohol is a better reinforcer when organisms are physiologically dependent. This is also true of many other drugs. The relationships of physiological dependence and tolerance to drug intake, including alcohol intake, have been reviewed (Cappell and LeBlanc, 1981).

Positive Reinforcement of Concurrent Behavior: Schedule-Induced Polydipsia.   Schedule-induced polydipsia has been used to generate high levels of sustained alcohol drinking. Sufficient intake occurs to disrupt the pattern of concurrent food-reinforced responding (Meisch and Thompson, 1972). When multiple polydipsia sessions are scheduled at regular intervals within 24-hr cycles, blood alcohol levels are sustained and physiological dependence can result (Falk et al., 1972; McMillan et al., 1976; Samson and Falk, 1974a,b). An important feature of schedule-induced drinking is that after rats have had experience drinking alcohol, they will prefer solutions of alcohol at concentrations of 4–5.6% (w/v) to water when both liquids are available (Freed, 1974; McMillan, 1978; Samson and Falk, 1974a). Thus, the high alcohol intakes cannot be interpreted as simply being secondary to high intake of the vehicle, water. In future studies of schedule-induced drinking the alcohol concentration should be varied when water is concurrently available so that the range of concentrations preferred to water can be determined. Schedule-induced polydipsia can also be used to obtain drinking of other drugs (for a review see Gilbert, 1978b).

Schedule of Alcohol Access.   Schedule of alcohol access describes the time periods when alcohol is available. Two values are required to specify the schedule of access: the duration of the access period, and the interval between access periods (Meisch, 1977).

Schedule of access has been most often studied with the two-bottle water–alcohol preference technique. When rats have access to alcohol for a period of time and subsequently lose access to it, an alcohol-deprivation effect occurs. When alcohol is reintroduced there is a marked but transient increase in drinking (Dember and Kristofferson, 1955; LeMagnen, 1960; Nichols and Hsiao, 1967; Sinclair and Senter, 1967, 1968). This deprivation effect has been systematically explored in a series of experiments (Sinclair, 1971, 1972, 1979; Sinclair et al., 1973a,b), and the results have been reviewed (Meisch, 1977).

Schedule of access has also been studied with the intravenous self-injection procedure. When alcohol was continuously available and each lever press by a rhesus monkey was reinforced by an alcohol injection of 0.1 g/kg, cycles of persistent responding that lasted 1 or more weeks alternated with periods of low responding that lasted for several days (Deneau et al., 1969; Winger and Woods, 1973; Woods et al., 1971). However, when alcohol was available for just 3 hr each day, a stable pattern of responding resulted, and this could be maintained over many weeks (Winger and Woods, 1973; Woods et al., 1971). Stability was indicated by the small range in the number of injections across sessions. With other drugs that can serve as reinforcers, stable patterns of responding can also be maintained when access to the drug is limited to short periods each day, such as 3 hr (Spealman and Goldberg, 1978).

Response Consequences

Schedules of Reinforcement.   Schedules of reinforcement specify the relationship between responses and the events that follow responding. An important characteristic of reinforcers is that they maintain behavior under conditions of intermittent reinforcement. Behavior is maintained under intermittent schedules when responses only occasionally produce access to a reinforcer. Schedules of reinforcement are important determinants of operant behavior (Zeiler, 1977), and the role of reinforcement schedules in controlling drug reinforced behavior has been reviewed (Spealman and Goldberg, 1978).

Intermittent schedules have been examined with alcohol in only a few studies. Alcohol can maintain responding under both interval and ratio schedules. Under fixed-interval schedules the delivery of a reinforcer follows the first response emitted after the lapse of a fixed interval of time. Alcohol can maintain fixed-interval performances in rats (Anderson and Thompson, 1974; Meisch and Thompson, 1974c). With variable-interval schedules the interval between reinforcement varies around a fixed value. Under such a schedule, intravenous alcohol injections maintained the lever-pressing behavior of rhesus monkeys (Carney et al., 1976).

Alcohol reinforcement has also been studied with fixed-ratio schedules. Such schedules require emission of a fixed number of responses for each unit of alcohol delivered. Under fixed-ratio schedules of alcohol, reinforcement has been demonstrated with both rats (Meisch and Thompson, 1973; Roehrs and Samson, 1981, 1982; Samson et al., 1982) and rhesus monkeys (Henningfield and Meisch, 1978). In one experiment with rats the fixed-ratio size was studied over a range of values (Meisch and Thompson, 1973). As the size of the fixed ratio was increased to a value of eight, the number of dipper presentations of 8% (w/v) alcohol remained constant. Further increases in the size of the ratio resulted in systematic decreases in the number of dipper presentations. These findings are consistent with results obtained with alcoholics studied on a clinical research unit (Babor et al., 1978; Bigelow and Liebson, 1972; Mello et al., 1968). In these human studies alcohol intake decreased with increases in the response requirement or response cost for alcohol.

Punishment of Alcohol Drinking.   The effects of punishment on alcohol drinking have been examined in several sophisticated studies. Food pellets were available to rats under a fixed-interval 26-sec schedule; the first lever press after 26 sec had elapsed resulted in the delivery of a food pellet. Each press on a second lever resulted in the presentation of a dipper cup containing either water or an alcohol solution. Alcohol concentrations of 0% (water), 4, 8, 16, and 32% (v/v) were studied. After a baseline of responding was obtained, a change in the contingencies was made such that presses on the lever that resulted in liquid delivery had the additional consequence of delaying food availability for 8 sec. This contingency resulted in a suppression of alcohol intake. This suppression did not occur in a yoked control group that received food coincidentally with the experimental rats (Poling and Thompson, 1977a).

In a related experiment the effects of presses on the lever for alcohol were studied when each press, or every second or fourth response, delayed food availability for 8 sec. This contingency decreased alcohol responding, and the degree of the decrease was greatest when each lever press delayed food availability and was least when every fourth press delayed food availability. These experimental conditions were studied with alcohol concentrations of 4, 8, and 16% (v/v), and the degree of the response rate decrease was inversely related to the alcohol concentration. As in the earlier experiment, yoked control rats received food pellets coincidentally with the experimental rats. The control rats did not display decreases in alcohol responding (Poling and Thompson, 1977b).

In a third experiment two food-deprived rats responded under a concurrent fixed-ratio 12 schedule of food reinforcement and a fixed-ratio 1 schedule of alcohol (8% v/v) reinforcement. During every second or fourth interpellet interval, a tone occurred. In the presence of this tone, presses on the lever that led to alcohol delivery also altered the schedule of food reinforcement such that subsequent presses on the food lever were punished by electric shock. After a food pellet was obtained, the food reinforcement schedule reverted back to the original nonpunishment condition. This arrangement of conditions produced a decrease in the frequency of alcohol-maintained responding which was directly related to shock voltage from 25 to 100 V (Poling and Thompson, 1977c). These experiments elegantly show that alcohol-reinforced responding is sensitive to punishment contingencies.

Alcohol Dosage and Concentration.   The dosage or concentration of alcohol is a determinant of alcohol-reinforced behavior (Karoly et al., 1978). Alcohol dosage has been studied in an intravenous self-administration experiment with rhesus monkeys. When the alcohol dose was varied from 0.05 to 0.2 g/kg, the number of injections during 3-hr sessions was inversely related to dosage. The total alcohol intake per session increased slightly with increases in the dosage; the intake over the 3-hr session ranged from 3.9 g/kg at the 0.05 g/kg dose to 4.7 g/kg at the 0.2 g/kg dose.

Alcohol concentration acts in a similar fashion to alcohol dosage. When the concentration was varied from 8 to 32% (w/v), the volume consumed by monkeys decreased with increases in concentration; however, at all alcohol concentrations volume consumed exceeded water values (Henningfield and Meisch, 1978). In this study, total intake per 3-hr session also increased slightly with increases in concentration. Intake ranged from 2.56 g/kg at 8% to 3.08 g/kg at 32%. These intakes are less than those found in the previous study when alcohol was taken intravenously (Karoly et al., 1978). In both the oral and intravenous studies the temporal pattern of intake was negatively accelerated; that is, the rate was highest at the beginning of the session and gradually decreased over the course of the session. This pattern of intake leads to the maximum effects within a limited time.

Alcohol concentration has also been studied in an experiment with two baboons (Henningfield et al., 1981). The volume consumed was inversely related to the alcohol concentration from 8 to 32% (w/v). Total intake within the 3-hr sessions was relatively constant across concentrations. In one baboon, blood alcohol levels regularly exceeded 200 mg/dl, and in the other blood alcohol levels were greater than 100 mg/dl.

In a study with rats the alcohol concentration was varied from 2 to 32%

(w/v). Number of alcohol deliveries was an inverted U-shaped function of concentration, and the amount of alcohol consumed (g/kg) increased with increases in the concentration (Meisch and Thompson, 1974a). In this study and in the studies with the rhesus monkeys and baboons, the relations between the independent variables of alcohol concentration or alcohol dosage and measures of reinforced performance are similar to those found in studies where other drugs functioned as reinforcers (Johanson and Schuster, 1981; Meisch and Carroll, 1981).

4.   ANIMAL MODELS OF ALCOHOLISM

Several investigators have specified criteria for an animal model of alcoholism (Cicero, 1979, 1980; Falk and Tang, 1977; Lester and Freed, 1973;

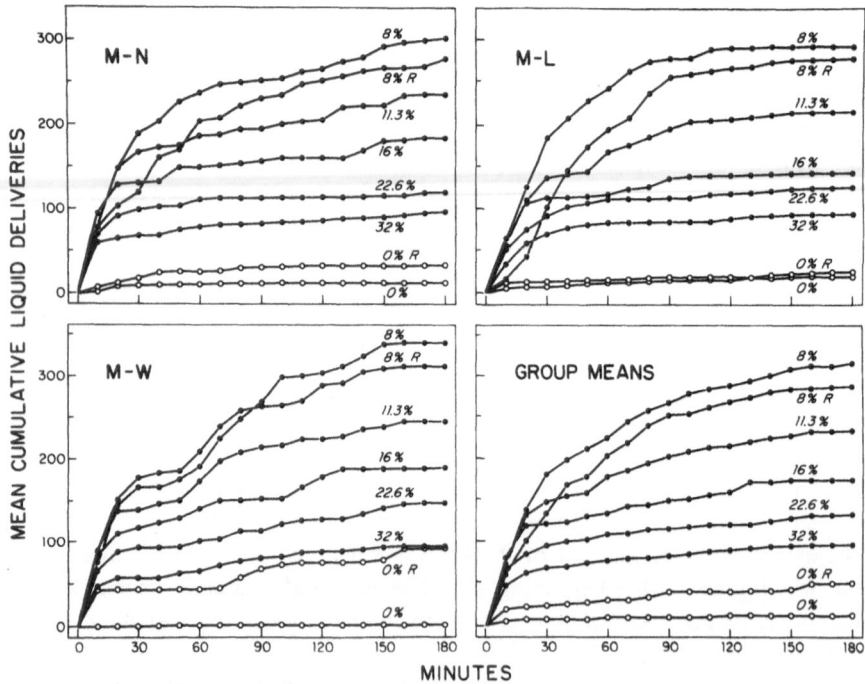

Figure 1. Mean cumulative liquid deliveries obtained by rhesus monkeys during 3-hr sessions as a function of alcohol concentration ($n = 5$ for individual graphs, and $n = 15$ for the group graph). For monkey M-W, eight lip contacts with the drinking spout were required for each liquid delivery (FR 8 reinforcement schedule), and for monkeys M-N and M-L, 16 lip contacts were required (FR 16). Filled circles, alcohol solutions; unfilled circles, water. R, retest values of 0 or 8% (w/v) alcohol. (From Henningfield and Meisch, 1978, with permission.)

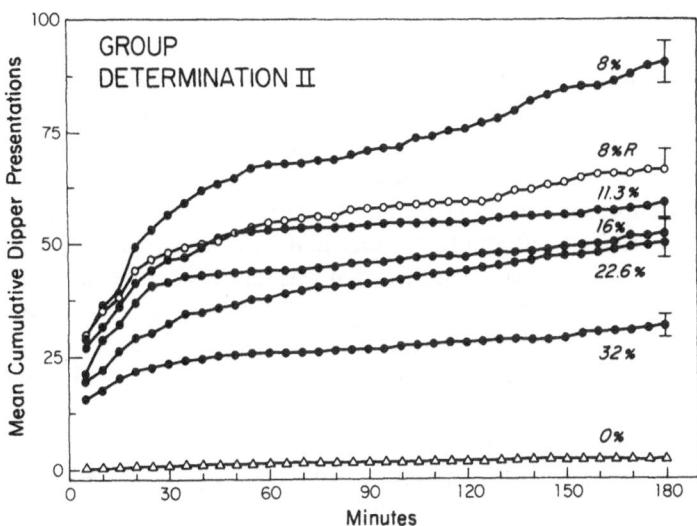

Figure 2. Mean cumulative dipper presentations obtained by rats as a function of alcohol concentration. Each press on a lever resulted in a dipper presentation (FR 1 reinforcement schedule). Filled circles, test values. Unfilled circles, 8% (w/v) retest values. Each point is the mean of 40 sessions (4 rats × 10 sessions each), except for those points at 0% (water) which are means of 240 sessions (4 rats × 6 concentrations × 10 sessions each). Brackets, mean standard error of the mean ($n = 4$; 4 rats × 1 SE each). Absence of brackets indicates that values fell within the area occupied by a plotted point. (From Beardsley et al., 1978, with permission.)

Mello, 1976). Ideally such a model should also work with pharmacologically related drugs such as the barbiturates, but this has not been explicitly considered by those who have listed criteria. The specification of such criteria presupposes that we have a good understanding of alcoholism. This may not be the case. For example, concepts such as craving have been found to be logically and empirically inadequate as explanations of alcoholic drinking (Mello, 1975). The frequency with which alcoholism is observed may lead us to confuse familiarity with understanding.

The specification of criteria for an animal model of alcoholism presupposes that we have solved the more general problem of identifying equivalent behavior in different species (Meisch, 1980). Little attention has been given to this problem. An exception is Skinner's (1959) comment that "one moves from the experimental analysis of behavior at the lower level to the human level, not by pointing out plausible analogues, but by constructing an experimental situation in which the same kinds of variables are manipulated and the same changes in behavior are measured." This statement should also hold true for the more general case of comparing behaviors of different animal species. Thus, a necessary condition for concluding that behaviors in two different species are equivalent is the demonstration that the same independent variables control a particular behavior in a similar manner.

An example of similar control in two species by an independent variable is illustrated by the results shown in Figs. 1 and 2. These figures show the time course of alcohol drinking over a range of concentrations for rhesus monkeys (Fig. 1) and rats (Fig. 2). The number of liquid deliveries for monkeys and the number of dipper presentations for rats are cumulated over the 3-hr experimental sessions. There are some similar features in the alcohol-drinking behavior of the two species. As the alcohol concentration was increased in steps from 8 to 32% (w/v), the volume consumed decreased but remained greater than water control values. Also, both species showed negatively accelerated patterns of alcohol intake. Details of the experiments with rats (Beardsley et al., 1978) and monkeys (Henningfield and Meisch, 1978) have been reported. The time course plots in these figures are similar to plots of the time course of alcohol intake when rhesus monkeys intravenously self-administer alcohol (Karoly et al., 1978). One would predict similar patterns of intake if human alcoholics were tested in a comparable situation.

5. CONCLUDING REMARKS

Over the last 20 years systematic progress has been made in studies of drug self-administration. In general, drugs that humans abuse serve as reinforcers for animals, and drugs that humans do not abuse do not serve as reinforcers for animals (Johanson and Balster, 1978). Drug intake and drug-reinforced behavior vary in an orderly way as a function of a number of variables. There has been a progressive increase in the complexity of the behavior and in the range of variables studied. Findings from any one laboratory have been replicated in other laboratories, and the findings with one species have usually been replicated in other species. Conditions have been identified that reliably produce high drug intake, such as 24-hr drug access, high drug dosages, and low response requirements. Conditions have also been identified that result in high rates of drug-reinforced responding under conditions where animals receive small dosages delivered infrequently.

In contrast to advances in studies of drug self-administration, less progress has been made in alcohol self-administration research. Several factors probably account for this lack of progress. One factor has been the difficulty in identifying conditions that lead to high elective alcohol intake. A second factor has been the absence of a conceptual framework within which to view alcohol intake, and a third factor has been a misplaced emphasis on tolerance and physiological dependence as determinants of intake.

Procedures are now available for establishing alcohol as an effective reinforcer when taken by mouth. These procedures work with rats, rhesus monkeys, and baboons. Importantly these procedures also work with other drugs. With rhesus monkeys alcohol can serve as a reinforcer when taken intragastrically or intravenously under conditions that are effective when other drugs are studied. As with other drugs the methods and principles of operant conditioning provide a conceptual framework for designing and interpreting studies. The methods and concepts used in studies of drug-reinforced behavior can be used in the particular case of alcohol. The outcome should be the rapid progress that has characterized the study of other abused drugs.

ACKNOWLEDGMENTS

Thanks to Frank R. George, Gregory A. Lemaire, and Sandra M. Schrader for their helpful comments on the manuscript. Work cited as being done by the author was supported by an NIDA Research Scientist Development Award DA 00007, by NIDA grant DA 00944, and by NIAAA grant 00299.

REFERENCES

Altshuler, H. L., and Talley, L., 1977, Intragastric self-administration of ethanol by the rhesus monkey: An animal model of alcoholism, in: *Currents in Alcoholism,* Vol. I (F. A. Seixas, ed.), pp. 243–253, Grune and Stratton, New York.

Altshuler, H., Weaver, S., and Phillips, P., 1975, Intragastric self-administration of psychoactive drugs by the rhesus monkey, *Life Sci.* **17**:883.

Amit, Z., Amir, S., and Corcoran, M. E., 1973, A possible artifact in studies of alcohol consumption in rats, *Q. J. Stud. Alcohol* **34**:524.

Anderson, S. M., McClearn, G. E., and Erwin, V. G., 1979, Ethanol consumption and hepatic enzyme activity, *Pharmacol. Biochem. Behav.* **11**:83.

Anderson, W. W., and Thompson, T., 1974, Ethanol self-administration in water satiated rats, *Pharmacol. Biochem. Behav.* **2**:447.

Babor, T. F., Mendelson, J. H., Greenberg, I., and Kuehnle, J., 1978, Experimental analysis of the "happy hour": Effects of purchase price on alcohol consumption, *Psychopharmacology* **58**:35.

Beardsley, P. M., Lemaire, G. A., and Meisch, R. A., 1978, Ethanol-reinforced behavior of rats with concurrent access to food and water, *Psychopharmacology* **59**:7.

Begleiter, H., 1975, Alcohol consumption subsequent to physical dependence, in: *Alcohol Intoxication and Withdrawal: Experimental Studies,* Vol. II (M. M. Gross, ed.), pp. 373–378, Plenum Press, New York.

Bigelow, G., and Liebson, I., 1972, Cost factors controlling alcoholic drinking, *Psychol. Rec.* **22**:305.

Branchey, M., Rauscher, G., and Kissin, B., 1971, Modifications in the response to alcohol following the establishment of physical dependence, *Psychopharmacologia* **22**:314.

Cappell, H., and LeBlanc, A. E., 1981, Tolerance and physical dependence: Do they play a role in alcohol and drug self-administration? in: *Research Advances in Alcohol and Drug Problems,* Vol. 6 (Y. Israel, F. B. Glaser, H. Kalant, R. E. Popham, W. Schmidt, and R. G. Smart, eds.), pp. 159–196, Plenum Press, New York.

Carney, J. M., Llewellyn, M. E., and Woods, J. H., 1976, Variable interval responding maintained by intravenous codeine and ethanol injections in the rhesus monkey, *Pharmacol. Biochem. Behav.* **5**:577.

Carroll, M. E., 1982, Rapid acquisition of oral phencyclidine self-administration in food-deprived and food-satiated rhesus monkeys: Concurrent phencyclidine and water choice, *Pharmacol. Biochem. Behav.* **17**:341.

Carroll, M. E., and Meisch, R. A., 1978, Etonitazene as a reinforcer: Oral intake of etonitazene by rhesus monkeys, *Psychopharmacology* **59**:225.

Carroll, M. E., and Meisch, R. A., 1979, Effects of food deprivation on etonitazene consumption in rats, *Pharmacol. Biochem. Behav.* **10**:155.

Carroll, M. E., and Meisch, R. A., 1980, Oral phencyclidine (PCP) self-administration in rhesus monkey: Effects of feeding conditions, *J. Pharmacol. Exp. Ther.* **214**:339.

Carroll, M. E., France, C. P., and Meisch, R. A., 1979, Food deprivation increases oral and intravenous drug intake in rats, *Science* **205**:319.

Carroll, M. E., France, C. P., and Meisch, R. A., 1981, Intravenous self-administration of etonitazene, cocaine and phencyclidine in rats during food deprivation and satiation, *J. Pharmacol. Exp. Ther.* **217**:241.

Cicero, T. J., 1979, A critique of animal analogs of alcoholism, in: *Biochemistry and Pharmacology of Ethanol,* Vol. 2 (E. Majchrowicz, and E. P. Noble, eds.), pp. 533–560, Plenum Press, New York.

Cicero, T. J., 1980, Animal models of alcoholism? in: *Animal Models in Alcohol Research* (K. Eriksson, J. D. Sinclair, and K. Kiianmaa, eds.), pp. 99–117, Academic Press, New York.

Crabbe, J. C., and Belknap, J. K., 1980, Pharmacogenetic tools in the study of drug tolerance and dependence, *Subst. Alcohol Actions Misuse* **1**:385.

Davis, W. M., Smith, S. G., and Werner, T. E., 1976, Intragastric alcohol: Effects of unit dosage on self-administration and on conditioned reinforcement, *Proc. West. Pharmacol. Soc.* **19**:346.

Davis, W. M., Smith, S. G., and Werner, T. E., 1978, Noradrenergic role in self-administration of ethanol, *Pharmacol. Biochem. Behav.* **9**:369.

Davis, W. M., Werner, T. E., and Smith, S. G., 1979, Reinforcement with intragastric infusions of ethanol: Blocking effect of FLA 57, *Pharmacol. Biochem. Behav.* **11**:545.

de la Garza, R., Bergman, J., and Hartel, C. R., 1981, Food deprivation and cocaine self-administration, *Pharmacol. Biochem. Behav.* **15**:141.

Dember, W. N., and Kristofferson, A. B., 1955, The relation between free-choice alcohol consumption and susceptibility to audiogenic seizures, *Q. J. Stud. Alcohol* **16**:86.

Deneau, G., Yanagita, T., and Seevers, M. H., 1969, Self-administration of psychoactive substances by the monkey, *Psychopharmacologia* **16**:30.

DeNoble, V. J., and Begleiter, H., 1978, Alcohol self-administration in monkeys (*Macaca radiata*): The effects of prior alcohol exposure, *Pharmacol. Biochem. Behav.* **8**:391.

Ellis, F. W., and Pick, J. R., 1970, Experimentally induced ethanol dependence in rhesus monkeys, *J. Pharmacol. Exp. Ther.* **175**:88.

Eriksson, K., 1969, Factors affecting voluntary alcohol consumption in the albino rat, *Ann. Zool. Fennici* **6**:227.

Eriksson, K., and Ruse, M., 1981, Finnish selection studies on alcohol-related behaviors: General outline, in: *Development of Animal Models as Pharmacogenetic Tools* (G. E. McClearn, R. A. Deitrich, and V. G. Erwin, eds.), pp. 87–117, National Institute on Alcohol Abuse and Alcoholism, Research Monograph No. 6, DHHS Publication No. (ADM) 81-1133, Rockville, Md.

Eriksson, K., Sinclair, J. D., and Kiianmaa, K. (eds.), 1980, *Animal Models in Alcohol Research,* Academic Press, New York.

Essig, C. F., and Lam, R. C., 1968, Convulsions and hallucinatory behavior following alcohol withdrawal in the dog, *Arch. Neurol.* **18**:626.

Falk, J. L., 1961, Production of polydipsia in normal rats by an intermittent food schedule, *Science* **133**:195.

Falk, J. L., 1969, Conditions producing psychogenic polydipsia in animals, *Ann. N.Y. Acad. Sci.* **157**:569.

Falk, J. L., and Tang, M., 1977, Animal model of alcoholism: Critique and progress, in: *Alcohol Intoxication and Withdrawal*, Vol. 3B (M. M. Gross, ed.), pp. 465–493, Plenum Press, New York.

Falk, J. L., Samson, H. H., and Winger, G., 1972, Behavioral maintenance of high concentrations of blood ethanol and physical dependence in the rat, *Science* **177**:811.

Freed, E. X., 1974, Fluid selection by rats during schedule-induced polydipsia, *Q. J. Stud. Alcohol* **35**:1035.

Freed, E. X., and Lester, D., 1970, Schedule-induced consumption of ethanol: Calories or chemotherapy? *Physiol Behav.* **5**:555.

Freed, E. X., Carpenter, J. A., and Hymowitz, N., 1970, Acquisition and extinction of schedule-induced polydipsic consumption of alcohol and water, *Psychol. Rep.* **26**:915.

Freund, G., 1969, Alcohol withdrawal syndrome in mice, *Arch. Neurol.* **21**:315.

Gilbert, R. M., 1974, Effects of food deprivation and fluid sweetening on alcohol consumption by rats, *Q. J. Stud. Alcohol* **35**:42.

Gilbert, R. M., 1978a, Schedule induction and sweetness as factors in ethanol consumption and preference by rats, *Pharmacol. Biochem. Behav.* **8**:739.

Gilbert, R. M., 1978b, Schedule-induced self-administration of drugs, in: *Contemporary Research in Behavioral Pharmacology* (D. E. Blackman, and D. J. Sanger, eds.), pp. 289–323, Plenum Press, New York.

Gillespie, R. J. G., and Lucas, C. C., 1958, An unexpected factor affecting the alcohol intake of rats, *Can. J. Biochem. Physiol.* **36**:37.

Goldstein, D. B., 1978, Animal studies of alcohol withdrawal reactions, in: *Research Advances in Alcohol and Drug Problems*, Vol. 4 (Y. Israel, F. B. Glaser, H. Kalant, R. E. Popham, W. Schmidt, and R. G. Smart, eds.), pp. 77–109, Plenum Press, New York.

Goldstein, D. B., and Pal, N., 1971, Alcohol dependence produced in mice by inhalation of ethanol: Grading the withdrawal reaction, *Science* **172**:288.

Goodrick, C. L., 1972, End bottle preferences of inbred mice during alcohol preference and fluid intake multiple-bottle test procedures, *Psychon. Sci.* **28**:185.

Henningfield, J. E., and Meisch, R. A., 1978, Ethanol drinking by rhesus monkeys as a function of concentration, *Psychopharmacology* **57**:133.

Henningfield, J. E., and Meisch, R. A., 1981, Ethanol and water drinking by rhesus monkeys: Effects of nutritive preloading, *J. Stud. Alcohol* **42**:192.

Henningfield, J. E., Ator, N. A., and Griffiths, R. R., 1981, Establishment and maintenance of oral ethanol self-administration in the baboon, *Drug Alcohol Depend.* **7**:13.

Hunter, B. E., Walker, D. W., and Riley, J. N., 1974, Dissociation between physical dependence and volitional ethanol consumption: Role of multiple withdrawal episodes, *Pharmacol. Biochem. Behav.* **2**:523.

Johanson, C. E., and Balster, R. L., 1978, A summary of the results of a drug self-administration study using substitution procedures in rhesus monkeys, *Bull. Narc.* **30**:43.

Johanson, C. E., and Schuster, C. R., 1981, Animal models of drug self-administration, in: *Advances in Substance Abuse*, Vol. 2, (N. K. Mello, ed.), pp. 219–297, JAI Press Inc., Greenwich, Connecticut.

Karoly, A. J., Winger, G. D., Ikomi, F., and Woods, J. H., 1978, The reinforcing property of ethanol in the rhesus monkey, *Psychopharmacology* **58**:19.

Kiianmaa, K., 1980, *Role of Brain Monoamines in Ethanol Consumption and Acute Intoxication of Rats*, Research Laboratories of the State Alcohol Monopoly (Alko), Helsinki, Finland.

Kliner, D. J., and Meisch, R. A., 1982, The effects of food deprivation and satiation on oral pentobarbital self-administration in rhesus monkeys, *Pharmacol. Biochem. Behav.* **16:**579.

Korn, S. J., 1960, The relationship between individual differences in the responsivity of rats to stress and intake of alcohol, *Q. J. Stud. Alcohol* **21:**605.

LeMagnen, J., 1960, Étude de quelques facteurs associés à des modifications de la consommation spontanée d'alcool ethylique par le rat, *J. Physiol. (Paris)* **52:**873.

Lester, D., and Freed, E. X., 1973, Criteria for an animal model of alcoholism, *Pharmacol. Biochem. Behav.* **1:**103.

Lester, D., and Greenberg, L. A., 1952, Nutrition and the etiology of alcoholism: The effect of sucrose, saccharin and fat on the self-selection of ethyl alcohol by rats, *Q. J. Stud. Alcohol* **13:**553.

Li, T.-K., Lumeng, L., McBride, W. J., Waller, M. B., Hawkins, T. D., 1979, Progress toward a voluntary oral consumption model of alcoholism, *Drug Alcohol Depend.* **4:**45.

Lieber, C. S., and DeCarli, L. M., 1974, An experimental model of alcohol feeding and liver injury in the baboon, *J. Med. Primatol.* **3:**153.

Mardones, J. R., Segovia, N. M., and Hederra, A. D., 1953, Heredity of experimental alcohol preference in rats. II. Coefficient of heredity, *Q. J. Stud. Alcohol* **14:**1.

Mardones, J. R., Segovia-Riquelme, N., Hederra, A. D., and Alcaino, F. G., 1955, Effect of some self-selection conditions on the voluntary alcohol intake of rats, *Q. J. Stud. Alcohol* **16:**425.

Marfaing-Jallat, P., Pruvost, M., and LeMagnen, J., 1974, La consommation d'ethanol par auto-administration intragastrique chez le rat, *J. Physiol. (Paris)* **68:**81.

McClearn, G. E., 1981, Animal models of genetic factors in alcoholism, in: *Advances in Substance Abuse Behavioral and Biological Research* (N. K. Mello, ed.), pp. 185–217, JAI Press Inc., Greenwich, Connecticut.

McClearn, G. E., Deitrich, R. A., and Erwin, V. G. (eds.), 1981, Development of Animal Models as Pharmacogenetic Tools, National Institute on Alcohol Abuse and Alcoholism, Research Monograph No. 6, DHHS Publication No. (ADM) 81-1133, Rockville, Md.

McClearn, G. E., and Rodgers, D. A., 1959, Differences in alcohol preference among inbred strains of mice, *Q. J. Stud. Alcohol* **20:**691.

McClearn, G. E., and Rodgers, D. A., 1961, Genetic factors in alcohol preference of laboratory mice, *J. Comp. Physiol. Psychol.* **54:**116.

McMillan, D. E., 1978, Effects of access to a running wheel on ethanol intake in rats under schedule-induced polydipsia, in: *Currents in Alcoholism*, Vol. III (F. A. Seixas, ed.), pp. 221–235, Grune & Stratton, New York.

McMillan, D. E., Lenader, J. D., Ellis, F. W., Lucot, J. B., and Frye, G. D., 1976, Characteristics of ethanol drinking patterns under schedule-induced polydipsia, *Psychopharmacology* **49:**49.

Meisch, R. A., 1975, The function of schedule-induced polydipsia in establishing ethanol as a positive reinforcer, *Pharmacol. Rev.* **27:**465.

Meisch, R. A., 1977, Ethanol self-administration: Infrahuman studies, in: *Advances in Behavioral Pharmacology* (T. Thompson, and P. B. Dews, eds.), pp. 35–84, Academic Press, New York.

Meisch, R. A., 1980, Ethanol as a reinforcer for rats, monkeys and humans, in: *Animal Models in Animal Research* (K. Eriksson, J. D. Sinclair, and K. Kiianmaa, eds.), pp. 153–158, Academic Press, New York.

Meisch, R. A., 1981, Animal studies of alcohol intake, *Br. J. Psychiat.* **141:**113.

Meisch, R. A., 1984, The relationship between physical dependence on ethanol and reinforcing properties of ethanol in animals, in: *Endocrinological Aspects of Ethanol Tolerance and Dependence* (T. J. Cicero, ed.), National Institute on Alcohol Abuse and Alcoholism, Research Monograph, Rockville, Md., in press.

Meisch, R. A., and Beardsley, P., 1975, Ethanol as a reinforcer for rats: Effects of concurrent access to water and alternate positions of water and ethanol, *Psychopharmacologia* **43:**19.

Meisch, R. A., and Carroll, M. E., 1981, Establishment of orally delivered drugs as reinforcers for rhesus monkeys: Some relations to human drug dependence, in: Behavioral Pharmacology of Human Drug Dependence (T. Thompson, and C. E. Johanson, eds.), pp. 197–209, National Institute on Drug Abuse, Research Monograph No. 37. DHHS Publication No. (ADM) 81-1137, Rockville, Md.

Meisch, R. A., and Henningfield, J. E., 1977, Drinking of ethanol by rhesus monkeys: Experimental strategies for establishing ethanol as a reinforcer, in: Alcohol Intoxication and Withdrawal, Vol. 3B (M. M. Gross, ed.), pp. 443–463, Plenum Press, New York.

Meisch, R. A., and Kliner, D. J., 1979, Etonitazene as a reinforcer for rats: Increased etonitazene-reinforced behavior due to food deprivation, Psychopharmacology 63:97.

Meisch, R. A., and Stark, L. J., 1977, Establishment of etonitazene as a reinforcer for rats by use of schedule-induced drinking, Pharmacol. Biochem. Behav. 7:195.

Meisch, R. A., and Thompson, T., 1971, Ethanol intake in the absence of concurrent food reinforcement, Psychopharmacologia, 22:72.

Meisch, R. A., and Thompson, T., 1972, Ethanol intake during schedule-induced polydipsia, Physiol. Behav. 8:41.

Meisch, R. A., and Thompson, T., 1973, Ethanol as a reinforcer: Effects of fixed-ratio size and food deprivation, Psychopharmacologia 28:171.

Meisch, R. A., and Thompson, T., 1974a, Ethanol intake as a function of concentration during food deprivation and satiation, Pharmacol. Biochem. Behav. 2:589.

Meisch, R. A., and Thompson, T., 1974b, Rapid establishment of ethanol as a reinforcer for rats, Psychopharmacologia 37:311.

Meisch, R. A., and Thompson, T., 1974c, Ethanol as a reinforcer: An operant analysis of ethanol dependence, in: Drug Addiction, Vol. 3, Neurobiology and Influences on Behavior (J. M. Singh and H. Lal, eds.), pp. 117–133, Stratton Intercontinental Medical Book Corporation, New York.

Meisch, R. A., Henningfield, J. E., and Thompson, T., 1975, Establishment of ethanol as a reinforcer for rhesus monkeys via the oral route: Initial results, in: Alcohol Intoxication and Withdrawal, Vol. II (M. M. Gross, ed.), pp. 323–342, Plenum Press, New York.

Mello, N. K., 1973, A review of methods to induce alcohol addiction in animals, Pharmacol. Biochem. Behav. 1:89.

Mello, N. K., 1975, A semantic aspect of alcoholism, in: Biological and Behavioral Approaches to Drug Dependence (H. D. Cappell, and A. E. LeBlanc, eds.), pp. 73–87, Addiction Research Foundation, Toronto.

Mello, N. K., 1976, Animal models for the study of alcohol addiction, Psychoneuroendocrinology 1:347.

Mello, N. K., and Mendelson, J. H., 1971a, The effects of drinking to avoid shock on alcohol intake in primates, in: Biological Aspects of Alcohol (M. K. Roach, W. M. McIsaac, and P. J. Creaven, eds.), pp. 313–332, University of Texas Press, Austin.

Mello, N. K., and Mendelson, J. H., 1971b, Evaluation of polydipsia technique to induce alcohol consumption in monkeys, Physiol. Behav. 7:827.

Mello, N. K., McNamee, H. B., and Mendelson, J. H., 1968, Drinking patterns of chronic alcoholics: Gambling and motivation for alcohol, in: Clinical Research in Alcoholism (J. O. Cole, ed.), pp. 83–118, American Psychiatric Association, Psychiatric Research Report #24, Washington, D.C.

Myers, R. D., 1978, Psychopharmacology of alcohol, Annu. Rev. Pharmacol. Toxicol. 18:125.

Myers, R. D., Stoltman, W. P., and Martin, G. E., 1972, Effects of ethanol dependence induced artificially in the rhesus monkey on the subsequent preference for ethyl alcohol, Physiol. Behav. 9:43.

Nichols, J. R., and Hsiao, S., 1967, Addiction liability of albino rats: Breeding for quantitative differences in morphine drinking, Science 157:561.

Numan, R., 1979, Multiple exposures to ethanol facilitate intravenous self-administration of ethanol by rats, *Pharmacol. Biochem. Behav.* **10**:767.

Oei, T. P. S., and Singer, G., 1979, The effects of a fixed time schedule and body weight on ethanol self-administration, *Pharmacol. Biochem. Behav.* **10**:767.

Papasava, M., Oei, T. P. S., and Singer, G., 1981, Low dose cocaine self-administration by naive rats: Effects of body weights and a fixed-time one minute food delivery schedule, *Pharmacol. Biochem. Behav.* **15**:485.

Pickens, R., Meisch, R. A., and Thompson, T., 1978, Drug self-administration: An analysis of the reinforcing effects of drugs, in: *Handbook of Psychopharmacology*, Vol. 12 (L. L. Iverson, S. D. Iverson, and S. H. Snyder, eds.), pp. 1–37, Plenum Press, New York.

Pieper, W. A., and Skeen, M. J., 1972, Induction of physical dependence on ethanol in rhesus monkeys using an oral acceptance technique, *Life Sci.* **11**:989.

Pieper, W. A., Skeen, M. J., McClure, H. M., and Bourne, P. G., 1972, The chimpanzee as an animal model for investigating alcoholism, *Science* **176**:71.

Pohorecky, L. A., 1981, Animal analog of alcohol dependence, *Fed. Proc.* **40**:2056.

Poling, A., and Thompson, T., 1977a, Suppression of ethanol-reinforced lever pressing by delaying food availability, *J. Exp. Anal. Behav.* **28**:271.

Poling, A., and Thompson, T., 1977b, Effects of delaying food availability contingent on ethanol-maintained lever pressing, *Psychopharmacology* **51**:289.

Poling, A., and Thompson, T., 1977c, Attenuation of ethanol intake by contingent punishment of food-maintained responding, *Pharmacol. Biochem. Behav.* **7**:393.

Richter, C. P., and Campbell, K., 1940, Alcohol taste thresholds and concentration of solution preferred by rats, *Science* **91**:507.

Roehrs, T. A., and Samson, H. H., 1981, Ethanol reinforced behavior assessed with a concurrent schedule, *Pharmacol. Biochem. Behav.* **15**:539.

Roehrs, T. A., and Samson, H. H., 1982, Relative responding on concurrent schedules: Indexing ethanol's reinforcing efficacy, *Pharmacol. Biochem. Behav.* **16**:393.

Samson, H. H., and Falk, J. L., 1974a, Alteration of fluid preference in ethanol-dependent animals, *J. Pharmacol. Exp. Ther.* **190**:365.

Samson, H. H., and Falk, J. L., 1974b, Schedule-induced ethanol polydipsia: Enhancement by saccharin, *Pharmacol. Biochem. Behav.* **2**:835.

Samson, H. H., Roehrs, T. A., and Tolliver, G. A., 1982, Ethanol reinforced responding in the rat: A concurrent analysis using sucrose as the alternate choice, *Pharmacol. Biochem. Behav.* **17**:333.

Sinclair, J. D., 1971, The alcohol-deprivation effect in monkeys, *Psychon. Sci.* **25**:21.

Sinclair, J. D., 1972, The alcohol-deprivation effect: Influence of various factors, *Q. J. Stud. Alcohol* **33**:769.

Sinclair, J. D., 1979, Alcohol-deprivation effect in rats genetically selected for their ethanol preference, *Pharmacol. Biochem. Behav.* **10**:597.

Sinclair, J. D., and Senter, R. J., 1967, Increased preference for ethanol in rats following alcohol deprivation, *Psychon. Sci.* **8**:11.

Sinclair, J. D., and Senter, R. J., 1968, Development of an alcohol-deprivation effect in rats, *Q. J. Stud. Alcohol* **29**:863.

Sinclair, J. D., Walker, S., and Jordan, W., 1973a, Alcohol intubation and its effect on voluntary consumption by rats, *Q. J. Stud. Alcohol* **34**:726.

Sinclair, J. D., Walker, S., and Jordan, W., 1973b, Behavioral and physiological changes associated with various durations of alcohol deprivation in rats, *Q. J. Stud. Alcohol* **34**:744.

Skinner, B. F., 1959, Animal research in the pharmacotherapy of mental disease, in: *Psychopharmacology: Problems in Evaluation* (J. O. Cole, and R. W. Gerard, eds.), pp. 224–228, NAS-NRC Publication No. 583, National Academy of Sciences, Washington, D.C.

Smith, S. G., and Davis, W. M., 1974, Intravenous alcohol self-administration in the rat, *Pharmacol. Res. Commun.* **6**:397.

Smith, S. G., Werner, T. E., and Davis, W. M., 1975, Technique for intragastric delivery of solutions: Application for self-administration of morphine and alcohol by rats, *Physiol. Psychol.* **3**:220.

Smith, S. G., Werner, T. E., and Davis, W. M., 1976, Comparison between intravenous and intragastric alcohol self-administration, *Physiol. Psychol.* **4**:91.

Sohler, A., Burgio, P., and Pellerin, P., 1969, Changes in drinking behavior in rats in response to large doses of alcohol, *Q. J. Stud. Alcohol* **30**:161.

Spealman, R. D., and Goldberg, S. R., 1978, Drug self-administration by laboratory animals: Control by schedules of reinforcement, *Annu. Rev. Pharmacol. Toxicol.* **18**:313.

Thompson, T., and Schuster, C. R., 1964, Morphine self-administration, food-reinforced, and avoidance behaviors in rhesus monkeys, *Psychopharmacologia* **5**:87.

Walker, D. W., and Freund, G., 1971, Impairment of shuttlebox avoidance learning following prolonged alcohol consumption in rats, *Physiol. Behav.* **7**:773.

Waller, M. B., McBride, W. J., Lumeng, L., and Li, T.-K., 1982, Induction of dependence on ethanol by free-choice drinking in alcohol-preferring rats, *Pharmacol. Biochem. Behav.* **16**:501.

Weeks, J. R., 1962, Experimental morphine addiction: Method for automatic intravenous injections in unrestrained rats, *Science* **138**:143.

Werner, T. E., Smith, S. G., and Davis, W. M., 1977, Intragastric self-administration of alcohol by rats: A dose–effect comparison, *Physiol. Psychol.* **5**:453.

Winger, G. D., and Woods, J. H., 1973, The reinforcing property of ethanol in the rhesus monkey. I. Initiation, maintenance and termination of intravenous ethanol-reinforced responding, *Ann. N.Y. Acad. Sci.* **215**:162.

Woods, J. H., Ikomi, F., and Winger, G., 1971, The reinforcing property of ethanol, in: *Biological Aspects of Alcohol* (M. K. Roach, W. M. McIsaac, and P. J. Creavan, eds.), pp. 371–388, University of Texas Press, Austin, Texas.

Yanagita, T., and Takahashi, S., 1973, Dependence liability of several sedative-hypnotic agents evaluated in monkeys, *J. Pharmacol. Exp. Ther.* **185**:307.

York, J. L., 1981, Consumption of intoxicating beverages by rats and mice exhibiting high and low preferences for ethanol, *Pharmacol. Biochem. Behav.* **15**:207.

Young, A. M., Herling, S., and Woods, J. H., 1981, History of drug exposure as a determinant of drug self-administration, in: *Behavioral Pharmacology of Human Drug Dependence* (T. Thompson and C. E. Johanson, eds.), pp. 75–89, National Institute on Drug Abuse, Research Monograph No. 37, DHHS Publication No. (ADM) 81-1137, Rockville, Md.

Zeiler, M., 1977, Schedules of reinforcement: The controlling variables, in: *Handbook of Operant Behavior* (W. K. Honig and J. E. R. Staddon, eds.), pp. 201–232, Prentice-Hall, Englewood Cliffs, New Jersey.

3

Genetics of Alcoholism and Alcohol Actions

RICHARD A. DEITRICH and KAREN SPUHLER

1. INTRODUCTION

In this chapter we will first review the more recent data from twin, half-sibling, and adoption studies which indicate that humans inherit a risk for alcoholism. We will then consider the genetics of alcohol metabolism in humans.

From these human studies we will move to the large literature on the inheritance of alcohol-related behaviors in animals. In making the leap from humans to animals, researchers assume that behavioral reactions to alcohol in humans can be modeled in animals. There are a number of goals in animal studies. The most obvious long-term goal is to determine what factors are inherited that makes an individual at risk for developing alcoholism. At best, studies with this goal in mind will point the way to research that must eventually be done in humans. The more immediately achievable goal, however, is to use genetic differences in behavioral, physiological, and biochemical reactions to alcohol to study the molecular mechanism of action of alcohol. We will divide the animal genetic studies into those of alcohol preference, acute effects of alcohol, and finally, tolerance and physical dependence. Presumably, such a progression occurs in humans who become alcoholic. Such a progression is studied in a series of separate animal models whereas all elements can and do occur sequentially in a single human.

RICHARD A. DEITRICH and KAREN SPUHLER ● Department of Pharmacology and Alcohol Research Center, University of Colorado School of Medicine, Denver, Colorado.

2. HUMAN STUDIES ON ALCOHOLISM

The first step in any study of human alcoholism is to define the term or at least clearly set out the criteria by which a person may be classified as an alcoholic. In most human studies use is first made of objective criteria, such as liver cirrhosis, blackouts, withdrawal symptoms, or other alcohol-related physical disorders. For a person to manifest such obvious symptoms of excessive alcohol intake, abusive drinking must have occurred for several years. Other, more subtle criteria of alcohol abuse, such as marital, employment, or legal problems, are largely social in nature and vary widely from culture to culture. For example, serious legal problems in Saudi Arabia can result from a level of alcohol consumption that would be considered near abstinence in a Western country. Obviously one cannot "acquire" this disease unless one is exposed to the agent, irrespective of genetic transmission.

Genetic Risk Factors

While a genetic basis for alcoholism has been suspected for more than a hundred years (see Goodwin, 1971), serious, systematic studies have been carried out from the 1950s to the present (see also Omenn, 1975; Oakeshott and Gibson, 1981; Goodwin, 1980; Martin, 1981). These studies are of family members of alcoholic probands, monozygotic and dizygotic twins, half-siblings, and finally, adoptive families. The power of the methods increase roughly in the order given. Two approaches may be discerned. One method involves identification of alcoholic probands and study of their relatives. The papers by Winokur et al. (1970) and Amark (1951) are prototypes of this style. The other approach is to sample a large number of individuals randomly in a designated population without regard to the presence or absence of alcoholism and investigate the prevalence of alcohol abuse and related traits among individuals and their relatives. This method is used with twins, half-sibs, and adoptees.

Several large and comprehensive studies were carried out and published as monographs in Scandinavian journals. These studies did not receive the notice they deserved and only in recent years have larger numbers of researchers, medical professionals, and the lay public in the United States taken note of the genetic aspects of alcoholism. Still, to date, the more comprehensive studies have been carried out in Scandinavia. For example, Sweden has a system of recordkeeping for alcohol intake by the individual citizen by Temperance Boards, National Health Insurance, hospitals, and courts. Also, it is much easier to trace adoptees in Scandinavia than in the United States, and there are large twin registries that can be used in sampling.

A number of papers could be cited as supporting a genetic influence on alcoholism, however, we shall confine our discussion to a few of the larger studies and, in particular, to those in which environmental influences have been assessed and more specific risk factors hypothesized.

Family Studies.   The design of family studies has generally not allowed the separation of environmental and genetic effects. Taken with other evidence, however, they do add weight to the genetic hypothesis. The study by Amark (1951), using 644 alcoholic probands, concluded that their male first-degree relatives were at significantly greater risk than the general population. Winokur (1970) came to a similar conclusion studying both male and female probands. An interesting study by Kaij and Dock (1975) explored the popular idea that alcoholism in men skips a generation and that the sons of daughters of alcoholic men would be at greater risk than the general population, since an alcoholic man would transfer his X-linked gene(s) to his grandsons via his daughter. However, they could find no evidence for sex-linked inheritance. Another situation resulting in generation-skipping is that a son of an alcoholic who is adversely conditioned towards drinking because of the negative home environment refrains from alcohol abuse. The grandchildren may have an increased genetic risk, but lack the aversive conditioning and therefore would be more likely to become alcoholic. Apparently, controlled studies have not been carried out, although there is some evidence that extremes of alcohol consumption are not imitated by offspring (Harburg et al., 1982).

Twin Studies.   One classic technique for studying genetic risk is to compare monozygotic (MZ) and dizygotic (DZ) twins. Any observed difference between monozygotic twins is presumed to be due to environmental effects, while a difference between dizygotic twins is presumed to be due to both environment and genetic effects. From the studies shown in Table 1, the most striking conclusion is that drinking behavior, whether it is high or low, is heritable, but alcoholism *per se* is not. The studies vary widely in criteria for classifying phenotypes. Some (Loehlin, 1972; Jonsson and Nilsson, 1968) are based only on self reports, while others (e.g., Partanen et al., 1966) are based on extensive interviews. The implicit assumption is that the environment for identical twin pairs is no more alike than that for fraternal twin pairs, but this is arguable. Greater similarity of MZ twin-pairs' common environment than DZ twin-pairs' common environment inflates the estimate of genetic contribution to the trait. The study by Kaij (1960) limited the definition of alcoholism to a pathologic desire for alcohol, physical dependence on alcohol, and blackouts. He found that 54.2% of MZ probands had a concordant twin whereas the concordance rate for DZ twins was 31.5%. The greater concordance rate for MZ twins, compared to DZ twins, suggests an inherited component for alcoholism.

Half-Sib Studies.   This technique has been used less extensively, but is a more powerful method for assessing genetic risks because it controls common environmental effects. The one major study, that of Schuckit et al. (1972), which preceded a number of adoption studies, concluded that the risk for alcoholism was best predicted by the presence of an alcoholic biologic parent.

Table 1. Family Studies

Study	Sex	Alcoholics	Nonalcoholics	Major conclusions
Amark, 1951	Males	644	704	Brothers and father of alcoholic probands are at greater risk for alcoholism
Winokur, 1970	Both	259	507	Alcoholism in brothers of male and female alcoholic probands is 46–50%
Kaij and Dock, 1975	Grandsons	136	75	No sex-linked factors
Propping, 1978	Both	1,845		Summary of four older studies where 24% of first-degree relatives of alcoholics were also alcoholics
Cotton, 1979	Both	6,251	4,083	Summary of 39 studies showed that alcoholism in relatives of alcoholics is 46–50%; Substantially higher than in relatives of nonalcoholics

Twins

Study	Sex	No. Pairs		Major conclusions
Kaij, 1960	Males	174		54.2% of MZ twins were both alcoholic, 31.5% of DZ twins were. ". . . Drinking habits are influenced by genetic factors and . . . greatly determine . . . chronic alcoholism"
Partanen et al., 1966	Males	902		Abstaining, normal use and heavy use more heritable; "alcoholism" was not
Jonsson and Nilsson, 1968	Both	1,500		". . . genetic factors are relevant for the consumption or non-consumption . . . of large alcohol quantities"
Loehlin, 1972	Both	850		Drinking behavior heritable (h^2 from 0.16 to 0.62 on various measures)
Kaprio et al., 1978	Both	11,500		Heritability for intake; males: 0.37, females: 0.25

Half-Sib

		No. subjects		
		Probands	Half-sibs	
Schuckit et al., 1972	Both	69	164	Biological alcoholic parent was best predictor

Adoptees

		No. subjects		
		Alcoholic	Control	
Goodwin et al., 1973	Males	55	78	Risk for sons of alcoholics, three to four times higher
Goodwin et al., 1974 (siblings of adoptees)	Males	35	50	Same alcoholism in nonadopted siblings as in adopted group
Goodwin et al., 1977	Females	49	81	No inheritance found
Cadoret and Gath, 1978	Both	84	Matched	33% of adoptees with alcoholic biological parent were alcoholic; 1.39% of adoptees of nonalcoholic biological parent were alcoholic
Cadoret et al., 1980	Males	23	69	Significant association between adoptee alcoholism and alcoholic biological background
Cloninger et al., 1981	Males	151	711	Three types of alcoholism. One type highly heritable from father to son
Bohman et al., 1981	Females	31	882	Threefold increased risk for daughters of alcoholic mothers

Adoption Studies.   The most-quoted study is that of Goodwin et al. (1973). This was carried out in Denmark and included adoptees of 55 alcoholics and 78 control parents (mostly fathers). Nine of 55 individuals in the alcoholic group were actually treated for alcoholism, and only 1 of 78 controls had been treated. The analyses indicated that there was a three- to fourfold increased rate of alcoholism in the sons of alcoholic biological fathers over the control adoptee population. However, with a sample of 49 female adoptees of alcoholic biological parents, Goodwin et al. (1977) could not find any increased frequency of al-coholism over that of control adoptees. One rather startling conclusion was that being raised in a home with an alcoholic parent either had no effect or was somewhat protective. Further information on this point was also obtained from studies of the nonadopted sons of the alcoholic parents. There was no significant difference in the frequency of alcohol-related problems between the nonadopted sons and adopted sons, again pointing to the overriding effect of biologic in-heritance (Goodwin et al., 1974). Cloninger et al. (1981) came to a similar conclusion with a much larger sample of alcoholics. A study carried out in the United States by Cadoret and Gath (1978) came to similar conclusions with both male and female adopted offspring of 89 alcoholics.

Perhaps the most complete adoption studies were recently published in two parts by Cloninger et al. (1981) for adopted men and Bohman et al. (1981) for adopted women. The study was conducted on all persons born out of wedlock from 1930 through 1949 in Stockholm, Sweden, and subsequently adopted by nonrelatives before the age of 3 years (average age at placement was 8 months). The study is by far the largest adoption study carried out in the area of alcoholism. There were 862 men and 913 women who met all the criteria for inclusion as adoptees in the study. The statistical technique used for classification of at risk subjects was cross-fostering, whereby one can separately identify four groups: (1) those at risk due to the adoptive environment, (2) those at risk due to genetic inheritance, (3) those at risk due to both environment and genetics, and (4) those at risk for neither genetic nor environmental influence.

Because of the large number of male adoptees available for study, and consequently the relatively large number of those with alcohol problems, Clon-inger et al. (1981) were able to divide their sample according to severity of alcoholism. The mild-type category included those individuals who had one registration with the Temperance Board but who were never treated for alco-holism. The fathers of these individuals had higher occupational status and less criminality than the fathers of other abusers. The mothers of mild abusers were also found more likely to be alcohol abusers than were the mothers of other types of abusers. As can be seen in Fig. 1, there was a significant contribution from both genetic and environmental factors for this type of abuse. The interaction was not greater than expected from addition of the two factors, however. Of the 77 individuals at genetic risk for this type of alcoholism 10.4% or 8 males were alcoholics; of the 168 at environmental risk, 12 (7.1%) were abusers, whereas of the 30 individuals at risk for both genetic and environmental factors 26.7%,

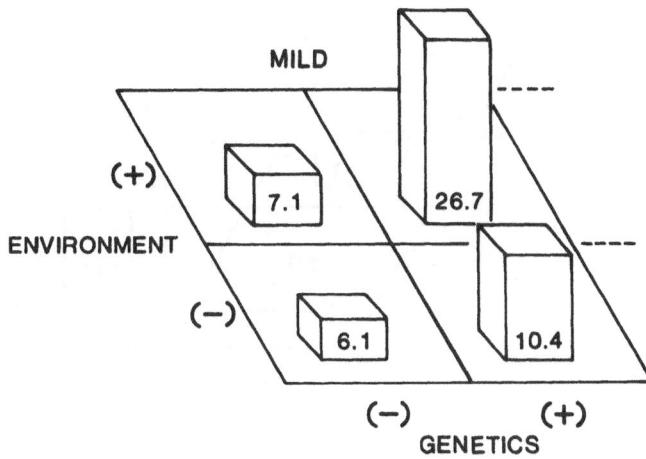

Figure 1. Percentage of men who develop mild alcoholism, of those at risk in each of four categories. (Drawn from data in Cloninger et al., 1981).

or 8 individuals, were mild alcohol abusers. The spontaneous rate (neither genetic nor environmental predisposition) of alcohol abuse in this category was significantly prevalent (36 of 587, 6.1% at risk). The upper-limit estimate of heritability of liability to mild alcohol abuse was 38% ± 15% (S.E.). Heritability is defined as the proportion of total phenotypic variance due to genetic effects. It gives the relative importance of genetic transmission in determining individual differences.

A strange finding was that severe alcohol abuse followed a similar pattern except that fathers of severe abusers had lower occupational status. Severe abusers were those who had four or more registrations with the Temperance Board and either voluntary or compulsory treatment with a diagnosis of alcoholism.

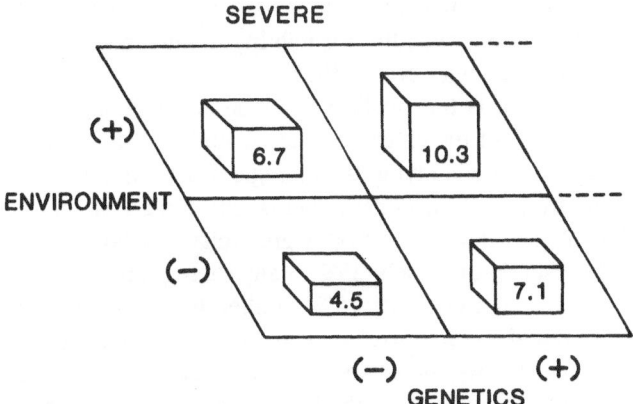

Figure 2. Percentage of men who develop severe alcoholism, of those at risk in each of four categories. (Drawn from data in Cloninger et al., 1981).

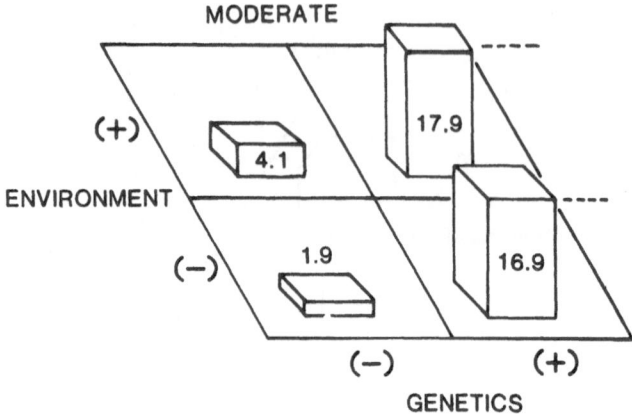

Figure 3. Percentage of men who develop moderate alcoholism, of those at risk in each of four categories. (Drawn from data in Cloninger et al., 1981).

As seen in Fig. 2, spontaneous severe alcoholism was a significant (4.5%) risk (21 of 465 persons), whereas neither environmental risk (6 of 90, 6.7%), genetic risk (17 of 239, 7.1%), nor the combined risk (7 of 68, 10.3%) produced a significant effect. The upper-limit heritability estimate was 25% ± 20% (S.E.). The mild and severe types are called "milieu-limited" by Cloninger et al. (1981).

Moderate abusers, however, present quite a different picture. They are defined as having two or more registrations with the Temperance Board but no treatment for alcoholism. Their fathers had low occupational status, criminality, and recurrent alcohol abuse. As seen in Fig. 3, spontaneous alcoholism (11 of 567, 1.9% at risk) was significant, but the most statistically significant and striking effect was the genetic risk: 12 of 71 (16.9%) were moderate abusers, a ninefold increased risk. The environmental risk alone was 4.1%. Almost no further risk could be attributed to both genetic and environmental contributions (5 of 28, 17.9%). The heritability for liability to moderate abuse was estimated to be about 90% (Cloninger et al., 1981).

While this group is the smallest (36 men accounting for only 24% of the abusers and only 4% of the entire population studied), it is obviously the group of most interest to those studying the biologic causes of alcoholism. Studies by Schuckit et al. are discussed in detail below but in these studies "at risk" males are defined as those who have a first-degree relative who is an alcohol abuser. Given that at most approximately 25% of any such group would have the highly heritable risk factor and the other 75% a milieu-limited type, such a population would have a great deal of "noise" in it. It would seem more profitable to select only those at risk for the moderate, highly heritable, form of alcoholism for initial studies and thereby reduce the heterogeneity for risk factors.

The other problem these results pose is that, even under the worst conditions of both increased genetic and environmental risk, only 26.7% of those at risk actually become alcoholic (for mild abuse). Perhaps a better question to ask is why so many who are at risk are protected from alcoholism rather than why the relatively few are subject to the disease.

Studies on 913 women were also completed by these researchers but published separately (Bohman et al., 1981). In contrast to men, where the overall prevalence for alcohol abuse was 17.5%, the frequency of alcoholism in women was only 3.4%. Thus, separate categories could not be distinguished. In Fig. 4 the genetic (or congenital as the authors prefer) contribution is now from the mother, rather than from the father. Genetic factors account for a fourfold increase in alcohol abuse ($p < 0.05$). The characteristics in parents that predispose their daughters to alcoholism included alcohol abuse, criminality, and low occupational status. Alcoholism alone in the mother was a poor predictor (only 5 of the 31 alcoholic women had alcoholic mothers). Thus, the situation with women is much less clear than with men.

Risk for alcoholism is a complex trait; both genetic and environmental factors play a role in its onset. However, research oriented towards major gene effects would be profitable. Perhaps one should be looking for protective factors present in the nonalcoholics but missing in the alcohol abusers, rather than for predisposing factors in the alcoholics.

Physiological Findings

Given at least the prerequisite that family history changes the likelihood of risk for alcoholism, several studies have been initiated with family-history-positive (FHP) and family-history-negative (FHN) individuals by Schuckit and his colleagues. Relatives are studied rather than the alcoholic proband because of the possibility that alcohol consumption *per se* may alter the system under investigation. Study of the alcoholic himself or herself is nearly fruitless if one is concerned with antecedent risk factors.

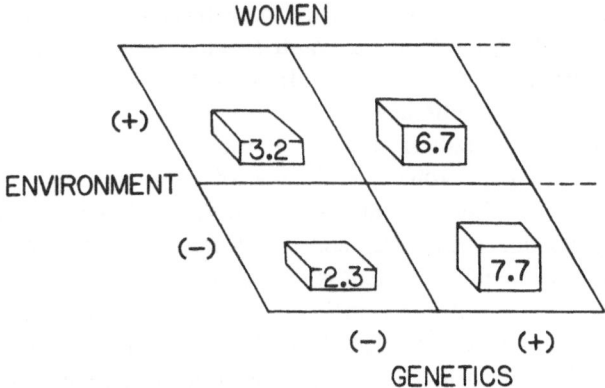

Figure 4. Percentage of women who develop alcoholism, of those at risk in each of four categories. (Drawn from data in Bohman et al., 1981).

In characterizing FHP individuals, a number of studies have been carried out. The first report was the finding that blood acetaldehyde levels were higher in FHP than FHN male subjects following a dose of ethanol (0.5 ml/kg) (Schuckit and Rayses, 1979). This study reported results similar to those of Truitt (1971) and Korsten et al. (1975) for alcoholic subjects. The studies have been criticized because of problems in the determination of blood acetaldehyde levels (Eriksson et al., 1982b; Eriksson, 1980). Eriksson and Peachy (1980) could not replicate these earlier findings. In fact the opinion of some workers in the field is that acetaldehyde is not detectable in the peripheral blood of normal human subjects following alcohol ingestion (Lindros and Eriksson, 1981). The problem of artifactual formation of acetaldehyde from ethanol during the assay procedure complicated all work with blood acetaldehyde. It would seem possible that the reason for the difference in blood acetaldehyde resided not in real *in vivo* differences of acetaldehyde metabolism but in genetic control of the mechanisms controlling artifactual formation of acetaldehyde. However, this appeared not to be the case, since in a study of blood from alcoholics or controls to which ethanol was added, there was no difference in artifactual acetaldehyde formation (Eriksson and Peachey, 1980; Eriksson, personal communication).

Further studies with FHP and FHN men are proceeding. So far it has been found that alcohol in FHP men (at 0.75 ml/kg of ethanol) causes a reduction in muscle tension while in FHN individuals no change in muscle tension was observed (Schuckit et al., 1981).

In another study Schuckit (1980) indicated that FHP subjects rated themselves less intoxicated than FHN subjects following a dose of 0.75 ml/kg ethanol. Blood alcohol levels (BALs) were measured every 15 min, and except at 60 min, when the level was higher in the FHN subjects, there were no differences in the BALs. There also was no difference in time to peak blood level or magnitude of that peak (Schuckit, 1981). Thus, three studies from Schuckit's laboratory all show a difference between FHP and FHN men in their responses to ethanol. Since one would predict, on the basis of the results from Cloninger et al. (1981), that no more than 25% of the FHP individuals would have highly heritable risk factors, the results are quite striking. However, Utne et al. (1977) had shown earlier that there was no difference between offspring of alcoholics and controls in alcohol elimination rates.

Two problems in experimental design should be pointed out in the Schuckit studies. The avoidable problem is the one of a placebo taste control. This is difficult to achieve when ethanol is given orally, but has been carried out successfully by floating a small amount of ethanol on the top of the fluid. The second, and unavoidable, problem is that while the FHP men were matched to FHN men on a variety of items, including drinking history, we have a paradox. The investigators hypothesize that alcoholism risk factors are inherited and can be identified by comparing these 2 groups. But most subjects drank, even if moderately, so that their varied response to subsequent alcohol doses might be

altered because of genetic make-up. In other words, if one proposes that physiological, biochemical, and psychological risk factors can be inherited, one is
forced to conclude that genetics may play a role in the development of behavioral
and physical tolerance to ethanol as well. Thus, while the overt pathology of
alcoholics has been avoided in these studies, the subjects have been exposed
previously to alcohol.

Other studies have involved the comparative physiology of identical and
fraternal twins. Perhaps the most extensive studies have been those of Propping
and his colleagues. While they have concentrated on the electroencephalogram
(EEG) to measure alcohol effects, they have also studied absorption, distribution,
and metabolism. It has been known for quite some time that the EEGs of MZ
twins are identical; it is not surprising that the effect of alcohol on the EEG of
MZ twins is the same while the effect of alcohol on the EEG of dizygotic twins
is dissimilar (Propping, 1977a, 1978).

Since the differential effect on DZ twins might have been due to a genetic
effect on absorption, degradation, and elimination, Kopun and Propping (1977)
studied these parameters in their subjects. They found that while there was genetic
variation in the kinetics, it did not account for the differential effect on the EEG.

Pollock et al. (1983) and Gabrielli et al. (1982) have studied the EEG in
high-risk (HR) subjects (sons of alcoholics) and controls. They found that alcohol
given to HR subjects caused greater increases of slow alpha activity and greater
decreases of fast alpha activity than in controls. Somewhat in accord with Schuckit's findings (1980), the HR subjects did not feel as intoxicated as controls.
Lipscomb and Nathan (1980) have obtained data showing that alcoholics are
less able to estimate blood alcohol levels than nonalcoholics after training. They
did not confirm that relatives of alcoholics shared this ability, however. Rather,
they found that individuals with low tolerance for alcohol, as measured by body
sway, were more accurate in estimating blood alcohol concentration than highly
tolerant individuals.

The EEG is a relatively crude way to eavesdrop on neuronal communication
and might be likened to monitoring all the telephone communication in the United
States simultaneously. While such a system certainly can detect patterns and
pathology, it is limited in its ability to pinpoint abnormalities more closely. The
technique of evoked potentials is a further refinement that allows one to monitor
one large sensory neuronal circuit, perhaps analogous to being able to narrow
down the telephone eavesdropping to a large city rather than the whole country.
While extensive use of this technique has not yet been made in studying the
genetics of alcohol effects, several groups have begun promising studies.

At a different level, numerous psychological tests have been shown to be
affected by alcohol. However, Propping (1977b) found no significant genetic
variation in twins for the effect of alcohol on tapping speed, simple reaction
time, discrimination reaction time, pursuit rotor, two-hand coordination, or results of tremometer test.

Metabolism

Rates of Alcohol Elimination.   The usual pharmacogenetic approach to drug metabolism is to compare MZ and DZ twins. This has been done by several individuals starting with Luth (1939), who studied 10 MZ and 10 DZ twins including a younger sample of 17- and 18-year-old pairs. Such a young sample would probably not be used now. The paper is useful since the data are presented in some detail. This allowed Kopun and Propping (1977) to calculate the heritability for the rate of ethanol elimination (β_{60}, in mg/ml per hr) as 0.63 and for ethanol degradation rate (EDR, in mg/kg per hr) as 0.67, which agree very closely.

Kopun and Propping (1977) also investigated 19 MZ and 21 DZ twin pairs for β_{60}, EDR, and absorption rate. They found heritabilities of 0.46, 0.41, and 0.57, respectively. These heritabilities are much lower than that calculated by Vesell et al. (1971) who found a heritability of 0.98 for β_{60}. Most authors now believe that this last estimate is impossibly high (see Kopun and Propping, 1977). However, both Kopun and Propping (1977) and Vesell et al. (1971) used the following equation for heritability:

$$h^2 = \frac{Vw(F) - Vw(I)}{Vw(F)}$$

where h^2 is the heritability, $Vw(F)$ is the within-twin-pair variance for fraternal twins, and $Vw(I)$ is the within-twin-pair variance for identical twin pairs. A more appropriate estimate of heritability is given by $h^2 = 2(t_{mz} - t_{dz})$, where t_{mz} is the intraclass correlation for monozygotic twin pairs and t_{dz} is the intraclass correlation for dizygotic twin pairs (see McClearn and DeFries, 1973). If one applies this equation to the data of Kopun and Propping (1977) for β_{60}, the estimate of h^2 is 0.76, for EDR it is 0.96, and for absorption rate it is 0.58. Likewise, using the data of Vesell et al. (1971) one obtains an estimate for h^2 of 0.68 instead of 0.98 for β_{60}. In this regard the estimates of h^2 for β_{60} fall in a moderately high range around 0.70 for both data sources, indicating a substantial genetic contribution to phenotypic variation. The estimate for EDR is very high. A possible complication in the data of Vesell et al. (1971) also arises in that the same blood ethanol levels were not achieved in all the dizygotic twin pairs. If first-order ethanol elimination rates were applicable, the calculated rates would be in error due to the different levels.

Ethnic Differences in Alcohol Metabolism.   In addition to individual differences, genetic variation in alcohol metabolism has been studied at the level of racial differences. The major groups studied have been North American Indians, Orientals, and whites. Schaefer (1978) presents a summary of the findings in the literature as well as his own data on south Indian Reddis. Most of the results favor the conclusion that Orientals metabolize ethanol faster than either

American Indians or whites. Comparisons of metabolic rates of ethanol in American Indians and whites show that the Indians metabolized ethanol more quickly in 5 of 20 comparisons, more slowly in 8 of 20 comparisons, and at the same rate in the remaining 7 comparisons. Given the great individual variability in ethanol metabolic rates, a large sample of individuals should be tested under identical conditions so that valid comparisons could be made. The largest number tested was by Reed et al. (1976): 58 whites, 20 Orientals, and 24 Indians. Even this may be too few subjects to allow extrapolation to entire racial populations.

Enzymes of Alcohol Metabolism.   The discovery of "atypical" alcohol dehydrogenase in human liver by von Wartburg et al. (1965) nearly 20 years ago and the subsequent discovery of a high proportion of the "atypical" enzyme among the Japanese (Fukui and Wakasugi, 1971) gave rise to the hope that at last a biochemical wedge had been found which would eventually provide an explanation for human alcoholism. While that optimistic prediction has not come to pass, similar work with human liver aldehyde dehydrogenase seems to provide an explanation for why some individuals are protected from alcoholism rather than why some become alcoholic.

Alcohol Dehydrogenase.   Alcohol dehydrogenase (ADH) in human liver contains two subunits, consisting of the possible monomers α, β_1, β_2, γ_1, and γ_2 (Smith et al., 1971). In addition, a newly discovered enzyme, the π enzyme also exists (Li et al., 1977; Bosron et al., 1979) that may account for as much as 40% of ADH activity at 60 mM ethanol. The monomers α, β, and γ are coded for by three separate autosomal loci: ADH_1, ADH_2, and ADH_3. Allelic variants are present at both the ADH_2 and ADH_3 loci. At ADH_2, the β_1 allele is at a high frequency in whites (Smith et al., 1971) and the β_2 allele is more frequent in Oriental populations (Fukui and Wakasugi, 1972). The alleles, γ_1 and γ_2 are present each at intermediate frequencies at the ADH_3 locus in whites (Harris et al., 1973). With 5 subunits the possible combinations to make up the active dimer is 15. Many of these can be distinguished by starch gel electrophoresis and isoelectric focusing (Harada et al., 1978) but some may not exist *in vivo.* The "atypical" or Oriental form of ADH in the homozygous condition would be $\beta_2\beta_2$.

The activity of typical ADH in liver as measured *in vitro* is about 2 IU/g liver, while that of the atypical ADH is about 8 IU/g liver (von Wartburg, 1980). Whether the presence of the atypical enzyme in Orientals is responsible for the more rapid rate of metabolism as discussed earlier, is open to question. Studies by Edwards and Evans (1967) of two individuals known to have the atypical enzyme and by von Wartburg and Schurch (1968) of one individual revealed that while the rate of alcohol metabolism was somewhat higher than normal, it was not strikingly so. Much has been made of these two studies, but given the very small number of individuals tested (three), more work is necessary before decisive conclusions can be proposed about the role of ADH isozymes in drinking behavior.

Aldehyde Dehydrogenase and Facial Flushing.   The situation with aldehyde dehydrogenase (ALDH) is complicated in animals by the presence of the
ALDH enzymes in nearly all tissues and in virtually every cell fraction. Beyond
that, both the mitochondria and cytosol have several enzymes of widely different
Km values and some cytosolic enzymes are inducible (see review by Li, 1977).

Human enzymes may prove to be as complicated, but the major interest
from a genetic point of view is the discovery in population studies that about
half of the Oriental individuals, but none of the Germans tested, lacked a low-
Km ALDH enzyme (Agarwal et al., 1981; Harada et al., 1980a,b). The lack of
the low-Km ALDH was discovered by using liver from autopsies. Analysis of
genetic variants of both the ADH and ALDH loci will be much easier since the
discovery that the enzymes are present in hair roots and can be analyzed by thin-
layer isoelectric focusing. Recently, Goedde et al. (1982) have reported that
some individuals among Japanese, Chinese, Vietnamese, Indonesian, Korean,
and Equadorean Indian populations are missing the low-Km ALDH enzyme
while few Europeans, Egyptians, Sudanese, or Liberians are missing the enzyme.
It may also be possible to use red blood cell ALDH activity as a marker of this
condition (Inoue et al., 1980).

Facial flushing following alcohol ingestion was found only in those individuals who were missing the low-Km ALDH. Flushing was not found in those
individuals who were homozygous for atypical ADH but who had the low-Km
ALDH (Agarwal et al., 1981). Flushers have higher blood acetaldehyde levels
and higher urinary excretion of epinephrine and norepinephrine (Mizoi et al.,
1979), as well as metabolites of these catecholamines (Mizoi et al., 1980).

This finding brings into much clearer focus a relatively extensive literature
on facial flushing following alcohol ingestion (Wolff, 1972, 1973; Ewing et al.,
1974; Mizoi et al., 1979, 1980). Reed et al. (1976) have reported higher blood
acetaldehyde levels in Orientals than whites without studying the flushing response. The consensus seems to be that individuals lacking the low-Km ALDH
will have higher blood levels of acetaldehyde with consequent flushing, increased
heart rate, and increased pulse pressure following ingestion of alcohol.

The mechanisms by which these symptoms are produced remains to be
elucidated. However, studies have investigated the possible involvement of catecholamine release without conclusive results (Mizoi et al., 1980). Several reports
have noted that flushing can be prevented with antihistamine compounds. Tan
et al. (1982) reported that a combination of H_1 and H_2 histamine receptor antagonists would block the reaction.

Perhaps the most significant finding to come from these studies is the report
that among Japanese alcoholics with cirrhosis, only 2% are flushers (Harada et
al., 1982). Thus, the paradoxical situation arises where the lack of an enzyme
apparently indirectly protects one from alcoholic cirrhosis by decreasing alcohol
intake, whereas one would expect a greater than average alcoholic damage should
alcohol be ingested.

The genetics of the missing ALDH have not been elucidated but Schwitters et al. (1982) presented data which indicated that flushing is genetically dominant. Data from one Japanese pedigree does not agree, however (Agarwal, 1980). It also is not certain if an inactive enzyme is produced (Teng, 1981).

A number of other markers of alcoholism have been studied. Plasma, especially platelet monoamine oxidase (MAO), is one such marker that has been studied in a variety of conditions. Major and Murphy (1978) obtained evidence that first-degree relatives of low-MAO alcoholic and control individuals had a higher incidence of alcoholism than the first-degree relatives of high-MAO alcoholic and control individuals. Schuckit et al. (1982) measured platelet MAO in FHP and FHN subjects. They were able to identify correctly only 8 of 15 FHP subjects and 12 of 15 FHN subjects on this basis.

Genetics of Tissue Damage due to Ethanol

It has been suspected from clinical observations for quite some time that the risk for alcoholic cirrhosis has a genetic component. Only recently, however, has this notion received solid empirical support. Hrubec and Omenn (1981) using data from 15,924 male twin pairs in the National Academy of Sciences, National Research Council Twin Registry, found a twin concordance rate for alcoholic cirrhosis of 14.6% for MZ twins and 5.4% for DZ twins. The biochemical mechanism underlying the finding is unknown although the possibility of an autoimmune disease is attractive.

At a more biochemical level, an attempt has been made to explain the occurrence of Wernicke–Korsakoff's syndrome from a genetic basis (Blass and Gibson, 1977). These investigators studied the Km value for thiamine pyrophosphate as a cofactor of transketolase in fibroblasts from four patients with Wernicke–Korsakoff's syndrome. They found that the apparent Km value was 40 μM for the enzyme from the affected patients and 16 μM in six control lines. This could mean that, in the face of decreased thiamine levels due to a poor diet, these patients had a more severe deficiency in the enzyme activity than unaffected individuals. These results have not received independent confirmation to date. The most convincing evidence would be to demonstrate such an increased Km value in the nonalcoholic relatives of alcoholics with Wernicke-Korsakoff's syndrome.

Conclusions

There is little room for doubt concerning the influence of genetics on the risk for alcoholism in humans. However, there are a great number of questions concerning interactions of genes and environment and the relationship of the inheritance of drinking behavior and alcoholism.

At the biochemical level of investigating alcohol metabolizing enzymes, researchers have had more success explaining how some genotypes are protected from alcoholism rather than how other genotypes have a propensity toward excessive alcohol intake. At the physiological level, the study of evoked potentials in offspring of alcoholics is a potentially powerful tool in understanding neuronal effects of ethanol.

3. ANIMAL STUDIES

There are no complete models of human alcoholism. One can take that as a discouraging pronouncement or, as the animal researchers have, as an opportunity to divide the components of human alcoholism into numerous animal models each reproducing only one aspect of the human condition. Models have been developed which at least superficially resemble the progress of alcohol abuse in humans. There are numerous studies of genetic variation for animals' preference for alcohol. The implicit assumption is that an animal's choice to drink alcohol rather than water would simulate the human drive to consume alcohol. This idea has been vigorously disputed, however (Cicero, 1980). The next step in the progress of alcoholism is the acute effect of the drug. This aspect has been studied in great detail using genetics as a tool. The last step to be approached genetically is the chronic effect of ethanol in producing tolerance and physical dependence.

While techniques are available to produce tissue damage in animals chronically exposed to alcohol, the use of genetics in this area has not been extensive.

Inbred Strain Differences

Animals are considered to be inbred after 20 successive generations of full brother–sister mating. At this point the inbreeding coefficient approaches 99%, which indicates that animals are nearly identical homozygotes (Falconer, 1960). Thus, except for sex differences they are essentially identical twins. Any differences in response between animals of a given strain is due to environmental factors.

Advantages.   Such animals provide a very stable background upon which to impose some drug or other environmental perturbation. The signal-to-noise ratio is usually better and the variance related to sampling error is smaller. Replications can be carried out straightforwardly, since all the animals within a strain are identical. With the exception of genetic differences introduced into the breeding stock due to mutation, the animals can be considered stable in their response over time and location. In the presence of genetic drift, in conjunction with mutation and possible environmental modification, that could change selection pressure in the laboratory, substrain differences can occur over a number of generations when a secondary breeding stock is initiated from a primary source of animals. However, phenotypic responses of inbred strains are remarkably reproducible.

One of the most important uses of inbred strains in this field is in establishing the correlation between two behavioral responses or between a behavioral response to alcohol and underlying physiological or biochemical mechanisms. In order to do this with reasonable confidence a study should use a minimum of three strains that differ in the behavioral response. Even then the correlation must be quite high to achieve significance ($r = 0.997$ for a probability of rejection of the null hypothesis at the 95% level). For as many as eight strains, a correlation of 0.707 is required for a p value of 0.05. Ideally, a sufficient number of inbred strains should be tested to sample uniformly the range of the total phenotypic response distribution.

Disadvantages. The major disadvantage in the use of inbred strains is that, except for mice, not a large number of strains of other species are available for study. While a great deal of valuable information has been obtained using inbred strains of mice, the possibility that a finding is species-specific always remains. The need to use multiple strains to test hypotheses about phenotypic associations can present problems in either the purchase and/or housing of the animals, as well as the concurrent testing of animals.

Heterogeneous Animal Stocks

Heterogeneous animals are systematically outbred in order to introduce a large amount of genetic variation into the population. The larger the number of mating pairs per generation, the greater the level of heterozygosity. In practice, for research laboratories with adequate housing facilities, a stock containing 40–60 breeding pairs, is a reasonable size so that first-cousin matings can be avoided. Thus, individual animals are genetically unique, with no two alike at all gene loci.

Heterogeneous stocks should be reproducible, like any other scientific tool. This can be accomplished by deriving the stock from the intercrossing of a number of inbred strains. Since the inbred strains are relatively stable, the heterogeneous stock can be reproduced at any time. Most commercial animal suppliers do have outbred animal stocks. Generally, these are not systematically derived and maintained as described above. Often animals of the same outbred strain purchased from two different suppliers will react differently, which scientists would not tolerate when purchasing reagent-grade chemicals from two different suppliers. The problem lies in the fact that there is not a defined "reagent-grade" stock. However, there needs to be greater attention directed towards the genotype of animals used in pharmacologic research.

Advantages. These animals are useful for hypothesis-generating experiments and for correlations of behavioral traits with biochemical or physiological traits. The more precise the measures, the lower the variance due to experimental error, and the fewer the number of animals required for testing. Since each animal is genetically different, each provides a single data point in a continuous distribution composed of genetic and environmental differences. With inbred strains each strain provides a single data point as a mean strain response, which is restricted to only a part of the range of the theoretical genotypic distribution. In contrast, the heterogeneous stock (HS) represents a broader cross-section of the species–response distribution, and measures of behavioral and correlated traits in HS should provide a picture of this response.

Disadvantages.   Unless the measures are precise, large numbers of animals will be required to sample the response distribution adequately. For multiple traits it is necessary to measure both phenotypes in a single animal. If two or more behavioral tasks are to be correlated, a Latin square design could be used to vary the order in which the tests are applied. This is not possible when a test requires the sacrifice of the animal. In this case one is faced with the question of whether the exposure of the animal to the drug or behavioral testing procedure alters the subsequent measure of the physiological or biochemical trait. Obtaining the physiological or biochemical measure on animals not previously exposed provides some of the data needed for a complete analysis of variance. Moreover, since each animal is genetically different, behavioral and biochemical measures cannot be carried out in different animals.

Selection Studies

The process of bidirectional selection for a behavioral or a related trait requires that one start with a base population in which there is sufficient genetic variation available for selection to act upon. Thus, in contrast to other research, one requires large variances for the drug response. If there are substantial and consistent differences in the phenotypes of interest when they are assessed in the inbred strains used to derive the heterogeneous stock, the likelihood of a selection study being successful is relatively high.

Selection for a given trait is started by testing a designated number of animals from the base population. Breeding pairs are chosen at random from the population to establish control lines. The highest responding animals are chosen to initiate "high" lines and the lowest responding animals are chosen for the "low" lines. From that time on the animals are bred within line in consecutive generations. Selection of the breeding pairs for production of the succeeding generation is based upon the test criteria for the phenotype. Ideally, one should establish replicate lines in each direction of selection and for the control in order to increase the power of determining the response to selection, the estimation of additive genetic variation for the trait (realized heritability), and the magnitude of genetic correlation of traits hypothesized to be associated with the selected trait. The control lines, since they are maintained under the same laboratory conditions, are used to estimate the effect of the environment during the course of the selection experiment (see DeFries, 1981).

Advantages. The advantages of selection include those of the use of a heterogeneous stock given a minimum amount of inbreeding. In addition, selection experiments have the advantage that the high and low lines should theoretically differ only in those traits that are related, to a greater or lesser degree, to the trait upon which selection is operating. Thus, the existence of a genetic correlation between the primary trait (e.g., behavioral response) and another secondary trait (e.g., biochemical, not used as an index for selection) can be determined. In the case of a drug such as ethanol, which alters so many systems, this is a distinct advantage.

A second advantage is that after the lines in each direction have been separated by 10–20 generations of breeding, such that their response distributions are minimally overlapping, animals can be used for study of associated traits without prior exposure to ethanol. This avoids the uncertainty of alteration of the system under study by the drug test environment.

Disadvantages. The major disadvantage to the use of selected lines is the time, space, and expense that must be devoted to their development. As mentioned above (see DeFries, 1981), an ideal selection study consists of two replicate high lines, two low lines, and two unselected control lines. Each of these should have a minimum of 10 breeding pairs and all the offspring must be tested. This results in 300–600 animals to be tested at each generation. For mice and rats only three to four generations per year can be obtained since normally adults are tested. Thus, for a trait in which directional response is slow, 4 to 5 years might be necessary before the animals are maximally useful. During the course of selection, the animals are genetically unique and temporally limited to a specific laboratory environment. Thus, it is essential that they be protected, preferably in a specific pathogen-free colony.

Recombinant Inbred Strains

Advantages. Recombinant inbred strains combine the advantages of inbred strains with some of the advantages of selection. Recombinant inbred strains are developed from two inbred strains or well-separated selected lines that differ widely on the trait of interest (see Bailey, 1971; Eleftheriou and Elias, 1975). The two strains are crossed to produce an F_1 population, and from the F_1 animals a large number (30–40) of mating pairs are chosen at random to initiate a series of separate inbred strains, employing a brother–sister mating system. This results in a number (e.g., 10–20) of inbred strains that should exhibit a response gradient between the extremes shown by the parental strains. Whereas the response distribution present for the set of recombinant inbred strains would be more discrete than that of the continuously varying heterogeneous stock, there are essentially unlimited animals representing a portion of the genotypic distribution. This is limited only by the extreme response of the two parental strains. Recombinant inbred strains are extremely valuable for studies of genetic linkage and major gene effects on a trait.

Disadvantages.   The major disadvantage to the use of recombinant inbred strains is the time and expense necessary to generate them. There is usually inbreeding depression, due to decrease in reproductive fitness, resulting in loss of a certain proportion of the original recombinant strains. Thus, the experimenter should initiate more strains than are required for future use.

Preference

Preference for alcohol in mice and rats has long been a favorite subject of behavioral geneticists, perhaps because it requires no special instrumentation. It can involve the testing of a large number of animals simultaneously and it is subject to several different experimental paradigms. The most common preference test situation is one in which the animal is provided with a choice of either a 10% (usually weight/volume) ethanol solution or water in standard drinking tubes (with a ball bearing in the tube to reduce leakage and evaporation) or in Richter tubes. The position of the drinking tubes is changed daily. Many variations on this basic paradigm have been introduced. Thus, experimental designs have employed an empty bottle, an increasing concentration of alcohol presented every few days, lick counters, fluid deprivation of the animals, etc. The question of which factors are most important in accounting for differences in preference is controversial. Taste, acetaldehyde levels, stress, and tetrahydroisoquinoline or carboline formation have all been implicated at one time or another in alcohol preference. However, the presence of genetic variation for preference is well-established from data employing inbred strains, heterogeneous stocks, and selective breeding (Erwin and McClearn, 1981).

Inbred Strains.   A number of papers by McClearn and Rodgers (1959), Rodgers and McClearn (1962), Rodgers et al. (1963), and Fuller (1964) sparked a great deal of interest in the possibility of a genetic component to preference in mice. Anderson et al. (1979) and Anderson and McClearn (1981) have selectively bred mice for high and low ethanol acceptance. Erwin et al. (1980) also carried out a study of preference and associated traits in the heterogeneous stock of mice (HS/Ibg). Surprisingly, in the HS mice ethanol preference was significantly correlated only with acquisition of acute tolerance, and not initial behavioral sensitivity.

Studies by Brewster (1968) and Satinder (1970), using inbred strains of rats, provided evidence for genetic differences in ethanol consumption. Subsequently K. Eriksson (1968, 1969, 1971) at ALKO in Finland selectively bred rats for alcohol acceptance (AA) and alcohol nonacceptance (ANA). Li et al. (1981) also were successful in selectively breeding rats for preference (P) and nonpreference (NP). The P animals will consume sufficient alcohol to become tolerant and dependent and will actively work for alcohol solutions.

An enormous number of studies have been carried out on associated traits using either inbred strains or the selected lines of rats and mice that differ in their ethanol preference. A great many characteristics associated with preference have been found. Generally speaking, however, none of these has served to define a specific set of conditions that, when all present in the same animal, would account for either high or low preference. Broadhurst (1978) discusses a great number of these studies. The recently published reviews by C. J. P. Eriksson (1981), K. Eriksson and Rusi (1981), and Li et al. (1981) also provide summaries that are most useful in attempting to approach the mechanism by which preference operates.

The relative importance, in the determination of alcohol preference, of factors such as sensory inputs of taste and odor of ethanol, which exist prior to absorption of the drug, has been emphasized by Belknap et al. (1977) in a comparative study of high-preferring C57BL/6J and low-preferring DBA/2J mouse strains. DBA mice quickly developed ethanol avoidance in a choice drinking situation at a much lower blood ethanol concentration than C57BL mice. C57BL mice appeared less able than DBA to distinguish ethanol from water at a 2% concentration. Thus, strain differences in ethanol preference could be partially due to aversion at the initial time of exposure before any substantial ingestion of alcohol.

In addition, several studies have dealt with the question of the relative importance of postingestion toxicity by acetaldehyde leading to a conditioned aversion for the drug in accounting for differential preference between C57BL and DBA mouse strains. Sheppard et al. (1968, 1970) found that C57BL not only had a 30% higher liver ADH activity than DBA, but a 300% higher ALDH activity with an *in vitro* assay. Schneider et al. (1973) found that DBA do have higher blood acetaldehyde levels for 3 hr after injection. However, ethanol elimination rate of the two strains was essentially identical. The fact that the two strains also differed in preference for propylene glycol suggested that conditioned aversion to acetaldehyde might not account for differential preference. Conditioned aversion to saccharin, using a paradigm in which ethanol administered intraperitoneally was the unconditioned stimulus (UCS), developed in DBA but not in C57BL mice (Horowitz and Whitney, 1975). Thus, given the much lower ALDH activity in DBA than in C57BL, it could be argued that acetaldehyde is a UCS stimulus for aversion to ethanol. In fact, Dudek and Fuller (1978) showed this to be the case for the saccharin-conditioned aversion test, where DBA developed greater aversion following intraperitoneal acetaldehyde than C57BL, and the F_1 hybrid was intermediate. Again, the relevance of these findings of differences in ethanol preference of the two strains needs to be investigated further, preferably using a large number of inbred strains in the design, since DBA mice do not voluntarily consume enough ethanol to reach detectable blood acetaldehyde levels.

A paper by Ho et al. (1975) illustrates the usefulness of inbred strain comparisons and the difficulty in hypothesis testing when using only two such strains. They found that there were no differences in the uptake of [H^3]norepinephrine and [H^3]dopamine nor in serotonin levels between high-preferring C57BL/6J and low-preferring DBA/2J mice. Thus, these factors may be ruled out as contributing to alcohol preference differences in these strains. However, they also found that brain content of acetylcholine and whole brain uptake of [^{14}C]choline was higher in the C57BL mice, but brain acetylcholinesterase was higher in the DBA mice. This finding in only two strains could well be fortuitous. Without further experiments including either additional inbred strains, a classic mendelian analysis, resemblance between F_2 and F_3 generations, or the heterogeneous stock of mice, little can be said about the genetic relationship between preference and these putatively associated factors. This paper nicely illustrates that an adequate hypothesis test is obtained only when there is *no* difference between two inbred strains for the secondary trait. In addition, Ho et al. (1975) found that an inhibitor of choline acetyltransferase decreased preference for ethanol in the C57BL mice, suggesting cholinergic factors in preference. Since the DBA mice already exhibit low preference, it is not surprising that there was no effect of choline acetyltransferase inhibition on their preference. Durkin et al. (1982) found differences between C57BL and BALB/c in the inhibition by acute ethanol *in vivo* of high-affinity Na-dependent choline uptake in striatal synaptosomes. The V_{max} of uptake in striatum was inhibited in C57BL, but not in BALB/c; whereas, hippocampal uptake was similar in both strains. Following acute ethanol administration, choline acetyltransferase activity in striatum also was increased in C57BL, but not in BALB/c. These findings suggested genetic differences in ethanol interaction with cholinergic neurons in altering their activity, which could involve postsynaptic dopaminergic input.

Using rat inbred strains, a strain-dependent effect for excess dietary tryptophan-induced increase in ethanol intake (Myers and Melchior, 1975), and for central catecholaminergic and serotonergic lesions on ethanol consumption (Melchior and Myers, 1976) was found. Three rat strains showed an increased ethanol preference after serotonergic lesion by 5,6-DHT (Melchior and Myers, 1976); however, Pickett and Collins (1975) found no association of central serotonin content and ethanol preference in a genetic analysis of crosses between C57BL and DBA/2 mouse strains.

Selected Lines.   C. J. P. Eriksson (1981) has reviewed the data from the AA and ANA rats, which indicate an interrelationship between food utilization, metabolism, and ethanol preference. The high-alcohol-preferring animals (AA) have a more active basal metabolic rate, higher food intake, and a higher energy requirement. Linkola (1976, 1982) and Linkola et al. (1977, 1980) have carried out extensive studies on fluid and electrolyte balance in these selected lines. Differences between the lines were indicated for water intake and conservation, thirst mechanisms, electrolyte metabolism, sensitivity of the renin–angiotensin system to ethanol, and vasopressin excretion and diuresis with ethanol intake. The differences in renal function of electrolyte balance and hormonal response might be related to the genetic selection for ethanol preference, although this association is difficult to discern since the experimental groups are limited to two comparisons (Linkola, 1982).

As with studies on mice for alcohol preference, it has been found that AA rats have lower blood acetaldehyde levels and also higher acetaldehyde metabolism in liver mitochondria (C. J. P. Eriksson, 1981). In studies of alcohol and acetaldehyde metabolism in perfused livers of AA and ANA rats (Eriksson, 1973), it was found that female AA rats had a higher rate of ethanol metabolism and higher oxygen consumption that those of the ANA line. The AA line also had lower acetaldehyde levels and higher 3-hydroxybutyrate/acetoacetate ratios indicating a higher NADH/NAD ratio in mitochondria. On the other hand, Inoue et al. (1981) found no relationship of brain ALDH to preference in AA and ANA rats. These results contrast with those of Amir (1977, 1978) and Amir and Stern (1978) who found a correlation between brain ALDH and preference.

It is important to stress that all data showing a lower acetaldehyde level in the preferring strain or line of animals (mice or rats) come from measures taken after administration of acute doses of alcohol (or acetaldehyde) and not after free-choice alcohol intake (the alcohol-preference paradigm). As reviewed before, since nonpreferring inbred strains or outbred stocks rarely consume enough ethanol to achieve measurable blood ethanol, let alone measurable blood acetaldehyde levels, it is difficult to argue that the acetaldehyde level is the cue to stop drinking. Lower blood acetaldehyde levels in preferring versus nonpreferring animals may well be a strongly associated trait but causality is not likely.

In spite of the foregoing caveat, C. J. P. Eriksson (1981) presents a rational scheme for development of preference based on the higher blood acetaldehyde levels. Indeed, since some investigators start preference studies with *forced* alcohol intake, such mechanisms may operate to establish the aversion to ethanol by association with the increased blood acetaldehyde levels. This theory, of course, is at odds with the theory proposed by Myers and Melchior (1977) of increased blood acetaldehyde leading to increased preference by the production of TIQs. Eriksson discusses this point by proposing a biphasic action of acetaldehyde, whereby it is reinforcing at low levels but aversive at high levels. In a study of ethanol as a stimulus cue in an operant conditioning task, the performance level of ANA rats was higher than that of AA rats; acetaldehyde also enhanced the ethanol-conditioned response to a greater extent in ANA than in AA rats. Pentobarbital also was effective as a drug stimulus, and the degree of motor impairment due to ethanol was similar in the two lines. These findings indicated that the sitmulus cues could be generalized to other sedative agents, and that ethanol-induced ataxia was independent of its value as a conditioned stimulus in the preference lines.

In contrast to the studies with AA and ANA rats, Li and Lumeng (1977) found no difference in blood acetaldehyde levels between P and NP rats. The P and NP rats, selectively bred by Li and co-workers, have also been useful in studying correlated traits and thus in obtaining leads regarding the mechanism of preference. Their results have been reviewed recently (Li et al., 1981). As with many other animal models, including the AA and ANA rats (Rusi et al., 1977), the preferring animals had a shorter sleep time than the nonpreferring animals, following a dose of 2.5 g/kg of ethanol (Lumeng et al., 1982). An analysis of 10 different putative neurotransmitters in brain areas of the P and NP rats was conducted by Penn et al. (1978). From a number of comparisons of neurotransmitter levels, the most useful ones were between P and NP rats on water or on forced ethanol for 1 week. These comparisons should determine whether there are existing neurotransmitter level differences between the lines and/or whether there are differences which are revealed only after ethanol is present. In the diencephalon–mesencephalon area, both P and NP rats had higher levels of tyrosine in the ethanol group; serotonin was higher in the telencephalon of the NP-ethanol group and GABA was higher in the telencephalon of the P-ethanol group. No differences in the cerebellum between the two lines were observed. It is extremely difficult, if not impossible, to weave a pattern from these many results which would explain preference. The major advantage of such data is that they point the way for more detailed genetic studies of correlated traits in these or other lines or strains which differ in preference. Thus, a major advantage from the use of two bidirectionally selected lines of animals is the exclusion of factors not related to preference differences.

Li and co-workers (1981) were able to show that ethanol administered intravenously would substitute for oral intake of ethanol and that ethanol was reinforcing for alcohol-preferring rats (Penn et al., 1978). Most recently, Waller et al. (1982) have shown that P rats will consume sufficient ethanol to become dependent as assessed by hyperactivity in open field and in a head-poke apparatus upon withdrawal.

A selection experiment in mice for ethanol acceptance has been carried out by Anderson and McClearn (1981). Measuring ethanol acceptance requires less time than measuring preference and consists of measuring the amount of water consumed on 2 consecutive days and then the amount of 10% ethanol solution consumed after 24 hr of fluid deprivation. The operational phenotype is the ratio of the 10% ethanol consumed on the test day to the mean fluid amount consumed on days 1 and 2. The results of 10 generations of selection indicated a realized heritability estimate of 0.21 ± 0.04. The correlation between ethanol acceptance and preference was a somewhat low value of 0.25 in a sample of 110 male and female mice of the F_2 generation of a cross of inbred strains, C57BL $\times$ C3H. Liver ADH and ALDH activity have been measured in these lines. The high-accepting selected line has a greater total liver alcohol dehydrogenase activity, which is consistent with the observations on the high-preferring C57BL strain. The correlation between total liver ADH and acceptance was only 0.25 in the F_2 animals (C57BL $\times$ C3H), however, this value could be an underestimation of the initial covariance present in the base population selected from the HS mice (Anderson et al., 1979).

Several studies have employed classic mendelian crosses of inbred strains to compare means and variances among parental, F_1, F_2, and backcross generations for the alcohol preference phenotype. This approach can be used to estimate the amount of phenotypic variation, which is contributed by additive genetic variation in the population sampled. The heritability index is estimated as the ratio of additive genetic to phenotypic variance and is employed in predicting the response to selection by the population. Brewster (1968) initially estimated a high heritability for alcohol preference in a reanalysis of mouse data from Rodgers and McClearn (1962; $h^2 = 0.82$) and Fuller (1964; $h^2 = 0.86$), and in his rat data ($h^2 = 0.72$). Essentially no dominance was indicated for the preference phenotype in the data from mice or rats, but complete dominance towards high ethanol intake was indicated in the data from rats. This inconsistency with regard to the presence of dominance for the two measures is of concern, since one would expect some degree of correlation between ethanol intake and preference.

Whitney et al. (1970) reanalyzed data from Brewster (1968), Rodgers and McClearn (1962), and Fuller (1964), as well as presenting an analysis of their data from mouse strain crosses, heterogeneous stock, and lines selected for high and low open field activity. Their estimates from the various data sources were considerably lower than those of Brewster (1968); they ranged from approximately 0.20 to 0.46. These estimates were similar to that of Fuller (1964; $h^2 = 0.39$) estimated from the intraclass correlation coefficient. Emphasizing that their method yields an upper-limit estimate, they conclude that heritability for alcohol preference is probably less than 0.20. K. Eriksson (1969), using offspring–midparent resemblance in an F_1 cross from the rat AA and ANA selected lines, estimated heritability as 0.14 for ethanol intake and found evidence of dominance for low intake, in contrast to Brewster's (1968) finding in the rat. Drewek and Broadhurst (1981) have pointed out that Eriksson's estimation procedure can yield an underestimate of heritability.

Unfortunately, as observed occasionally with behavioral measures, these data have presented difficulty in the scale used to measure phenotypic distributions because the phenotypic variance is smaller in the F_2 segregating generation than in isogenic (parental and F_1) generations. Thus, heritability has to be estimated from generation means (first degree statistics) rather than variances (second degree statistics). This method of estimation is less efficient and requires the more restrictive assumption that the inbred strains are drawn at random from a large population in linkage equilibrium (Mather and Jinks, 1971). The methods of estimating heritability in these various animal studies on alcohol preference have elicited an important discussion on the appropriate measure to employ, and its application in behavioral genetic research of alcohol preference with respect to interpretation and generalizing across different populations (see Drewek, 1980; Drewek and Broadhurst, 1981, 1983a; Whitney et al., 1982). Fuller and Collins (1972) analyzed mendelian crosses from C57BL × DBA/2 which did not exhibit sufficient variance in the segregating generations. Using a nonparametric method of genetic analysis they found that the observed generation distributions did not deviate widely from that expected with a system of two gene loci or two independent gene blocks accounting for phenotypic variation in ethanol intake and preference.

Recently, Drewek and Broadhurst (1979, 1981, 1983b) have initiated a complete biometrical analysis of genetic variation for alcohol preference in the rat. Significant additive genetic variation for alcohol intake and preference ratio in a set of seven inbred strains (Drewek and Broadhurst, 1979) was found. From these observations, Drewek and Broadhurst (1981, 1983b) demonstrated that phenotypic variation for alcohol preference ratio, alcohol intake, and alcohol calorie contribution ratio fit a simple additive–dominance model with no interaction between gene loci using a triple-test cross design with six inbred rat strains and two or three tester strains. Estimating heritability from second-degree statistics, where scale of distribution criteria were met, high values for all three phenotypes were calculated ($h^2 = 0.84$ for ethanol intake, $h^2 = 0.70$ for preference ratio, $h^2 = 0.77$ for calorie contribution ratio, averaged over sexes). Directional dominance for the high-end phenotype was indicated for all three phenotypes, suggesting that selection for high preference, either directly for ethanol or for an associated trait, might have operated during the evolution of the rodent population (Drewek and Broadhurst, 1981, 1983b). The findings among the three measures were highly consistent, where the estimates of genetic correlation between them were greater than 0.80. Again, discussion has arisen that these heritability estimates might be overestimates (Whitney et al., 1982), since they were derived with a formula that would provide an estimate of the maximal amount of additive genetic variation (Drewek and Broadhurst, 1983a), rather than the formula more often used in biometric genetic analysis. Regardless of the most appropriate estimate for heritability, the existence of significant genetic variation in alcohol intake and preference is obvious from experiments using rat and mouse inbred strains, heterogeneous stocks, mendelian crosses, and response to directional selection.

In a study of four mouse inbred strains, Goodrick (1978) found significant strain differences in ethanol intake. Interestingly, there were significant dose–dependent differences among the strains, such that while intake decreased with increasing ethanol concentration, C57BL consumed at all concentrations, DBA/2 abstained at all concentrations, and BALB/c mice consumed at only less than 10–15% concentrations. Sex effects were present in the F_1 hybrid crosses and dominance for high intake was suggested in the C57BL × BALB/c hybrid [consistent with Drewek and Broadhurst's findings in the rat strains (1981, 1983b)]. The heritability estimate was 0.38 for intake. Especially noteworthy is Goodrick's (1978) evidence for significant age-related changes in ethanol consumption that appear to have a complex genetic basis. Consumption, in general, decreased with age through maturity, but then increased through old age and senescence. Goodrick (1978) stressed the need to direct research towards assessment of physiological changes that might underlie these developmental changes in the behavioral trait of ethanol intake.

Initial Sensitivity

As with preference studies, a large literature exists concerning initial sensitivity to ethanol. Perhaps there are even more ways to measure initial sensitivity than to measure preference. The most common measure is that of "sleep time" which is assessed by administering an anesthetic dose of ethanol and then recording the duration of the loss of the righting reflex. Various tasks measuring ataxia, balance, and analgesia to assess acute ethanol effects have been used. These include measuring the animal's ability to remain on a moving belt, remain on a tilting plane without sliding, ambulate on an alley-way grid, remain on a rotating or stationary rod (rotorod performance), right itself after being dropped upside down (aerial righting reflex), as well as acquisition and retention of passive avoidance tasks and hyperactivity at low doses. While it is not our purpose to review the advantages and disadvantages of these methods, it is important to point out that often the different measures of initial sensitivity do not intercorrelate very highly and are definitely task-dependent. The rank order of strains for a series of initial sensitivity measures is not necessarily uniform (Belknap and Deutsch, 1982). It is also important to measure blood, or preferably, brain ethanol levels at the behavioral end point to enable distinction between metabolic and neural effects, and in addition, measure at a time beyond rapid fluctuations, that is, 30 min after intraperitoneal injection of ethanol.

Inbred Strains.   Strain differences in initial sensitivity have been a subject of many studies. The results of these studies have been reviewed by Broadhurst (1978) and more recently, by Erwin and McClearn (1981). Again, the most popular strains for work in this aspect of ethanol action are C57BL, BALB/c, and DBA/2. Consistently, it has been found that C57BL mice are less impaired by a >3.0 g/kg hypnotic dose of ethanol than are mice of the DBA or BALB strains (Kakihana et al., 1966; Belknap et al., 1972; Damjanovich and MacInnes, 1973; Randall and Lester, 1974). While a great number of studies have been carried out to demonstrate correlated traits using these, and other inbred strains, little has been elucidated about the *mechanism* by which the differences were manifest. The study of the mechanism by which alcohol depresses the central nervous system has been advanced more dramatically by groups of studies that have combined the use of mouse selected lines, heterogeneous stock, and inbred strains.

One of the most common correlated traits associated with sleep time measures is hypothermia, or perhaps more correctly, poikilothermia, since the effect seems to be one of causing an inability to regulate temperature rather than true loss of heat (Myers, 1981). In general, hypothermia and sleep time are positively correlated in inbred strains as well as in selected animals (Moore and Kakihana,

1978). However, Eriksson et al. (1982*a*), in a study using the heterogeneous stock of mice, found that sleep time and hypothermia were negatively correlated ($r = -0.3$, $p < 0.01$), when a correction for room temperature fluctuations was made.

Crabbe et al. (1980*b*) conducted a genetic analysis of the hypothermic effect of ethanol and the effect of a vasopressin analog on development of hypothermic tolerance. They used C57BL/6By and BALB/cBy mice as parental strains, the F_1 hybrid and seven recombinant inbred strains derived from the F_2 generation. Significant strain differences were found for baseline temperature and initial sensitivity to ethanol with hypothermia as the index. It is interesting, however, that they found no significant strain differences in tolerance to ethanol-induced hypothermia, in agreement with Belknap and Deutsch (1982) in a study of C57BL and DBA/2 strains. Vasopressin altered the strain differences in initial sensitivity and blocked tolerance development in a strain pattern compatible with additive genetic inheritance. In contrast, Moore and Kakihana (1978) found genotypic differences for tolerance to ethanol-induced hypothermia: C57BL and BALB/c expressed it and DBA/2 did not. Tolerance for ethanol-induced hypothermia was independent of metabolism with respect to the hypothermia index.

Ethanol effect on locomotor activity was studied in C57BL and BALB/c strains by Randall et al. (1975). From 0 to 2.25 g/kg doses, C57BL exhibited decreasing activity and BALB showed increasing activity. This contrasting dosage effect with regard to ethanol's hypnotic strain effect suggested that the dosage-dependent biphasic action of ethanol might be controlled by separate mechanisms. In support of these findings, Crabbe et al. (1980*a*), in testing C57BL and DBA/2 strain responses to low dosages of ethanol in the open field, found a dosage- and strain-dependent effect. Depressant and stimulatory effects of ethanol appeared to be under separate genetic control. In a comparison of 20 inbred mouse strains (Crabbe, 1983), three separate major groups of ethanol-related traits were distinguished on the basis of their intercorrelations. For hypothermic sensitivity (1), strains with greater sensitivity to ethanol-induced hypothermia also exhibited greater ataxic response and acute anesthetic sensitivity to ethanol and less baseline activity. For ethanol-induced increase in activity (2), strains with increased activity also showed more rapid loss of balance from a dowel with ethanol. For increased basal activity (3), strains also had increased open-field activity and were less sensitive to ethanol-induced ambulatory ataxia. This study emphasizes the complexity of the genetic systems underlying ethanol modification of behavior.

Using the Bailey recombinant inbred strains of mice, Oliverio and Eleftheriou (1976) carried out a genetic analysis of the effect of alcohol on motor activity. They concluded that the effect of alcohol on basal activity was controlled by at least one major gene, EAM (ethanol activity modifier), and designated the C57BL-like locus as EAM^h and the BALB-like locus as EAM^l. The EAM locus was assigned to chromosome 4 (LG VIII). In a study assessing a large number of ethanol-related traits (open-field activity, ambulatory ataxia, ethanol acceptance and severity of withdrawal), Crabbe et al. (1983c) used recombinant inbred strains (BXD/Ty) from the cross of C57BL/6J × DBA/2J to evaluate major gene effects by the strain distribution pattern. Polygenic control was indicated for the motor impairment traits; substantial major gene control was indicated for ethanol acceptance and less so for severity of withdrawal.

The question of the influence of alcohol and aldehyde dehydrogenase activity on either initial effects or preference is intriguing. While it is generally believed that levels of ADH are not rate-limiting for alcohol disappearance rates, nevertheless, Belknap et al. (1972) found a significant correlation between liver ADH and sleep time, (-0.3 to -0.57) and liver ALDH and sleep time (-0.29 to -0.52) at 90 min after ethanol injection, following the sleep time test, in the genetically heterogeneous mouse stock (HS/Ibg). No correlation of sleep time and enzyme activity, assessed at 42 days after ethanol injection, was seen. McClearn and Anderson (1979) performed a factor analysis of the intercorrelations of a set of ethanol-related phenotypes measured in the HS mice. Their study indicated that sleep time and liver ADH measures were correlated poorly and loaded on completely independent factors, indicating a lack of association.

Estimates of genetic variation for sleep time, plus open-field activity and heart rate following ethanol administration, were determined in the HS mice by Reed (1977). Both open-field activity (OFA) and heart rate were found to increase after a low dosage of ethanol, indicating the excitatory action of the drug. Significant, but low heritability (additive genetic variation) was demonstrated for all three phenotypes ($h^2 = 0.12$, OFA; $h^2 = 0.18$, heart rate; $h^2 = 0.17$, sleep time) by midparent–midoffspring regression analysis.

Selected Lines.   To date, there have been three selection programs in which direct selection for a phenotype related to initial sensitivity has been applied. The oldest is that of McClearn and Kakihana (1981) originally started in 1962, which resulted in short-sleep (SS) and long-sleep (LS) mice. Studies in rats by Bass and Lester (1981a) produced the most-affected (MA) and least-affected (LA) rats. Studies by Eriksson provided alcohol-tolerant (AT) and alcohol-nontolerant (ANT) rats (K. Eriksson and Rusi, 1981). These experiments have suffered from various deficiencies that are more apparent in hindsight. For example, the selection experiment of McClearn and his colleagues, while starting with a genetically defined heterogeneous population, has only a single high line (LS), a single low line (SS), and no control line; also, a severe reduction in breeding pairs was experienced early in the selection program (McClearn and Kakihana, 1981).

The study by Lester and his co-workers also has only a single high and a single low line but started from a genetically undefined group of animals that would be difficult to reproduce systematically. Similar problems exist with the

AT and ANT lines. Nevertheless, the aim of these and other breeding studies has been to provide animals that are useful in determining the mechanism of action of ethanol and in this respect they have been very informative.

SS and LS Mice. The phenotypic response which is the basis of selection of these animals is the length of time from loss to recovery of the righting reflex following a hypnotic dose of ethanol. The animals are now in the 35th generation of breeding which encompasses 25 generations of effective selection. Selection has been relaxed several times and then terminated at generation 34, since no further response to selection has been apparent.

The realized heritability of the "sleep time" response is estimated as 0.20 (McClearn and Kakihana, 1981), and the difference in sleep time has reached the point where a lethal dose to some LS animals will not cause some SS mice even to lose their righting reflex. Sleep time differences between the high and low line continued to separate with applied selection, but the apparent ED_{50} values have remained essentially the same with about a twofold difference between LS and SS lines since generation 14. This might indicate that different mechanisms are responsible for the initial loss of the righting reflex and the eventual regain of the reflex.

Because of the design of the selection experiment, the animals are probably more useful in excluding postulated mechanisms. Studies on differences between the two SS and LS lines for traits hypothesized to underlie the selected behavioral sensitivity difference can only be suggestive of association. Test of genetic correlation and models for causal analysis must be approached using biochemical and physiological studies in the genetically heterogeneous stock of mice, sets of inbred strains, recombinant inbred strains, or mendelian crosses between SS and LS lines. Thus, some of the most useful data come from experiments where no differences in a trait between the lines are found, since it allows one to exclude that trait as being important in accounting for why the animals differ in central nervous system sensitivity to ethanol (see Collins, 1981, for earlier review of LS and SS mice). In this way, it has been shown that metabolism of ethanol is not of critical importance to the sleep time differences (Heston and Erwin, 1974; Hjelle et al., 1981; Kakihana, 1976; Tabakoff et al., 1980). In more recent studies it appears that there are differences in the metabolic rate of ethanol if nonlinear kinetics are employed, such that the SS animals metabolize somewhat faster, perhaps as a result of the higher dose administered to this line, compared to LS, in selecting the breeding pairs during the later generations (Gilliam and Collins, 1982*a, b;* Gilliam et al., 1983). Gilliam and Collins (1983) have argued that, given that the selection index was based upon an anesthetic dosage of ethanol delivered from a 30% (volume/volume) solution of ethanol in isotonic saline, the LS mice might have been selected for toxic effects of ethanol in peripheral organs, as well as greater CNS sensitivity, compared to SS mice. Decline in blood ethanol level was found to be concentration-dependent in LS mice, but not in SS mice. In the LS and SS of earlier selected generations, it has been found that CNS depression was the same in both lines when animals were given barbiturates (Erwin et al., 1976; Siemens and Chan, 1976), chloral hydrate, paraldehyde and ether (Erwin et al., 1976), or halothane (Baker et al., 1980). This calls into question the long-held belief that the mechanism of action of these sedative agents and alcohol were the same. Moreover, in support of a different mechanism of action of these classes of sedative agents, O'Connor et al. (1982) and Howerton et al. (1983) have found further line differences. SS mice are more sensitive to pentobarbital and other more lipid-soluble sedatives, due to differential elimination rate and whole-body lipid content, while LS mice are more sensitive to the more water-soluble sedatives (short-chain alcohols and methyprylon), due to CNS sensitivity differences. Lipid solubility of the agent appears to be a major factor in the hypnotic sensitivity order of the two lines. CNS depression caused by salsolinol, a condensation product of dopamine and acetaldehyde, was greater in the LS than in the SS mice (Church et al., 1977).

Control of ethanol intake has been studied in LS and SS mice to determine the association of this phenotype with neural sensitivity to an acute ethanol dose. Church et al. (1979) observed that SS mice consume more ethanol (4% in glucose–saccharin solution) than LS mice regardless of previous exposure to the drug. In follow-up studies Fuller (1980) noted that the daily ethanol intake decreased for both lines as the ethanol concentration increased in the two-bottle, free-choice situation (consistent with inbred strain findings). On the average, although both genotypes consume ethanol, intake was higher for SS than LS mice; however, sex differences ($\female > \male$) and dependence upon ethanol concentration during prior exposure were apparent. These findings supported the idea that SS mice have a higher threshold for ethanol intake than LS mice but that the threshold adjusts upward to allow greater intake when animals have been exposed previously to higher levels of ethanol.

Conditioned taste aversion induced by ethanol has been postulated to play a role in ethanol intake (see section on alcohol preference). Dudek (1982) found that LS and SS mice do not differ. Thus, genetic systems involved in CNS sensitivity for ataxic effects of ethanol apparently differ from those that mediate development of conditioned taste aversion paired to ethanol.

At generation 12 of selective breeding Baer and Crumpacker (1977) observed that LS mice were about half as fertile as SS mice. In addition, the survival of LS progeny was decreased by maternal consumption of ethanol. LS lactating mothers consumed much less liquid than SS lactating mothers in a forced ethanol consumption condition and exhibited increased cannibalism of pups. Thus, LS mothers responded more stressfully with consequent reduced ability to provide nutritional care to their young than SS mothers. Higher fitness of SS mice compared to LS mice was observed also by Swanberg and Wilson (1979).

General CNS excitability and seizure susceptibility seems to be greater in SS than LS mice, since latency to myoclonic seizure in response to flurothyl is shorter in SS mice (Sanders and Sharpless, 1978; Greer and Alpern, 1977; Sanders, 1980), and low dosages of ethanol cause a greater increase in locomotor activity in SS than in LS mice (Sanders, 1976; Church et al., 1977). However, no differences were found between LS and SS mice in flurothyl-induced clonic seizures (Greer and Alpern, 1977; Sanders and Sharpless, 1978; Sanders, 1980). SS mice were more active in the open field than LS mice without ethanol administration. Both lines were less active in the open field test and more susceptible to clonic convulsions to the same degree 7–9 hr after ethanol administration (Sanders, 1980). However, LS mice showed a decrease in latency to myoclonic seizures with ethanol (SS decreased no further) and greater ethanol-induced hypothermia 2 hr after ethanol injection with a slower rate of recovery. SS mice exhibited more severe withdrawal from morphine (naloxone-precipitated) than did LS mice (Horowitz and Allan, 1982). This finding is consistent with the greater seizure severity (elicited by handling) of SS mice than of LS mice induced by alcohol withdrawal (Goldstein and Kakihana, 1974).

Studies using SS and LS mice, aimed at a molecular understanding of the action of ethanol, have been undertaken by several investigators in recent years. Goldstein and her colleagues have found that ethanol causes a greater disordering of synaptic plasma membranes isolated from LS than from SS mice (Goldstein et al., 1982). Similar results were obtained with HS mice that fell at the extremes of the ethanol-induced sleep time distribution (Goldstein et al., 1982).

Various neurotransmitter and receptor systems in the CNS of LS and SS mice have been studied extensively; however, to date none of these have accounted sufficiently for sleep time differences between the two lines. Initially, Collins et al. (1976) reported that whole-brain turnover rate of norepinephrine (NE) was not affected by ethanol, whereas whole-brain turnover rate of dopamine (DA) was reduced in both lines with LS mice exhibiting a larger decrease (65%) than SS mice (35%).

Recently, Howerton and Collins (1984) reported that ethanol inhibited K^+-stimulated release of NE from slices of frontal cortex at lower concentrations in LS mice than in SS mice; increased spontaneous overflow of NE from cortical slices was apparent at lower concentrations of ethanol in the SS mice. The response of cerebellar slices to ethanol was similar, and PGE_2 inhibited NE release to the same extent in the two lines of mice. Dibner et al. (1980) reported that β-adrenergic receptor density was lower in the cerebral cortex but not in the cerebellum of LS mice compared to SS mice. However, cyclic AMP accumulation in response to isoproterenol in the cortex was similar in LS and SS mice. No difference between the two lines was found in ligand binding to DA receptors in the striatum or to muscarinic cholinergic receptors in the cortex, striatum, or hippocampus. DA-stimulated adenylate cyclase in striatum was found to be slightly greater in LS mice than SS mice. However, recently Rabin and Molinoff (personal communication) have found no differences between LS and SS mice in stimulation of adenylate cyclase by ethanol in the presence or absence of DA or isoproterenol, in striatum, cerebral cortex or cerebellum.

A significant increase in tyrosine hydroxylase activity in the striatum, locus coeruleus, and frontal cortex in both LS and SS mice about 25 min after a 4.1 g/kg ethanol dose was reported by Baizer et al. (1981). Most interesting is the reported different time course for hypothalamic tyrosine hydroxylase activation, which coincided with regain of the righting reflex (SS: 25 min; LS: 125 min), and therefore, might be associated with alteration in arousal state from ethanol sedation. Dopaminergic involvement in the hypnotic effect of ethanol in these selected lines has been approached in a study by Dudek and Fanelli (1980). They found that gamma-butyrolactone (GBL), converted to the metabolically active form, gamma-hydroxybutyrate (GHB) *in vivo*, is more potent in lengthening sleep time of LS than SS mice. It also induced a threefold greater increase in whole-brain DA levels in LS than SS mice. It was argued that GBL's anesthetic effects at high dosages might serve as a model for ethanol depression of DA neurons since it has been found to depress DA firing rate in substantia nigra neurons and inhibit DA release (Roth et al., 1973). However, its usefulness as a means to elucidate genetic differences in the interaction of ethanol with dopaminergic systems can be questioned. GHB has been noted to induce epileptiform activity, consistent with a CNS excitatory state, rather than depressant anesthetic effects. Moreover, GHB has been reported to increase brain GABA levels and competitively inhibit GABA-transaminase. Most noteworthy is the finding that lesioning of substantia nigra and areas receiving dopaminergic innervation has no effect on GHB-induced behavioral depression (see Snead, 1977 for review). Chan (1976) found no evidence for GABA involvement in LS–SS ethanol sensitivity differences.

Recently, further investigations into the role of catecholaminergic systems, as well as cholinergic systems, in the differential neural sensitivity to ethanol have been initiated by Masserano and Weiner (1982), who studied the effects of neurotransmitter agonist and antagonists injected intraventricularly on LS and SS ethanol-induced sleep times. Nicotinic antagonists decreased LS sleep time, but did not alter SS sleep time. This nicotinic effect was blocked by a muscarinic antagonist. It was postulated that nicotinic receptor blockade might result in a net increase in acetylcholine (ACh) release (via reduction in negative feedback inhibition), followed by ACh acting at muscarinic receptors to reduce sleep time in the LS mice. However, cholinergic agonists did not modify either LS or SS ethanol-induced sleep time. Despite the uncertainty of the interaction of these agents with ethanol, SS mice appeared to have a less responsive cholinergic system. The effect of catecholamine agonists was less straightforward. The trend observed was that LS sleep time was decreased by NE and DA, while SS sleep time was increased by NE and DA. Amphetamine and GABA showed no effect in either LS or SS. The blocking of the DA and NE agonist effects by various α- and β-adrenergic and DA antagonists, as well as their effects alone, on ethanol-induced sleep time were inconsistent between the two lines. This indicates the extreme complexity of noradrenergic and dopaminergic mechanisms interacting with ethanol's central depressant action. Masserano and Weiner (1982) found that LS and SS expressed differential responses to catecholamine depletion by reserpine. The dosage administered was lethal to SS probably because of prolonged hypothermia, while LS ethanol-induced sleep time was decreased at a time after reserpine treatment when catecholamine turnover would be elevated (tyrosine hydroxylase activity increased). This might point towards a more deficient catecholamine mechanism in LS mice than SS which could be related to ethanol sensitivity differences. Horowitz et al. (1982) recently found that plasma levels of dopamine-β-hydroxylase (DBH) were higher in LS than SS mice. The DBH inhibitor, fusaric acid, decreased the ethanol-induced sleep time in a sample of mice from the heterogeneous stock. Horowitz et al. (1982) suggested that if greater adrenergic activity in LS is associated with longer sleep time, then inhibition of DBH should lead to a decrease in sleep time.

Kakihana (1976) reported that plasma corticosterone levels 1 hr after ethanol administration were higher in LS than SS mice. SS adrenal weight was 20–30% higher than LS, and SS showed a greater initial increase in corticosteroids following the control saline injection. More research on genetic differences for control regulation of adrenal response is required to assess the association of alcohol activation of corticosterone production with neuronal sensitivity to ethanol.

The recent electrophysiological experiments of Hoffer, Palmer, and their colleagues indicate that the behavioral responses of LS and SS mice to soporific dosages of ethanol correlate with ethanol sensitivities of cerebellar Purkinje neurons in these two mouse lines (see Seiger et al., 1983). Thus, the spontaneous discharge of cerebellar Purkinje cells *in situ* is depressed by ethanol locally applied from multibarreled micropipettes at dosages 30-fold lower in LS mice than in SS mice (Sorensen et al., 1980). In addition, the mean ethanol dosage needed to elicit equivalent depressions of Purkinje cell activity in HS mice is between the mean LS and SS values (Sorensen et al., 1981). This observation agrees well with the intermediate ethanol-induced sleep time in HS animals (Erwin et al., 1976, 1980). Furthermore, isolated cerebellar preparations from LS and SS mice, such as *in vitro* slices (Basile et al., 1983) or *in oculo* transplants (Palmer et al., 1982), manifest the same differential Purkinje cell sensitivity as is seen *in situ*. The *in oculo* transplant study indicated that this LS–SS Purkinje cell sensitivity difference appears to be an intrinsic genetic property of the cerebellum of the donor line and not directly influenced by input from the host animal. The pharmacologic specificity of these alcohol effects is suggested by the observation that no differential effects on cerebellum physiology were seen with local halothane application in LS and SS lines (Sorensen et al., 1981). This agrees with behavioral data showing that halothane-induced sleep time is similar in the LS and SS lines (Baker et al., 1980). However, this differential ethanol sensitivity is probably not expressed by neurons in all brain areas of LS and SS mice, since Sorensen et al. (1981) have found that locally applied ethanol is equipotent in depressing the activity of hippocampal pyramidal neurons from these two lines of mice.

Considering these data, one might hypothesize that cerebellar ethanol sensitivities in LS and SS mice might directly mediate the soporific behavioral sensitivities in these two mouse lines. This question has been recently assessed by Palmer et al. (1983), who investigated ethanol-induced sleep time in neonatally cerebellectomized LS and SS mice. While these authors did find that the presence of the cerebellum has an important influence on sleep time in SS mice (i.e., SS sleep time was longer in cerebellectomized SS mice than control SS), they also found that cerebellectomy did not alter the sleep time of LS mice. Indeed, a difference in sleep time between LS and SS mice remained after neonatal cerebellectomy. Palmer et al. (1983) suggest that populations of neurons also exist in noncerebellar brain areas, perhaps in the brain stem, which would exhibit markedly different sensitivites to ethanol in LS and SS mice and, therefore, also might play a more important role in mediating differences in behavioral sensitivity to ethanol. It is intriguing to hypothesize that perhaps the LS cerebellum already has reached the minimal threshold level of functional circuitry for ethanol effects as a result of genetic selection, while cerebellectomy simulated this change in threshold level in the SS mice. Similar findings by Northrup (1976) indicated that homozygous recessive "nervous" mutant mice (Sidman and Green, 1970), which have lost a significant proportion of their cerebellar Purkinje neurons, are more sensitive to the ataxic effects of alcohol than are wild-type mice of the same strain.

Recently, in order to test for genetic correlation of behavioral and cerebellar electrophysiological sensitivity to ethanol, Spuhler et al. (1982) compared the eight inbred strains of mice that were intercrossed to make up HS/Ibg from which LS and SS lines were derived. It was found that sleep time and the inhibition of cerebellar Purkinje neuron by locally applied ethanol were very highly correlated genetically ($r_A = 0.95 \pm 0.12$) among the strains, suggesting a common genetic mechanism or very close linkage of genes controlling expression of behavioral and neurophysiological sensitivity to ethanol. Cerebellar ethanol level upon recovery of righting reflex and metabolism accounted for much less variance in sleep time than cerebellar neuronal sensitivity. These findings in the inbred strains support the hypothesis that the two-line LS–SS cerebellar ethanol sensitivity differences are correlated with genotypic differences for the selected trait, sleep time.

This work has prompted several investigations of the biochemical differences between SS and LS cerebellum. Analysis of whole-brain phospholipids and cholesterol revealed no differences between SS and LS mice (Koblin and Deady, 1981). Two-dimensional gel electrophoresis studies of high-molecular-weight polypeptides (16–200 K) in the cerebellum of SS and LS mice also revealed a number of differences (Deitrich et al., 1982). Which of these are critical to the sleep time differences is the subject of current investigations.

MA and LA Rats. These selected lines of rats were originated at Rutgers by Riley et al. (1976, 1977). The two foundation strains were Sprague–Dawley and Long–Evans rats. The selection index was based upon the impairment of motor activity by ethanol. The breeding protocol was such that the most affected (MA) Sprague–Dawley rats were bred with the most affected MA Long–Evans rats to produce the first selected generation. Least affected lines were bred similarly.

In selecting parents for subsequent generations, a hooded and an albino rat were always crossed, presumably on the assumption that the albino rats were more Sprague–Dawley-like and the hooded rats more Long–Evans-like. Only four breeding pairs were chosen as parents for the subsequent generation within each line, thus a substantial amount of inbreeding could have affected the response to selection in the lines. There was no control line. The animals in the MA line did become progressively more affected with each generation, until at generation five there was a significant mean difference between the LA and MA lines. The LA line remained at about the same level as in the first generation until the ninth generation when there was a deviation (Bass and Lester, 1981a). Whether or not either of these lines differs significantly from a control foundation population is unknown, as there is no population of genetically similar rats. Again, hindsight affords an opportunity to design a more satisfying breeding project from a genetic point of view.

A number of possibly correlated behavioral measures have been reported, but very few biochemical and no neurophysiological measures have been reported using the MA and LA rats. The animals do differ in sleep time response to a hypnotic dosage of ethanol, to barbiturate, and chloral hydrate (Riley et al., 1979). The response to ethanol is different from that of alcohol-tolerant (AT) and alcohol-nontolerant (ANT) rats (discussed below), which showed no difference in sleep time response to ethanol.

The MA animals differ from LA animals with respect to ethanol effect on: reduction of startle response, changes in ambulation with amphetamine administration, incidence of vocalization on foot shock, escape avoidance tasks, wheel running, and swimming ability. The two lines do not differ in alcohol preference, ethanol-induced conditioned aversion, acquisition or loss of tolerance, or rates of ethanol clearance (Shapiro and Riley, 1980; Shapiro et al., 1979; Bass and Lester, 1981b). In a recent study, comparing ethanol-induced depression of motor activity and swimming impairment by analysis of the F_2 generation from the LA × MA cross, evidence for genetic independence of these two ethanol-related phenotypes was presented (Bass and Lester, 1983).

AT and ANT Rats. These animals were selectively bred for differential response to ethanol impairment on the tilting plane (K. Eriksson and Rusi, 1981; Eriksson et al., 1982a). The base stock was a mixture of Sprague–Dawley, Wistar, and Long–Evans strains. After eight generations of selection the ANT (alcohol nontolerant) animals became more sensitive to ethanol but the AT line did not become less affected. Thus, in these two rat studies (i.e., LA and MA lines; AT and ANT lines) it has apparently been possible to increase the sensitivity to alcohol but not to decrease sensitivity to ethanol. The AT and ANT lines are also without a control line. Relatively few correlated traits have been measured with the AT and ANT lines. As mentioned earlier, these lines do not differ with respect to ethanol-induced sleep time. This is somewhat surprising and suggests that in these rat strains the mechanisms underlying ethanol impairment of behavior as measured by the anesthetic effect and the depression of motor coordination are different. In addition, no difference in response to barbiturates is seen between these lines, which is similar to that found in earlier studies of the SS and LS mice (Erwin et al., 1976) but different from that observed in the MA and LA lines. This reiterates that ethanol-related phenotypes appear to be very much task-dependent and should be recognized as such in selective breeding studies. However, the relatively fine degree of selection that can be achieved is illustrated as well.

Tolerance and Dependence

For many years it was assumed that tolerance and dependence were on a continuum, and that if tolerance developed to ethanol then a degree of dependence also occurred, and vice versa. Numerous studies in the past several years have been carried out showing that this is not necessarily the case (Tabakoff and Ritzman, 1977). It is true that procedures for production of tolerance and dependence are usually similar if not identical. These procedures consist of ethanol administration by one of several methods: orally by forced intubation, orally in the drinking water with no choice, in sweetened water, in a complete liquid diet, by intraperitoneal injection, by inhalation, by slow release from silastic tubes, from silastic pillows placed subcutaneously, and others. Again, it is not our purpose to review the variety of methods, but point out that there may well be an interaction between the mode of ethanol administration and the development of tolerance and dependence.

Inbred Strains. A number of studies have been carried out using inbred strains and selected lines of mice and rats to assess the genetic influence on the development of tolerance. Perhaps the most comprehensive study published recently is that of Crabbe et al. (1982), in which 20 inbred strains of mice were tested for tolerance, as measured by reduction in hypothermia with chronic ethanol administration. They injected 10% w/v ethanol once daily at a dose of 3 g/kg intraperitoneally for 8 days. DBA/1J, DBA/2N, MA/MyJ, and PL/J did not develop tolerance; however, for the other strains, tolerance development exhibited a positive genetic correlation with initial sensitivity to the hypothermic effect of ethanol. Strain differences for ethanol-induced changes in activity and ataxia were found as well, but these traits did not correlate strongly with tolerance development. Recently Crabbe et al. (1983d) have reported that severity of withdrawal from ethanol was negatively genetically correlated with initial sensitivity and tolerance development. The responses of these traits in the lines selected for differences in withdrawal severity (discussed below) also support this genetic association of ethanol disruption of thermoregulation and predisposition to strong physical dependence.

Studies by Grieve et al. (1979) and Grieve and Littleton (1979) used three inbred strains, C57BL, DBA/2 and TO Swiss mice. Tolerance was assessed by measurement of ethanol levels at repeated losses of righting reflex following repeated exposure to ethanol vapor. As soon as the animals regained a response, they were returned to the vapor chamber and the process repeated over a 5-hr period. Grieve et al. (1979) found that the order of tolerance development was C57BL > TO Swiss > DBA/2, and that the order for severity of withdrawal was DBA/2 > TO Swiss > C57BL following prolonged ethanol exposure. Development of acute tolerance (assessed by comparison of blood ethanol levels with loss and recovery of righting reflex) to hypnotic doses of ethanol by C57BL and C3H, but not by DBA/2 was reported by Tabakoff and Ritzmann (1979), which is consistent with the findings of Grieve and co-workers. Neither LS nor SS mice developed acute tolerance to the hypnotic effect of ethanol, but they both developed tolerance with chronic exposure (Tabakoff et al., 1980). Similar studies have been conducted by Goldstein and her colleagues. In earlier investigations, Goldstein and Kakihana (1974) also found that C57BL were more tolerant than DBA/2, as measured by blood ethanol levels with chronic ethanol vapor exposure. C57BL exhibited less severity of withdrawal (i.e., handling-induced convulsions) than DBA/2 or Swiss–Webster mice. However, in later investigations (employing measurement of tolerance development by changes in brain ethanol level during repeated rotorod performance with ataxic dosages of ethanol), the order of tolerance was found to be DBA > C57BL > LS = SS (Parsons et al., 1982). The erythrocyte membrane cholesterol level was higher in the ethanol-tolerant mice than in the controls. Kakihana (1979) reported the order of withdrawal severity after chronic exposure with a liquid alcohol diet as DBA/2 > BALB/c > C57BL, consistent with the findings of Grieve et al. (1979) and Goldstein and Kakihana (1974). DBA/2 mice also had elevated levels of plasma corticosterone during ethanol consumption and withdrawal, whereas C57BL showed no significant increase in levels during consumption or withdrawal. Grieve et al. (1979) proposed that the more rapidly that tolerance develops, the more rapidly it is lost. Thus, C57BL mice that develop tolerance rapidly would lose it rapidly and undergo *milder* withdrawal. This assumes that tolerance and dependence (at least as measured by severity of withdrawal) share a common physiological mechanism, which might not be the case. This relationship is not the experience with opiate dependence, where the more rapidly tolerance is lost, the more *severe* the withdrawal syndrome is: heroin withdrawal is more severe than methadone withdrawal and the most severe withdrawal is that precipitated by a narcotic antagonist, such as naloxone (Jaffe, 1980).

Several studies correlating tolerance development with other behaviors have been carried out. Tampier et al. (1981) studied rats originally selectively bred for alcohol preference (UChB) or aversion (UChA) by Mardones and his colleagues in Chile (Segonia-Riguelme et al., 1971). These investigators found that tolerance developed in the UChA (low-preference) rats and that very little tolerance developed in UChB (high-preference) rats when sleep time was used as the measure of tolerance. The method of chronic alcohol administration was forced-choice 10% ethanol in the drinking water. The preferring rats consumed more alcohol but nevertheless developed less tolerance. Using a group of HS mice, Allen et al. (1982) found no correlation between ethanol preference and subsequent severity of withdrawal following a liquid alcohol-containing diet for 8 days.

Selective Breeding.   Goldstein (1973) carried out an abbreviated selective breeding experiment on alcohol withdrawal using Swiss–Webster mice. Ethanol was administered by the vapor phase. Selection was applied for two generations only, but did show a small separation of severe and mild response to ethanol withdrawal between the lines. Subsequently, two selection experiments have started, using the HS mice from the Institute for Behavioral Genetics (IBG). The method of alcohol administration in the selection program at IBG (McClearn et al., 1982; Allen et al., 1983) is with a liquid diet. The selection index is a composite of phenotypes derived from severity of withdrawal and ethanol intake measures. The phenotypes were chosen in a pilot experiment as those that best distinguished withdrawing from nonwithdrawing animals. There are replicate high, low, and control lines, each with 10 mating pairs each generation. After five generations of selection the realized heritability calculated for the severe withdrawal phenotype is 0.10 in line 1 and 0.21 in line 2. For the mild withdrawal phenotype it is not significant in line 1 and 0.16 in line 2 (Allen et al., 1983).

The other selection experiment being conducted by Crabbe (1983a) uses essentially the same breeding paradigm, but a different method for development of physical dependence (inhalation) and a different measure of severity of withdrawal (degree of intensity of convulsions on handling during the withdrawal phase). After five generations of selective breeding, the seizure-prone and seizure-resistant lines have responded to selection, but a greater response has been observed in the seizure-prone direction (higher realized heritability). The prone and resistant lines also differ in peak seizure intensity during withdrawal (Crabbe et al., 1983b). It will be of interest to compare animals from these two selection experiments in the future.

A major problem in studies of tolerance and dependence is devising an efficient measure. Criteria for tolerance have depended mostly upon measures of acute effects with a decreased response at a given dosage, or an increased dosage to elicit the same response. The criteria for dependence, however, are many and varied. Most investigators are cautious in asserting that any one measure is an actual measure of dependence. Instead, dependence is assessed indirectly through measures of withdrawal, which (e.g., convulsive behavior) are presumed to correlate highly with dependence. A better measure of dependence *per se* might be the blood level of alcohol required to prevent the withdrawal phase. Such an experiment would be technically difficult and expensive, however, since a large number of animals would be needed and carefully controlled repeated monitoring blood alcohol levels would be tedious. Presently, the relationship between measures of withdrawal severity and *cellular* dependence is not known, simply because exactly what constitutes cellular dependence is unknown. Possible mechanisms could involve cell membrane fluidity, altered neurotransmitter receptor density or affinity, altered enzyme content or substrate affinity, or a combination of factors. Of course the alcohol investigator is not alone in attempting to define cellular tolerance; the same problem exists for all drugs that produce dependence.

Metabolism of Ethanol and Acetaldehyde

With regard to the evidence for genetic variation in the enzymes of ethanol and acetaldehyde metabolism and the subsequent influence of such variation on the behavioral response to ethanol, relatively few studies have been done. The first is that of Burnett and Felder (1980). These investigators have discovered that one line of deer mice, *Peromyscus maniculatus,* lacks liver alcohol dehydrogenase activity (ADH−) and another line expresses a functional ADH enzyme. Apparently the ADH− animals utilize a microsomal ethanol-oxidizing system. These animals promise to be of considerable use in metabolic and perhaps behavioral research in the future. In the LS and SS selected lines French et al. (1979) have found similar total basal microsomal ethanol-oxidizing activities and cytosolic ADH and ALDH activities in the liver. However, the SS mice showed higher inducibility by 3-methylcholanthrene and chronic ethanol ingestion of microsomal ethanol oxidation than did LS mice (French et al., 1979; Hjelle et al., 1981).

Major genetic abnormalities in ALDH discovered in animals to date are confined to the induction of a cytosolic ALDH in livers of selected rat lines (Deitrich, 1971; Deitrich et al., 1977). The enzyme is induced by phenobarbital, as well as other inducing agents to a maximum 30-fold increase in total cytosolic ALDH. Since the induced enzyme does not account for all of the cytosolic activity normally, the actual total induction is much larger. The enzyme has a relatively high Km for acetaldehyde, so that even in the induced state it exhibits relatively little influence on blood acetaldehyde levels following alcohol administration (Petersen et al., 1977).

Thurman (1980) has presented evidence that the rate of ethanol metabolism and the "swift increase in alcohol metabolism" (SIAM, 2–3 hr increase following ethanol metabolism) is inherited in the rat. In the mouse, while ethanol elimination rate was increased in four inbred strains (DBA/2, C3H, AKR, C57BL), after 4 hr of ethanol vapor there were ethanol dosage–dependent strain differences with respect to the maximal rate achieved. DBA/2 SIAM occurred at a lower blood ethanol concentration than C57BL, C3H, and AKR SIAMs (Thurman et al., 1982).

Relationship between Human and Animal Genetic Studies

To date, human genetic studies have been most useful in providing solid evidence that genotype plays a role in the development of alcoholism. As such, we now have at least one biologic risk factor for alcoholism: the presence of alcoholism in a first-degree relative (father, mother, full sibling). It is crucial that the scientific community educate the lay public that inheritance does not account for all risk factors, and that the providers of alcohol treatment not overinterpret the current evidence to the point of claiming that all forms of alcoholism are inherited, leaving no room for the importance of environmental conditions. Many individuals with no known genetic predisposition become alcoholic, whereas even more individuals with strong genetic predisposition do not become alcoholic.

The role of animals genetic studies is twofold. The first is to provide guidance for human studies, narrowing down the myriad possibilities to those most likely to prove fruitful. The second and independent role is the use of animal genetics to develop feasible models that will lead to the determination of the molecular mechanisms by which alcohol acts. The variety of techniques available eventually must be able to solve this difficult problem. The ultimate goal of such research is to be able to prevent, or at least alleviate, the damaging action of alcohol, whether it be on the cell (hepatocyte, or neuron) or, more globally, on an individual's judgment and contribution to society.

ACKNOWLEDGMENT

The original research reported was supported by grant AA03527 from the National Institute on Alcohol Abuse and Alcoholism.

REFERENCES

Agarwal, D. P., Harada, S., and Goedde, H. W., 1981, Racial differences in biological sensitivity to ethanol: The role of alcohol dehydrogenase and aldehyde dehydrogenase isozymes, *Alcohol Clin. Exp. Res.* **5**:12–16.

Allen, D. L., Fantom, H. J., and Wilson, J. R., 1982, Lack of association between preference for and dependence on ethanol, *Drug Alcohol Depend.* **9**:119–125.

Allen, D. L., Petersen, D. R., Wilson, J. R., McClearn, G. E., and Nishimoto, T. K., 1983, Selective breeding for a multivariate index of ethanol dependence in mice: Results from the first five generations, *Alcohol Clin. Exp. Res.* **7**:443–447.

Amark, C., 1951, A study in alcoholism, *Acta Psychiatr. Neurol. Scand.* [*Suppl*] **70**:107–120.

Amir, S., 1977, Brain and liver aldehyde dehydrogenase: relations to ethanol consumption in Wistar rats, *Neuropharmacology* **16**:781–784.

Amir, S., 1978, Brain and liver aldehyde dehydrogenase activity and voluntary ethanol consumption by rats: Relations to strains, sex and age, *Psychopharmacology* **57**:97–102.

Amir, S., and Stern, M. H., 1978, Electrical stimulation and lesions of the medial forebrain bundle of the rat: Changes in voluntary ethanol consumption, *Psychopharmacology* **57**:167–174.

Anderson, S. M., and McClearn, G. E., 1981, Ethanol consumption: Selective breeding in mice, *Behav. Genet.* **11**:291–300.

Anderson, S. M., McClearn, G. E., and Erwin, V. G., 1979, Ethanol consumption and hepatic enzyme activity, *Pharmacol. Biochem. Behav.* **11**:83–88.

Baer, D. S., and Crumpacker, D. W., 1977, Fertility and offspring survival in mice selected for different sensitivities to alcohol, *Behav. Genet.* **7**:95–103.

Bailey, D. W., 1971, Recombinant-inbred strains: An aid to finding identity, linkage, and function of histocompatibility and other genes, *Transplantation* **11**:325–327.

Baizer, L., Masserano, J., and Weiner, N., 1981, Ethanol-induced changes in tyrosine hydroxylase activity in brains of mice selectively bred for differences in sensitivity to ethanol, *Pharmacol. Biochem. Behav.* **15**:945–949.

Baker, R., Melchior, C., and Deitrich, R., 1980, The effect of halothane on mice selectively bred for differential sensitivity to alcohol, *Pharmacol. Biochem. Behav.* **12**:691–695.

Basile, A., Hoffer, B., and Dunwiddie, T., 1983, Differential sensitivity of cerebellar Purkinje neurons to ethanol in selectively outbred lines of mice: Maintenance *in vitro* independent of synaptic transmission. *Brain Res.* **264**:69–78.

Bass, M. B., and Lester, D., 1981a, Selective breeding for ethanol sensitivity: Least affected and most affected rats, in: Development of Animal Models as Pharmacogenetic Tools (G. E., McClearn, R. A. Deitrich, and V. G. Erwin, eds.), pp. 193–202, U.S. Dept. of Health and Human Services Research Monograph-6, Publication No. (ADM) 81-1133.

Bass, M. B., and Lester, D., 1981b, Task-dependent ethanol effects on escape in rats bred for ethanol sensitivity, *Pharmacol. Biochem. Behav.* **15**:33–36.

Bass, M. B., and Lester, D., 1983, Genetic analysis of sensitivity to ethanol-induced depression of motor activity and impairment of swimming in rats, *Behav. Genet.* **13**:77–89.

Belknap, J. K., and Deutsch, C. K., 1982, Differential neurosensitivity to three alcohols and phenobarbital in C57BL/6J and DBA/2J mice, *Behav. Genet.* **12**:309–317.

Belknap, J. K., MacInnes, J. W., and McClearn, G. E., 1972, Ethanol sleep times and hepatic alcohol and aldehyde dehydrogenase activities in mice, *Physiol. Behav.* **9**:453–457.

Belknap, J.K., Belknap, N.D., Berg, J.H., and Coleman, R., 1977, Preabsorptive vs. postabsorptive control of ethanol intake in C57BL/6J and DBA/2J mice, *Behav. Genet.* **7**:413–425.

Blass, J. P., and Gibson, G. E., 1977, Abnormality of a thiamine-requiring enzyme in patients with Wernicke-Korsakoff syndrome, *N. Engl. J. Med.* **297**:1367–1370.

Bohman, M., Sigvardsson, S., and Cloninger, C. R., 1981, Maternal inheritance of alcohol abuse, *Arch. Gen. Psychiatry* **38**:965–969.

Bosron, W. F., Li, T.-K., Dafeldecker, W. P., Vallee, B. L., 1979, Human liver π-alcohol dehydrogenase: Kinetic and molecular properties, *Am. Chem. Soc.* **6**:1101–1105.

Brewster, D. J., 1968, Genetic analysis of ethanol preference in rats selected for emotional reactivity, *J. Hered.* **59**:283–286.

Broadhurst, P. L., 1978, *Drugs and the Inheritance of Behavior*, Plenum Press, New York.

Burnett, K. G., and Felder, M. R., 1980, Ethanol metabolism in *Peromyscus* genetically deficient in alcohol dehydrogenase, *Biochem. Pharmacol.* **29**:125–130.

Cadoret, R. J., and Gath, A., 1978, Inheritance of alcoholism in adoptees, *Br. J. Psychiatry* **132**:252–258.

Cadoret, R. J., Cain, C. A., and Grove, W. M., 1980, Development of alcoholism in adoptees raised apart from alcoholic biologic relatives, *Arch. Gen. Psychiatry* **37**:561–563.

Chan, A. W. K., 1976, Gamma aminobutyric acid in different strains of mice: Effect of ethanol, *Life Sci.* **19**:597–604.

Church, A. C., Fuller, J. L., and Dudek, B. C., 1977, Behavioral effects of salsolinol and ethanol on mice selected for sensitivity to alcohol-induced sleep time, *Drug Alcohol Depend.* **2**:443–452.

Church, A. C., Fuller, J. L., and Dann, L., 1979, Alcohol intake in selected lines of mice: Importance of sex and genotype, *J. Comp. Physiol.* **93**:242–246.

Cicero, T. J., 1980, Animal models of alcoholism? in: *Animal Models in Alcohol Research* (K. Eriksson, J. D. Sinclair, and K. Kiianmaa, eds.), pp. 99–117, Academic Press, New York.

Cloninger, C. R., Bohman, M., and Sigvardsson, S., 1981, Inheritance of alcohol abuse, *Arch. Gen. Psychiatry* **38**:861–868.

Collins, A. C., 1981, A review of research using short-sleep and long-sleep mice, in *Development of Animal Models as Pharmacogenetic Tools* (G. E. McClearn, R. A. Deitrich, and V. G. Erwin, eds.), pp. 161–170, U.S. DHHS Monograph 6, Publ. No. (ADM) 81-1133.

Collins, A. C., Lebsack, M. E., and Yeager, T. N., 1976, Mechanisms that underlie sex-linked and genotypically determined differences in the depressant actions of alcohol, *Ann. N.Y. Acad. Sci.* **273**:303–316.

Cotton, N. S., 1979, The familial incidence of alcoholism, *J. Stud. Alcohol* **40**:89–115.

Crabbe, J. C., 1983, Sensitivity to ethanol in inbred mice: Genotypic correlations among several behavioral responses, *Behav. Neurosci.* **97**:280–289.

Crabbe, J. C., Janowsky, J. S., Young, E. R., and Rigter, H., 1980a, Strain-specific effects of ethanol on open field activity in inbred mice. *Subst. Alcohol Actions Misuse* **1**:537–543.

Crabbe, J. C., Rigter, H., and Kerbusch, S., 1980b, Genetic analysis of tolerance to ethanol hypothermia in recombinant inbred mice: Effect of desglycinamide (^{9})-arginine(^{8})-vasopressin, *Behav. Genet.* **10**:139–152.

Crabbe, J. C., Janowsky, J. S., Young, E. R., Kosobud, A., Stack, J., and Rigter, H., 1982, Tolerance to ethanol hypothermia in inbred mice: Genotypic correlations with behavioral responses, *Alcohol Clin. Exp. Res.* **6**:446–458.

Crabbe, J. C., Kosobud, A., and Young, E. R., 1983a, Genetic selection for ethanol withdrawal severity: Differences in replicate mouse lines. *Life Sci.* **33**:955–962.

Crabbe, J. C., Kosobud, A., and Young, E. R., 1983b, Peak ethanol withdrawal convulsions in genetically selected mice. *Proc. West. Pharmacol. Soc.* **26**:201–204.

Crabbe, J. C., Kosobud, A., Young, E. R., and Janowsky, J. S., 1983c, Polygenic and single-gene determination of responses to ethanol in BXD/Ty recombinant inbred mouse strains. *Neurobehav. Toxicol. Teratol.* **5**:181–187.

Crabbe, J. C., Young, E. R., and Kosobud, A., 1983d, Genetic correlations with ethanol withdrawal severity. *Pharmacol. Biochem. Behav.* **18**:541–547.

Damjanovich, R. P. and MacInnes, J. W., 1973, Factors involved in ethanol narcosis: Analysis in mice of three inbred strains, *Life Sci.* **13**:55–65.

DeFries, J. C., 1981, Current perspectives on selective breeding: Example and theory, in: *Development of Animal Models as Pharmacogenetic Tools*, (G. E. McClearn, R. A. Deitrich, and V. G. Erwin, eds.), pp. 11–35, U.S. Dept. of Health and Human Services Monograph-6, Publication No. (ADM) 81-1133.

Deitrich, R. A., 1971, Genetic aspects of increase in rat liver aldehyde dehydrogenase induced by phenobarbital, *Science* **173**:334–336.

Deitrich, R. A., Bludeau, P., and Roper, M., 1977, Induction of different rat liver supernatant aldehyde dehydrogenase by phenobarbital and tetrachlorodibenzo-p-dioxin, *J. Biol. Chem.* **252**:6169–6176.

Deitrich, R. A., Baker, R., Markley, H., Nelson, R., Ebert, M., and Merril, C., 1982, Polypeptide differences in brains of mice with different ethanol sensitivity, *Alcohol Clin. Exp. Res.* **6**:293.

Dibner, M. D., Zahniser, N. R., Wolfe, B. B., Rabin, R. A., and Molinoff, P. B., 1980, Brain neurotransmitter receptor systems in mice genetically selected for differences in sensitivity to ethanol, *Pharmacol. Biochem. Behav.* **12**:509–513.

Drewek, K. J., 1980, Inherited drinking and its behavioral correlates, in: *Animal Models in Alcohol Research* (K. Eriksson, J. D. Sinclair, and K. Kiianmaa, eds.), pp. 35–49, Academic Press, New York.

Drewek, K. J., and Broadhurst, P. L., 1979, Alcohol selection by strains of rats selectively bred for behavior, *J. Stud. Alcohol* **40:**723–728.

Drewek, K. J., and Broadhurst, P. L., 1981, A simplified triple-test cross analysis of alcohol preference in the rat, *Behav. Genet.* **11:**517–531.

Drewek, K. J., and Broadhurst, P. L., 1983*a*, More on the heritability of alcohol preference in laboratory mice and rats, *Behav. Genet.* **13:**123–125.

Drewek, K. J., and Broadhurst, P. L., 1983*b*, The genetics of alcohol preference in the female rat confirmed by a full triple test cross, *Behav. Genet.* **13:**107–116.

Dudek, B. C., 1982, Ethanol-induced conditioned taste aversions in mice that differ in neural sensitivity to ethanol, *J. Stud. Alcohol* **43:**129–136.

Dudek, B. C., and Fanelli, R. J., 1980, Effects of gamma-butyrolactone, amphetamine, and haloperidol in mice differing in sensitivity to alcohol, *Psychopharmacology* **68:**89–97.

Dudek, B. C., and Fuller, J. L., 1978, Task-dependent genetic influences on behavioral response of mice *(Mus musculus)* to acetaldehyde, *J. Comp. Physiol. Psychol.* **92:**749–758.

Durkin, T. P., Hashem-Zadeh, H., Mandel, P., and Ebel, A., 1982, A comparative study of the acute effects of ethanol on the cholinergic system in hippocampus and striatum of inbred mouse strains, *J. Pharmacol. Exp. Ther.* **220:**203–208.

Edwards, J. A., and Evans, D. A. P., 1967, Ethanol metabolism in subjects possessing typical and atypical alcohol dehydrogenase, *Clin. Pharmacol. Ther.* **8:**824.

Eleftheriou, B. E., and Elias, P. K., 1975, Recombinant inbred strains: A novel genetic approach for psychopharmacogeneticists, in: *Psychopharmacogenetics*, (B. E. Eleftheriou, ed.), pp. 43–71, Plenum Press, New York.

Eriksson, C. J. P., 1973, Ethanol and acetaldehyde metabolism in rat strains genetically selected for their ethanol preference, *Biochem. Pharmacol.* **22:**2283–2292.

Eriksson, C. J. P., 1980, Elevated blood acetaldehyde levels in alcoholics and their relatives: A reevaluation, *Science* **207:**1383.

Eriksson, C. J. P., 1981, Finnish selection studies on alcohol-related behaviors: Factors regulating voluntary alcohol consumption, in: *Development of Animal Models as Pharmacogenetic Tools* (G. E. McClearn, R. A. Deitrich, and V. G. Erwin, eds.), pp. 119–145, U.S. Department of Health and Human Services Research, Monograph-6, Publication No. (ADM) 81-1133.

Eriksson, C. J. P., and Peachey, J. E., 1980, Lack of difference in blood acetaldehyde of alcoholics and controls after ethanol ingestion, *Pharmacol. Biochem. Behav.* **13:**101–105.

Eriksson, C. J. P., Deitrich, R. A., Clay, K., and Petersen, D. A., 1982*a*, Dissociation of components of ethanol intoxication and tolerance, in: *Advances in Pharmacology and Therapeutics II*, Vol. 5, (H. Yoshida, Y. Hagihara, and S. Ebashi, eds.), pp. 245–251, Pergamon Press, Oxford and New York.

Eriksson, C. J. P., Mizoi, Y., and Fukunaga, T., 1982*b*, The determination of acetaldehyde in human blood by the perchloric acid precipitation method: The characterization and elimination of artefactual acetaldehyde formation, *Anal. Biochem.* **125:**259–263.

Eriksson, K., 1968, Genetic selection for voluntary alcohol consumption in the albino rat, *Science* **159:**739–741.

Eriksson, K., 1969, The estimation of heritability for the self-selection of alcohol in the albino rat, *Ann. Med. Exp. Biol. Fenn.* **47:**172–174.

Eriksson, K., 1971, Rat strains specially selected for their voluntary alcohol consumption, *Ann. Med. Exp. Biol. Fenn.* **49:**67–72.

Eriksson, K., and Rusi, M., 1981, Finnish selection studies on alcohol-related behaviors: General outline, in: *Development of Animal Models as Pharmacogenetic Tools*, (G. E. McClearn, R. A. Deitrich, and V. G. Erwin, eds.), pp. 87–117, U.S. Dept. of Health and Human Services Monograph-6, Publication No. (ADM) 81-1133.

Erwin, V. G., and McClearn, G. E., 1981, Genetic influences on alcohol consumption and actions of alcohol, in: *Currents in Alcoholism* Vol VIII (M. Galanter, ed.), pp. 405–420, Grune and Stratton, New York.

Erwin, V. G., Heston, W. D. W., McClearn, G. E., and Deitrich, R. A., 1976, Effect of hypnotics on mice genetically selected for sensitivity to ethanol, *Pharmacol. Biochem. Behav.* **4:**679–683.

Erwin, V. G., McClearn, G. E., and Kuse, A. R., 1980, Interrelationships of alcohol consumption, actions of alcohol, and biochemical traits, *Pharmacol. Biochem. Behav.* **13:**297–302.

Ewing, J. A., Rouse, B. A., and Pellizzari, E. D., 1974, Alcohol sensitivity and ethnic background, *Am. J. Psychiatry* **131:**206–210.

Falconer, D. S., 1960, *Introduction to Quantitative Genetics,* Ronald Press, New York.

French, T. A., Atkinson, N., Petersen, D. R., and Chung, L. W. K., 1979, Differential induction of hepatic microsomal ethanol and drug metabolism by 3-methylcholanthrene in LS and SS mice, *J. Pharmacol. Exp. Ther.* **209:**404–410.

Fukui, M. and Wakasugi, C., 1971, Liver alcohol dehydrogenase in a Japanese population, *Jpn. J. Legal Med.* **26:**46–51.

Fuller, J. L., 1964, Measurement of alcohol preferences in genetic experiments, *J. Comp. Physiol. Psychol.* **57:**85–88.

Fuller, J. L., 1980, Regulation of ethanol intake in long- and short-sleep mice, in: *Animal Models in Alcohol Research* (K. Eriksson, J. D. Sinclair, and K. Kiianmaa, eds.), pp. 57–62, Academic Press, New York.

Fuller, J. L., and Collins, R. L., 1972, Ethanol consumption and preference in mice: A genetic analysis, *Ann. N.Y. Acad. Sci.* **197:**42–48.

Gabrielli, W. F., Mednick, S. A., Volavka, J., Pollock, V. E., Schulsinger, F., and Itil, T. M., 1982, Electroencephalograms in children of alcoholic fathers, *Psychophysiology* **19:**404–407.

Gilliam, D. M., and Collins, A. C., 1982a, Acute ethanol effects on blood pH, PCO_2, and PO_2 in LS and SS mice, *Physiol. Behav.* **28:**879–883.

Gilliam, D. M., and Collins, A. C., 1982b, Circadian and genetic effects on ethanol elimination in LS and SS mice, *Alcohol Clin. Exp. Res.* **6:**344–348.

Gilliam, D. M., and Collins, A. C., 1983, Concentration-dependent effects of ethanol in long-sleep and short-sleep mice, *Alcohol Clin. Exp. Res.* **7:**337–342.

Gilliam, D. M., Bloedow, D. C., and Collins, A. C., 1983, Nonlinear pharmacokinetics of ethanol elimination in LS and SS mice, *Alcohol Clin. Exp. Res.* **7:**95–99.

Goedde, H. W., Agarwal, D. P., Harada, S., and Meier-Tackmann, D., 1982, ALDH polymorphism and alcohol sensitivity: Biochemical and population genetic studies, *Alcohol Clin. Exp. Res.* **6:**434.

Goldstein, D. B., 1973, Inherited differences in intensity of alcohol withdrawal reactions in mice, *Nature* **245:**154–156.

Goldstein, D. B., and Kakihana, R., 1974, Alcohol withdrawal reactions and reserpine effects in inbred strains of mice, *Life Sci.* **15:**415–425.

Goldstein, D. B., Chin, J. H., and Lyon, R. C., 1982, Ethanol disordering of spin-labeled mouse brain membranes: Correlation with genetically determined ethanol sensitivity of mice, *Proc. Natl. Acad. Sci. USA* **79:**4231–4233.

Goodrick, C. L., 1978, Ethanol selection by inbred mice: Mode of inheritance and the effect of age on the genetic system, *J. Stud. Alcohol* **39:**19–38.

Goodwin, D. W., 1971, Is alcoholism hereditary? A review and critique, *Arch. Gen. Psychiatry* **25:**545–549.

Goodwin, D. W., 1980, The genetics of alcoholism, *Subst. Alcohol Actions Misuse* **1:**101–117.

Goodwin, D. W., Schulsinger, F., Hermansen, L., Guze, S. B., and Winokur, G., 1973, Alcohol problems in adoptees raised apart from alcoholic biological parents, *Arch. Gen. Psychiatry* **28:**238–243.

Goodwin, D. W., Schulsinger, F., Moller, N., Hermansen, L., Winokur, G., and Guze, S. B., 1974, Drinking problems in adopted and nonadopted sons of alcoholics, *Arch. Gen. Psychiatry* **31:** 164–169.

Goodwin, D. W., Schulsinger, F., Knop, J., Mednick, S., and Guze, S., 1977, Psychopathology in adopted and nonadopted daughters of alcoholics, *Arch. Gen. Psychiatry* **34:**1005–1009.

Greer, C. A., and Alpern, H. P., 1977, Mediation of myoclonic seizures by dopamine and clonic seizures by acetylcholine and GABA, *Life Sci.* **21:**385–392.

Grieve, S. J., and Littleton, J. M., 1979, Age and strain differences in the rate of development of functional tolerance to ethanol by mice, *J. Pharm. Pharmacol.* **31:**696–700.

Grieve, S. J., Griffiths, P. J., and Littleton, J. M., 1979, Genetic influences on the rate of development of ethanol tolerance and the ethanol physical withdrawal syndrome in mice, *Drug Alcohol Depend.* **4:**77–86.

Harada, S., Agarwal, D. P., and Goedde, H. W., 1978, Human liver alochol dehydrogenase isoenzyme variations, *Hum. Genet.* **40:**215–220.

Harada, S., Agarwal, D. P., and Goedde, H. W., 1980a, Electrophoretic and biochemical studies of human aldehyde dehydrogenase isozymes in various tissues, *Life Sci.* **26:**1773–1780.

Harada, S., Misawa, S., Agarwal, D. P., and Goedde, H. W., 1980b, Liver alcohol dehydrogenase and aldehyde dehydrogenase in the Japanese: Isozyme variation and its possible role in alcohol intoxication, *Am. J. Hum. Genet.* **32:**8–15.

Harada, S., Takagi, S., Agarwal, D. P., and Goedde, H. W., 1982, Ethanol and aldehyde metabolism in alcoholics from Japan, *Alcohol Clin. Exp. Res.* **6:**299.

Harburg, E., Davis, D. P., and Caplan, R., 1982, Parent and offspring alcohol use. Imitative and aversive transmission, *J. Stud. Alcohol* **43:**497–516.

Harris, H., Hopkinson, D. A., and Robson, E. B., 1973, The incidence of rare alleles determining electrophoretic variants: Data on 43 enzyme loci in man, *Ann. Hum. Genet.* **37:**237–253.

Heston, W. D. W., and Erwin, V. G., 1974, A comparison of the effects of alcohol on mice selectively bred for differences in ethanol sleep-time, *Life Sci.* **14:**365–370.

Hjelle, J. J., Atkinson, N., and Petersen, D. R., 1981, The effects of chronic ethanol ingestion on ethanol binding to hepatic cytochrome P-450 and on certain hepatic and renal parameters in the "long sleep" and "short sleep" mouse, *Alcohol Clin. Exp. Res.* **5:**198–203.

Ho, A. K. S., Tsai, C. S., and Kissin, B., 1975, Neurochemical correlates of alcohol preference in inbred strains of mice, *Pharmacol. Biochem. Behav.* **3:**1073–1076.

Horowitz, G. P., and Allan, A. M., 1982, Morphine withdrawal in mice selectively bred for differential sensitivity to ethanol, *Pharmacol. Biochem. Behav.* **16:**35–39.

Horowitz, G. P., and Whitney, G., 1975, Alcohol-induced conditioned aversion: Genotypic specificity in mice *(Mus musculus)*, *J. Comp. Physiol. Psychol.* **89:**340–346.

Horowitz, G. P., Dendel, P. S., Allan, A. M., and Major, L. F., 1982, Dopamine-β-hydroxylase activity and ethanol-induced sleep time in selectively bred and heterogeneous stock mice, *Behav. Genet.* **12:**549–561.

Howerton, T. C., O'Connor, M. F., and Collins, A. C., 1983, Lipid solubility is correlated with hypnotic and hypothermic responses of long-sleep (LS) and short-sleep (SS) mice to various depressant drugs, *J. Pharmacol. Exp. Ther.* **227:**389–393.

Howerton, T. C., and Collins, A. C., 1984, Ethanol-induced inhibition of norepinephrine release from brain slices obtained from LS and SS mice. *Alcohol*, in press.

Hrubec, Z., and Omenn, G. S., 1981, Evidence of genetic predisposition to alcoholic cirrhosis and psychosis: Twin concordances for alcoholism and its biological end points by zygosity among male veterans, *Alcohol Clin. Exp. Res.* **5:**207–215.

Inoue, K., Fukunaga, M., and Yamasawa, K., 1980, Correlation between human erythrocyte aldehyde dehydrogenase activity and sensitivity to alcohol, *Pharmacol. Biochem. Behav.* **13:**295–297.

Inoue, K., Rusi, M., and Lindros, K. O., 1981, Brain aldehyde dehydrogenase activity in rat strains with high and low ethanol preferences, *Pharmacol. Biochem. Behav.* **14:**107–111.

Jaffe, T. H., 1980, Drug addiction and drug abuse, in: *The Pharmacological Basis of Therapeutics* (A. G. Gilman, L. S. Goodman, and A. Gilman, eds.), p. 539, MacMillan Press, New York.

Jonsson, E. and Nilsson, T., 1968, Alkohol konsumption hos monozygota och dizygota tvillinger, *Nord. Hyg. Tidskr.* **49:**21–25.

Kaij, L., 1960, *Alcoholism in Twins: Studies on the Etiology and Sequels of Abuse of Alcohol,* Almqvist and Wiksell, Stockholm.

Kaij, L., and Dock, J., 1975, Grandsons of alcoholics, *Arch. Gen. Psychiatry* **32:**1379–1381.

Kakihana, R., 1976, Adrenocortical function in mice selectively bred for different sensitivity to ethanol, *Life Sci.* **18:**1131–1138.

Kakihana, R., 1979, Alcohol intoxication and withdrawal in inbred strains of mice: Behavioral and endocrine studies, *Behav. Neurol. Biol.* **26:**97–105.

Kakihana, R., Brown, D. R., McClearn, G. E., and Tabershaw, I. R., 1966, Brain sensitivity to alcohol in inbred mouse strains, *Science* **154:**1574–1575.

Kaprio, J., Sarna, S., Koskenvuo, M., and Rantasalo, I., 1978, The Finnish Twin Registry: Baseline characteristics. Section II. History of symptoms and illness, use of drugs, physical characteristics, smoking, alcohol and physical activity, *Kansanterveystieteen Julkaisuja* m 37, Helsinki.

Koblin, D. D., and Deady, J. E., 1981, Anaesthetic requirement in mice selectively bred for differences in ethanol sensitivity, *Br. J. Anaesth.* **53:**5–10.

Kopun, M., and Propping, P., 1977, The kinetics of ethanol absorption and elimination in twins and supplementary repetitive experiments in singleton subjects, *Eur. J. Clin. Pharmacol.* **11:**337–344.

Korsten, M. A., Matsuzaki, S., Feinman, L., and Lieber, C. S., 1975, High blood acetaldehyde levels after ethanol administration. Difference between alcoholic and nonalcoholic subjects, *N. Engl. J. Med.* **292:**386–389.

Li, T.-K., 1977, Enzymology of human alcohol metabolism, *Adv. Enzymol.* **45:**427–481.

Li, T.-K., and Lumeng, L., 1977, Alcohol metabolism of inbred strains of rats with alcohol preference and nonpreference, in: *Alcohol and Aldehyde Metabolizing Systems* (R. G. Thurman, ed.), pp. 625–633, Academic Press, New York.

Li, T.-K., Bosron, W. F., Dafeldecker, W. P., Lange, L. G., Vallee, B. L., 1977, Isolation of π-alcohol dehydrogenase of human liver: Is it a determinant of alcoholism? *Proc. Natl. Acad. Sci. USA* **74:**4378–4381.

Li, T.-K., Lumeng, L., McBride, W. J., and Waller, M. B., 1981, Indiana selection studies on alcohol-related behaviors, in: *Development of Animal Models as Pharmacogenetic Tools* (G. E. McClearn, R. A. Deitrich, and V. G. Erwin, eds.), pp. 171–191, U.S. Dept. of Health and Human Services Research Monograph-6, Publication No. (ADM) 81-1133.

Lindros, K. O., and Eriksson, C. J. P., 1981, Artefactual acetaldehyde formation during human blood perchlorate deproteinization, *Subst. Alcohol Actions Misuse* **2:**341–347.

Linkola, J., 1976, Urine sodium, potassium and osmolality in two rat strains selected for their different ethanol preferences, *Med. Biol.* **54:**254–259.

Linkola, J., 1982, Strain differences in water and electrolyte metabolism between alcohol preferring (AA) and alcohol avoiding (ANA) rats, Academic dissertation, Helsinki, Finland.

Linkola, J., Fyhrquist, F., and Forsander, O., 1977, Effects of ethanol on urinary arginine vasopressin excretion in two rat strains selected for their different ethanol preferences, *Acta Physiol. Scand.* **101:**126–128.

Linkola, J., Tikkanen, I., Fyhrquist, F., and Rusi, M., 1980, Renin, water drinking, salt preference and blood pressure in alcohol preferring and alcohol avoiding rats, *Pharmacol. Biochem. Behav.* **12:**293–296.

Lipscomb, T. R., and Nathan, P. E., 1980, Blood alcohol level discrimination: The effects of family history of alcoholism, drinking pattern, and tolerance, *Arch. Gen. Psychiatry* **37:**571–576.

Loehlin, J. C., 1972, An analysis of alcohol-related questionnaire items from the National Merit twin study, *Ann. N.Y. Acad. Sci.* **197:**121–125.

Lumeng, L., Waller, M. B., McBride, W. J., and Li, T.-K., 1982, Different sensitivities to ethanol in alcohol preferring and nonpreferring rats, *Pharmacol. Biochem. Behav.* **16**:125–130.

Luth, K. F., 1939, Untersuchungen uber die alkoholblutkonzentration nach alkoholgaben bei 10 eineiigen und 10 zweieiigen zwillingspaaren, *Dtsch. Z. Gerichtl. Med.* **32**:145–164.

Major, L. F., and Murphy, D. L., 1978, Platelet and plasma amine oxidase activity in alcoholic individuals, *Br. J. Psychiatry* **132**:548–554.

Martin, P. R., 1981, The human genetics of alcoholism, *Subst. Alcohol Actions Misuse* **2**:389–406.

Masserano, J. M., and Weiner, N., 1982, Investigations into the neurochemical mechanisms mediating differences in ethanol sensitivity in two lines of mice, *J. Pharmacol. Exp. Ther.* **221**:404–409.

Mather, K., and Jinks, J. L., 1971, *Biometrical Genetics*, Cornell University Press, Ithaca, pp. 81–82, 125–126.

McClearn, G. E., and Anderson, S. M., 1979, Genetics and ethanol tolerance, *Drug Alcohol Depend.* **4**:61–76.

McClearn, G. E., and DeFries, J. C., 1973, *Introduction to Behavioral Genetics*, Freeman, San Francisco.

McClearn, G. E., and Kakihana, R., 1981, Selective breeding for ethanol sensitivity: Short-sleep and long-sleep mice, in: *Development of Animal Models as Pharmacogenetic Tools* (G. E. McClearn, R. A. Deitrich, and V. G. Erwin, eds.), pp. 147–159, U.S. Dept. of Health and Human Services Research Monograph-6, Publication No. (ADM) 81-1133.

McClearn, G. E., and Rodgers, D. A., 1959, Differences in alcohol preference among inbred strains of mice, *Q. J. Stud. Alcohol* **20**:691–695.

McClearn, G. E., Wilson, J. R., Petersen, D. R., and Allen, D. L., 1982, Selective breeding in mice for severity of the ethanol withdrawal syndrome, *Subst. Alcohol Actions Misuse* **3**:135.

Melchior, C. L., and Myers, R. D., 1976, Genetic differences in ethanol drinking of the rat following injection of 6-OHDA, 5,6-DHT or 5,7-DHT into the cerebral ventricles, *Pharmacol. Biochem. Behav.* **5**:63–72.

Mizoi, Y., Ijiri, I., Tatsuno, Y., Kijima, T., Fujiwara, S., and Adachi, J., 1979, Relationship between facial flushing and blood acetaldehyde levels after alcohol intake, *Pharmacol. Biochem. Behav.* **10**:303–311.

Mizoi, Y., Hishida, S., Ijiri, I., Maruyama, J., Asakura, S., Kijima, T., Okada, T., and Adachi, J., 1980, Individual differences in blood and breath acetaldehyde levels and urinary excretion of catecholamines after alcohol intake, *Alcohol Clin. Exp. Res.* **4**:354–360.

Moore, J. A., and Kakihana, R., 1978, Ethanol-induced hypothermia in mice: Influence of genotype on development of tolerance, *Life Sci.* **23**:2331–2338.

Myers, R. D., 1981, Alcohol's effect on body temperature: Hypothermia, hyperthermia, or poikilothermia, *Brain Res. Bull.* **7**:209–221.

Myers, R. D., and Melchior, C. L., 1975, Dietary tryptophan and the selection of ethyl alcohol in different strains of rats, *Psychopharmacology* **42**:109–115.

Myers, R. D., and Melchior, C. L., 1977, Differential action on voluntary alcohol intake of tetrahydroisoquinolines or a β-carboline infused chronically in the ventricle of the rat, *Pharmacol. Biochem. Behav.* **7**:381–392.

Northrup, L., 1976, Additive effects of ethanol and Purkinje cell loss in the production of ataxia in mice, *Psychopharmacology* **48**:189–192.

Oakeshott, J. G., and Gibson, J. B., 1981, The genetics of human alcoholism: A review, *Aust. N. Z. J. Med.* **11**:123–128.

O'Connor, M. F., Howerton, T. C., and Collins, A. C., 1982, Effects of pentobarbital in mice selected for differential sensitivity to ethanol, *Pharmacol. Biochem. Behav.* **17**:245–248.

Oliverio, A., and Eleftheriou, B. E., 1976, Motor activity and alcohol: Genetic analysis in the mouse, *Physiol. Behav.* **16**:577–581.

Omenn, G. S., 1975, Alcoholism: A pharmacogenetic disorder, in *Genetics and Psychopharmacology. Modern Problems of Pharmacopsychiatry* (J. Mendlewicz, ed.), pp. 12–22, S. Karger, Basel, Switzerland.

Palmer, M. R., Sorensen, S. M., Freedman, R., Olson, L., Hoffer, B., and Seiger, A., 1982, Differential ethanol sensitivity of intraocular cerebellar grafts in long-sleep and short-sleep mice, *J. Pharmacol. Exp. Ther.* **222**:480–487.

Palmer, M. R., Olson, L., Dunwiddie, T. V., Hoffer, B. J., and Seiger, A., 1984, The effect of neonatal cerebellectomy on ethanol-induced sleep time of short sleep and long sleep mice, *Pharmacol. Biochem. Behav.* **20**:153–159.

Parsons, L. M., Gallaher, E. J., and Goldstein, D. B., 1982, Rapidly developing functional tolerance to ethanol is accompanied by increased erythrocyte cholesterol in mice, *J. Pharmacol. Exp. Ther.* **223**:472–476.

Partanen, J., Bruun, K., and Markkanen, T., 1966, Inheritance of Drinking Behavior, A study of intelligence, personality and use of alcohol of adult twins. The Finnish Foundation for Alcohol Studies, Publication 14, Distributed by Rutgers University Center of Alcohol Studies, New Brunswick, New Jersey.

Penn, P. E., McBride, W. J., Lumeng, L., Gaff, T. M., and Li, T.-K., 1978, Neurochemical and operant behavioral studies of a strain of alcohol-preferring rats, *Pharmacol. Biochem. Behav.* **8**:475–481.

Petersen, D. R., Collins, A. C., and Deitrich, R. A., 1977, Role of cytosolic aldehyde dehydrogenase in control of blood acetaldehyde concentration, *J. Pharmacol. Exp. Ther.* **201**:471–481.

Pickett, R. A., and Collins, A. C., 1975, Use of genetic analysis to test the potential role of serotonin in alcohol preference, *Life Sci.* **17**:1291–1296.

Pollock, V. E., Volavka, J., Goodwin, D. W., Mednick, S. A., Gabrielli, W. F., Knop, J., Mikkelson, U. M., and Schulsinger, F., 1983, The EEG after alcohol in men at risk for alcoholism, *Arch. Gen. Psychiatry* **40**:857–861.

Propping, P., 1977*a*, Genetic control of ethanol action on the central nervous system, *Hum. Genet.* **35**:309–334.

Propping, P., 1977*b*, Psychophysiologic test performance in normal twins and in a pair of identical twins with essential tremor that is suppressed by alcohol, *Hum. Genet.* **36**:321–325.

Propping, P., 1978, Alcohol and alcoholism, *Hum. Genet.* **1**:91–99.

Randall, C. L., and Lester, D., 1974, Differential effects of ethanol and pentobarbital on sleep times in C57BL and BALB mice, *J. Pharmacol. Exp. Ther.* **188**:27–33.

Randall, C. L., Carpenter, J. A., Lester, D., and Friedman, H. J., 1975, Ethanol-induced mouse strain differences in locomotor activity, *Pharmacol. Biochem. Behav.* **3**:533–535.

Reed, T. E., 1977, Three heritable responses to alcohol in a heterogeneous randomly mated mouse strain, *J. Stud. Alcohol* **38**:618–632.

Reed, T. E., Kalant, H., Gibbins, R. J., Kapur, B. M., and Rankin, J. G., 1976, Alcohol and acetaldehyde metabolism in Caucasians, Chinese and Amerinds, *Can. Med. Assoc. J.* **6**:851–855.

Riley, E. P., Freed, E. X., and Lester, D., 1976, Selective breeding of rats for differences in reactivity to alcohol. An approach to an animal model of alcoholism. I. General procedures, *J. Stud. Alcohol* **37**:1535–1546.

Riley, E. P., Worsham, E. D., Lester, D., and Freed, E. X., 1977, Selective breeding of rats for differences in reactivity to alcohol. An approach to an animal model of alcoholism. II. Behavioral measures, *J. Stud. Alcohol* **38**:1705–1717.

Riley, E. P., Shapiro, N. R., and Lochry, E. A., 1979, Hypnotic susceptibility to various depressants in rats selected for differential ethanol sensitivity, *Psychopharmacology* **60**:311–312.

Rodgers, D. A., and McClearn, G. E., 1962, Mouse strain differences in preference for various concentrations of alcohol, *Q. J. Stud. Alcohol* **23**:26–33.

Rodgers, D. A., McClearn, G. E., Bennett, E. L., and Hebert, M., 1963, Alcohol preference as a function of its caloric utility in mice, *J. Comp. Physiol. Psychol.* **56**:663–672.

Roth, R. H., Walters, J. R., and Aghajanian, G. K., 1973, Effect of impulse flow on the release and synthesis of dopamine in the rat striatum, in: *Frontiers in Catecholamine Research* (E. Usdin and S. H. Snyder, eds.), pp. 567–574, Pergamon Press, New York.

Rusi, M., Eriksson, K., and Maki, J., 1977, Genetic differences in the susceptibility to acute ethanol intoxication in selected rat strains, in: *Alcohol Intoxication and Withdrawal*, Vol. 3A (M. M. Gross, ed.), pp. 97–109, Plenum Publishing Corp., New York.

Sanders, B., 1976, Sensitivity to low doses of ethanol and pentobarbital in mice selected for sensitivity to hypnotic doses of ethanol, *J. Comp. Physiol. Psychol.* **90:**394–398.

Sanders, B., 1980, Withdrawal-like signs induced by a single administration of ethanol in mice that differ in ethanol sensitivity, *Psychopharmacology* **68:**109–113.

Sanders, B. and Sharpless, S. K., 1978, Dissociation between the anticonvulsant action of alcohol and its depressant action in mice of different genotypes, *Life Sci.* **23:**2593–2600.

Satinder, K. P., 1970, Behavior-genetic-dependent self-selection of alcohol in rats, *J. Comp. Physiol. Psychol.* **80:**422–434.

Schaefer, J. M., 1978, Alcohol metabolism and sensitivity reactions among the Reddis of South India, *Alcohol Clin. Exp. Res.* **2:**61–69.

Schneider, C. W., Evans, S. K., Chenoweth, M. B., and Beman, F. L., 1973, Ethanol preference and behavioral tolerance in mice: Biochemical and neurophysiological mechanisms, *J. Comp. Physiol. Psychol.* **82:**466–474.

Schuckit, M. A., 1980, Self-rating of alcohol intoxication by young men with and without family histories of alcoholism, *J. Stud. Alcohol* **41:**242–249.

Schuckit, M. A., 1981, Peak blood alcohol levels in men at high risk for the future development of alcoholism, *Alcohol Clin. Exp. Res.* **5:**64–66.

Schuckit, M. A. and Rayses, V., 1978, Ethanol ingestion: Differences in blood acetaldehyde concentrations in relatives of alcoholics and controls, *Science* **203:**54–55.

Schuckit, M. A., Goodwin, D. A., and Winokur, G., 1972, A study of alcoholism in half siblings, *Am. J. Psychiatry* **128:**122–125.

Schuckit, M. A., Engstrom, D., Alpert, R., and Duby, J., 1981, Differences in muscle-tension response to ethanol in young men with and without family histories of alcoholism, *J. Stud. Alcohol* **42:**918–924.

Schuckit, M. A., Shaskan, E., Duby, J., and Moss, M., 1982, Platelet monoamine oxidase activity in relatives of alcoholics. Preliminary study with matched control subjects, *Arch. Gen. Psychiatry* **39:**137–140.

Schwitters, S. Y., Johnson, R. C., Johnson, S. B., and Ahern, F. M., 1982, Familial resemblances in flushing following alcohol use, *Behav. Genet.* **12:**349–352.

Segonia-Riquelme, N., Varela, A., and Mardones, J., 1971, Appetite for alcohol, in: *Biological Basis of Alcoholism*, p. 299, (Y. Israel and J. Mardones, eds.), Wiley, New York.

Seiger, A., Sorensen, S. M., and Palmer, M. R., 1983, Cerebellar role in the differential ethanol sensitivity of long sleep and short sleep mice, *Pharmacol. Biochem. Behav.*, **18:**495–499.

Shapiro, N. R., and Riley, E. P., 1980, Avoidance behavior in rats selectively bred for differential alcohol sensitivity, *Psychopharmacology* **72:**79–83.

Shapiro, N. R., Garg, A. P., and Riley, E. P., 1979, Genotypic-dependent amphetamine effects in rats bred for differences in alcohol sensitivity, *Physiol. Psychol.* **7:**403–406.

Sheppard, J. R., Albersheim, P., and McClearn, G. E., 1968, Enzyme activities and ethanol preference in mice, *Biochem. Genet.* **2:**205–212.

Sheppard, J. R., Albersheim. P., and McClearn, G. E., 1970, Aldehyde dehydrogenase and ethanol preference in mice, *J. Biol. Chem.* **245:**2876–2882.

Sidman, R. L., and Green, M. C., 1970, "Nervous," a new mutant mouse with cerebellar disease, *in* "Les Mutants Pathologiques Chez L'Animal," pp. 69–79, (M. Soboundy, ed.), Centre National de la Recherche Scientifique, Paris.

Siemens, A. J., and Chan, A. W. K., 1976, Differential effects of pentobarbital and ethanol in mice, *Life Sci.* **19:**581–590.

Smith, M., Hopkinson, D. A., and Harris, H., 1971, Developmental changes and polymorphism in human alcohol dehydrogenase, *Ann. Hum. Genet.* **34:**251–271.

Snead, O. C., 1977, Minireview: Gamma hydroxybutyrate, *Life Sci.* **20:**1935–1944.

Sorensen, S., Palmer, M., Dunwiddie, T., and Hoffer, B., 1980, Electrophysiological correlates of ethanol-induced sedation in differentially sensitive lines of mice, *Science* **210:**1143–1145.

Sorensen, S., Dunwiddie, T., McClearn, G., Freedman, R., and Hoffer, B., 1981, Ethanol-induced depressions in cerebellar and hippocampal neurons of mice selectively bred for differences in ethanol sensitivity: An electrophysiological study, *Pharmacol. Biochem. Behav.* **14:**277–234.

Spuhler, K., Hoffer, B., Weiner, N., and Palmer, M., 1982, Evidence for genetic correlation of hypnotic effects and cerebellar Purkinje neuron depression in response to ethanol in mice, *Pharmacol. Biochem. Behav.* **17:**569–578.

Swanberg, K. M., and Wilson, J. R., 1979, Genetic and ethanol-related differences in maternal behavior and offspring viability in mice, *Dev. Psychobiol.* **12:**61–66.

Tabakoff, B., and Ritzmann, R. F., 1977, The effects of 6-hydroxydopamine on tolerance to and dependence on ethanol, *J. Pharmacol. Exp. Ther.* **203:**319–331.

Tabakoff, B., and Ritzmann, R. F., 1979, Acute tolerance in inbred and selected lines of mice, *Drug Alcohol Depend.* **4:**87–90.

Tabakoff, B., Ritzmann, R. F., Raju, T. S., and Deitrich, R. A., 1980, Characterization of acute and chronic tolerance in mice selected for inherent differences in sensitivity to ethanol, *Alcohol Clin. Exp. Res.* **4:**70–73.

Tampier, L., Quintanilla, M. E., and Mardones, J., 1980, Genetic differences in tolerance to ethanol: A study in UChA and UChB rats, *Pharmacol. Biochem. Behav.* **14:**165–168.

Tan, O. T., Stafford, T. J., Sarkany, I., Gaylarde, P. M., Tilsey, C., and Payne, J. P., 1982, Supression of alcohol-induced flushing by a combination of H_1 and H_2 histamine antagonists, *Br. J. Dermatol.* **107:**647–652.

Teng, Y.-S., 1981, Human liver aldehyde dehydrogenase in Chinese and Asiatic Indians: Gene deletion and its possible implications in alcohol metabolism, *Biochem. Genet.* **19:**107–114.

Thurman, R. G., 1980, Ethanol elimination in the rat, *Adv. Exp. Med. Biol.* **132:**655–661.

Thurman, R. G., Paschal, D., Abu-Murad, C., Pakkanen, L., Bradford, B. U., Bullock, K., and Glassman, E., 1982, Swift increase in alcohol metabolism (SIAM) in the mouse: Comparison of the effect of short-term ethanol treatment on ethanol elimination in four inbred strains, *J. Pharmacol. Exp. Ther.* **223:**45–49.

Truitt, E. B., 1971, Blood acetaldehyde levels after alcohol consumption by alcoholic and nonalcoholic subjects, *Adv. Mental Sci.* **3:**212–232.

Utne, H. E., Hansen, F. V., Winkler, K., and Schulsinger, F., 1977, Alcohol elimination rates in adoptees with and without alcoholic parents, *J. Stud. Alcohol* **38:**1219–1223.

Vesell, E. S., Page, J. G., and Passananti, G. T., 1971, Genetic and environmental factors affecting ethanol metabolism in man, *Clin. Pharmacol. Ther.* **12:**192–201.

von Wartburg, J. P., 1980, Alcohol metabolism and alcoholism—pharmacogenetic considerations, *Acta Psychol. Scand.* **62:**179–188.

von Wartburg, J. P., and Schurch, P. M., 1968, Atypical human liver alcohol dehydrogenase, *Ann. N.Y. Acad. Sci.* **151:**936–946.

von Wartburg, J. P., Papenberg, J., and Aebi, H., 1965, An atypical human alcohol dehydrogenase, *Can. J. Biochem.* **43:**889–898.

Waller, M. B., McBride, W. J., Lumeng, L., and Li, T.-K., 1982, Induction of dependence on ethanol by free-choice drinking in alcohol-preferring rats, *Pharmacol. Biochem. Behav.* **16:**501–507.

Whitney, G., McClearn, G. E., and DeFries, J. C., 1970, Heritability of alcohol preference in laboratory mice and rats, *J. Hered.* **61:**165–169.

Whitney, G., McClearn, G. E., and DeFries, J. C., 1982, Heritability of alcohol preference in laboratory mice and rats: Erroneous estimates, *Behav. Genet.* **12:**543–546.

Winokur, G., Reich, T., Rimmer, J., and Pitts, F. N., 1970, Alcoholism: III. Diagnosis and familial psychiatric illness in 259 alcoholic probands, *Arch. Gen. Psychiatry* **23:**104–111.

Wolff, P. H., 1972, Ethnic differences in alcohol sensitivity, *Science* **175**:449–450.

Wolff, P. H., 1973, Vasomotor sensitivity to alcohol in diverse Mongoloid populations, *Am. J. Hum. Genet.* **25**:193–199.

York, J. L., 1981, The ethanol stimulus in rats with differing ethanol preferences, *Psychopharmacology* **74**:339–343.

4

Alcohol Consumption and Ischemic Heart Disease

The Epidemiologic Evidence

MARY JANE ASHLEY

1. INTRODUCTION

Ischemic heart disease* (IHD) is the leading cause of death in Canada, as it is in many developed countries. This complex disease entity accounted for 48,683 or 28.5% of the 171,029 deaths recorded in this country in 1981 (Statistics Canada, 1982). Epidemiologic studies of IHD conducted over the past several decades have clearly established the etiologic significance of various factors, such as smoking (e.g., Public Health Service, 1979b) and hypertension (e.g., Dawber, 1980). The risk or protective roles of certain other factors, however, remain unresolved, and some are the subjects of controversy. Alcohol consumption is one of these factors.

Some of the evidence concerning the relationship between alcohol consumption and IHD comes from studies of alcoholics and heavy drinkers. Most,

---

* Ischemic heart disease (IHD) comprises the rubrics 410–414 in Section VII, Diseases of the Circulatory System of the International Classification of Diseases, 8th and 9th Revisions (ICD-8, ICD-9) (Public Health Service, 1975). In the 6th and 7th Revisions of the ICD the rubrics 420 and 422 covered the entity "arteriosclerotic and degenerative heart disease," also called "coronary heart disease" (CHD) and "coronary artery disease" (CAD) (Moriyama et al., 1971; Public Health Service, 1975, 1979a). These rubrics correspond approximately to 410–413 in the ICD-8. The titles of these rubrics are as follows: 410, "acute myocardial infarction;" 411, "other acute and subacute forms" (of ischemic heart disease); 412, "chronic ischemic heart disease;" 413, "angina pectoris." The rubric 414 is entitled asymptomatic ischemic heart disease.

MARY JANE ASHLEY • Department of Preventive Medicine and Biostatistics, University of Toronto, Toronto, Ontario, Canada.

however, comes from observational studies of general or quasigeneral populations, involving several methodologic approaches. These methods permit definition of the association between alcohol consumption and IHD in either groups or individuals.

Covariation studies are carried out in an attempt to define the direction and measure the degree of the association between the postulated factor and the disease in populations over time, and in different places, as well as among groups within populations. As such, these studies are based on total population rather than on individual experiences. Prevalence, case–control, and cohort studies, on the other hand, permit testing of the association between the factor and the disease in individuals. In a prevalence study a representative sample of individuals from a population is selected and measurements of the factor and the disease are carried out at the same time in each person. In the case–control (retrospective) method individuals from a population who have the disease (cases) are compared with persons from the same population who do not have the disease (controls) with regard to prior exposure to the postulated factor. In the cohort (prospective) approach healthy individuals from the population are categorized according to their exposure to the factor and are observed over time for the development of the disease.

In this chapter evidence arising from each of these types of studies is reviewed and summarized. The limitations inherent in the epidemiologic approach to resolving the relationship between alcohol consumption and IHD are discussed, and the available data are evaluated against the criteria for judging causal relationships.

2.   STUDIES OF GENERAL POPULATIONS

Covariation Studies

Temporal Associations.   LaPorte et al. (1980a) undertook temporal analyses relating age-adjusted atherosclerotic heart disease* (ASHD) rates in the United States to *per capita* adult alcohol consumption (total and beverage-specific) in the years 1950–1975. The correlations were negative for beer, wine, and spirits, and for total alcohol consumption. The strongest and only significant association, however, was that between ASHD and beer consumption. In a time lag analysis using 0–10-year lag periods between the alcohol consumption variables and ASHD rates (4 alcohol variables, each with 11 lag periods), all but 2 of the 44 correlation coefficients obtained were negative. The strongest negative association was for beer consumption with a 5-year lag period ($r = -0.94$).

---

* Atherosclerotic heart disease (ASHD) was defined as the entity A-83 in the "A" list of the ICD-8, corresponding to the rubrics 410–414 (IHD) in the detailed list of the ICD-8.

Table 1. Comparison of the Sign and Significance of Temporal Correlations Based on Age-Standardized and Unstandardized[a] Death Rates from IHD and Alcohol Consumption[b] in 19 Countries[c]

	Total period (1950–1977) (number of coefficients of correlation)		Prerevision period (1950–1968) (number of coefficients of correlation)	
	+	−	+	−
IHD (standardized) and consumption	7 (6)[d]	12 (7)	13 (10)	6 (2)
IHD (unstandardized) and consumption	13 (9)	6 (2)	17 (14)	2 (0)
IHD (unstandardized) and cirrhosis	15 (11)	4 (1)	19 (15)	0 (0)

[a] Unstandardized mortality rates were based on population aged 25 and older.
[b] Alcohol consumption rates were based on the population aged 15 and older.
[c] Adapted from Schmidt and Popham (1981a, 1982).
[d] Bracketed numbers indicate coefficients that were statistically significant ($p \leq 0.05$).

For lag periods up to and including 7 years beer consumption was the best predictor of ASHD death rates, but for the longer lag periods (8–10 years) wine consumption was the best predictor. A multiple regression analysis using the highest lag association between the alcohol variables and ASHD, and including cigarette and fat consumption 5 and 4 years prior to the ASHD death rates, respectively, showed that beer consumption with a 5-year lag period and cigarette consumption with a 5-year lag period accounted for 92% of the variance associated with ASHD death rates. When the linear relationships of cigarette and fat consumption were controlled (Kuller et al., 1979), beer consumption was still highly related to ASHD death rates.

In a more extensive study, Schmidt and Popham (1981a, 1982) analyzed the relationships over time between *per capita* alcohol consumption and IHD* death rates in each of 19 countries (not including the USA) for which data for at least 20 years were available. The resulting correlation coefficients lacked consistency, varying considerably in both size and sign (Table 1). When age-standardized death rates were used, positive correlations were found in seven countries, of which six were large enough to be statistically significant. On the other hand, negative correlations were found in 12 countries, 7 of these being large enough to be statistically significant. A further analysis, confined to the period 1950–1968, to avoid artifacts associated with the change from the seventh to the eighth Revision of the ICD, increased the consistency of the findings

* The entities comprising IHD were B26 in the abbreviated list of the ICD-6, corresponding to the rubrics 420–422 in the detailed list, A-81 (ICD-7) corresponding to the rubrics 420–422 in the detailed list, and A-83 (ICD-8) corresponding to the rubrics 410–414 in the detailed list.

somewhat. Thirteen of the coefficients were positive, 10 being large enough to be statistically significant. However, six remained negative, of which two were statistically significant. In addition, Schmidt and Popham (1981a) analyzed their original temporal series for various time lags, with results that were difficult to interpret. However, it appeared that when time lags of 4 and 6 years were used, somewhat fewer positive significant correlations were obtained. In reference to this analysis, however, the authors pointed out the questionable appropriateness of time lag adjustments lacking *a priori* theoretical or empirical justification.

A further dimension to these analyses was provided by studying the relationship over time between deaths from IHD and liver cirrhosis, the latter being a valid index of the prevalence of heavy drinking in a population. Using data from the period prior to the eighth Revision (1950–1968), Schmidt and Popham (1981a) found positive temporal correlations between the two variables in each of the 19 countries studied, 15 of these being statistically significant (Table 1). LaPorte et al. (1980b) correlated age-adjusted death rates for ASHD in the United States between 1950 and 1975 with the corresponding cirrhosis mortality rates and found a weak positive correlation coefficient ($r = +0.10$), which was not statistically significant.

Spatial Associations.   Some years ago, Brummer (1969) studied the relationship between alcohol consumption and coronary mortality* in 20 countries. A negative correlation coefficient was found in each of the six age–sex-specific population groups in which the relationship was examined. None, however, was statistically significant.

In 1970, St. Leger et al. (1979) studied death rates from IHD (A-83, ICD-8)† in men and women ages 55–64 in 18 countries in relation to total alcohol consumption, specific beverage (wine, beer, spirits) consumption, and several other variables including gross national product, cigarette consumption, caloric intake, and consumption of various fats. Alcohol consumption was very strongly negatively associated with IHD mortality in both sexes, a finding not explained by any of the other variables examined. Of the three beverages, wine had the strongest association, the relationship being negative and of the same magnitude as that of total alcohol consumption. For spirits the relationship was weakly negative. In contrast, the relationship for beer was weakly positive. Regression analysis, which controlled for gross national product, cigarette consumption,

---

* Brummer defined "coronary mortality" as deaths due to "myocardial infarction." The ICD rubrics were not provided. The term "myocardial infarction" refers to the heart muscle damage which occurs in persons with IHD. Acute myocardial infarction (410, ICD-8) refers to the diagnostic entity associated with obstruction of blood flow through the coronary arteries to the heart muscle resulting in an inadequate supply of oxygen (ischemia). Symptoms may be none, mild, or severe, including chest pain (which may radiate to the shoulder, arm, neck, or jaw), nausea, cold sweat, and shortness of breath. When ischemia is severe and prolonged, heart muscle is damaged and death or disability may occur. Diagnostic serum enzyme and/or electrocardiographic changes may be found. This event is commonly called a "heart attack" (Public Health Service, 1979a).

† Corresponds to 410–414, ICD-8 (detailed list).

Table 2. Correlations between *Per Capita* Alcohol Consumption[a] and Rates of
Death from Ischemic Heart Disease[b] and between Ischemic Heart Disease and
Liver Cirrhosis in Various Countries for Each of Three Years[c]

Year	Number of countries	Per capita consumption and ischemic heart disease		Ischemic heart disease and liver cirrhosis	
		Correlation coefficient	p	Correlation coefficient	p
1955	17	−0.44	<0.08	−0.50	<0.04
1965	26	−0.44	<0.03	−0.54	<0.01
1975	24	−0.42	<0.05	−0.48	<0.02

[a] Liters of absolute alcohol *per capita* 15 years of age and older.
[b] Per 100,000 population 25 years of age and older.
[c] Adapted from Schmidt and Popham (1981a).

and diet (total calories and consumption of carbohydrates and fats) did not alter the strong negative association between wine consumption and IHD death rates in either sex. The exclusion of France from the analysis (because of concerns about the reporting of IHD deaths in that country) did not alter the findings.

Using 1972 data from 20 countries, LaPorte et al. (1980a)* essentially confirmed these findings. They found a negative correlation ($r = -0.54$) between total alcohol consumption and IHD (A-83, ICD-8) death rates in men ages 55–64. A strong negative correlation for wine ($r = -0.65$) and a very weak negative correlation for spirits ($r = -0.03$) were observed. On the other hand, the correlation for beer was moderately positive ($r = +0.51$). When meat consumption was controlled, total alcohol consumption had the highest relationship to IHD death rates (partial correlation coefficient $= -0.65$). Controlling for cigarette smoking did not diminish this negative relationship.

Schmidt and Popham (1981a)† also found similar associations. They studied the relationships between *per capita* alcohol consumption and IHD death rates among various countries in three years, 1955, 1965, and 1975. The correlation coefficients were consistently negative (Table 2). The data were further analyzed by type of beverage. Consistently higher negative correlations were found for wine than for total alcohol. The correlation for spirits, although negative, was not significant, while that for beer was positive.

Buck et al. (1982) conducted a multiple regression analysis of data from 15 countries, which included gross national product, and *per capita* consumption of garlic/onion oil, animal fat, wine, beer, spirits, and cigarettes. They failed to find statistically significant correlation coefficients between IHD mortality and any of the alcohol variables, including wine, in either men or women in two age groups (45–54 and 55–64). Furthermore, the directions of the relationships were mixed. The lack of a negative relationship between wine and IHD was not due to confounding of wine with garlic/onion oil consumption. When the regres-

* See footnote, page 100.
† See footnote, pate 101.

sion was repeated excluding garlic/onion oil, wine was still unrelated to IHD. Instead, gross national product appeared to be the variable most strongly related (negatively) to IHD mortality across the countries studied.

Werth (1980), in a comparison across 50 states and the District of Columbia of death rates (not age-adjusted) from acute myocardial infarction per 100,000 population with adult per capita wine consumption for each of 10 years (1969–1978), found moderately strong negative correlation coefficients in all years. However, LaPorte and Cauley (1981), in a response to Werth, reported a correlational analysis between age-adjusted CHD mortality rates in 1970 for 48 states of the United States and *per capita* consumption of spirits, wine, beer, and total alcohol. For "white" men and women all of the correlation coefficients were positive, but they were small and statistically insignificant. For "other" men and women the correlation coefficients were mostly negative in direction, but only one was statistically significant, that for beer ($r = -0.29$, $p < 0.05$). They concluded that there was little relationship between *per capita* alcohol consumption and heart disease death rates. Schmidt and Popham (1981a) also analyzed 1970 data from 50 states and the District of Columbia. The correlation coefficients between IHD death rates (per 100,000 population 20 years of age and older) and consumption of each of wine, beer, spirits, and total alcohol (liters *per capita* 15 years of age and older) were all negative, but all were small and statistically insignificant. Partial correlations between IHD and each beverage and total alcohol, holding cigarette consumption constant, were -0.06, -0.28, -0.26, and -0.25 for beer, wine, spirits, and total alcohol, respectively. The last three coefficients were statistically significant, but none explained more than 8% of the variance.

Again, a further dimension has been added to the spatial correlational studies by an attempt to relate heavy consumption to IHD mortality. Schmidt and Popham (1981a), in three sets of spatial analyses, found strong negative correlations between mortality from IHD* and liver cirrhosis, a valid index for the prevalence of heavy drinking in a population (Table 2). In an analysis of interstate variation in the United States, however, they found the correlation between IHD and cirrhosis to be virtually zero ($r = -0.03$).

Associations across Population Groups.   Using data from England and Wales (1949–1953), Mathews (1976) calculated standardized mortality ratios (SMRs) for cirrhosis and coronary heart disease (420, ICD-6) in men aged 20–64 in 13 socioeconomic groups. Those groups with high SMRs for cirrhosis also had high SMRs for CHD ($r = +0.87$, $p < 0.01$). As a single predictor variable in a regression equation using the natural logarithms of the SMRs, the SMR for cirrhosis accounted for 65% of the variance in the SMR for CHD.

Interpretation of Findings.   The results of covariance studies always must be interpreted with considerable caution. These studies examine consumption and mortality rates over total populations, not relationships based on individuals.

* See footnote, page 101.

While significant relationships between average alcohol consumption and average IHD mortality across total populations may suggest that certain levels of alcohol consumption may be associated with certain risks of disease, studies based on individuals are needed to test the relationship. Fallacious associations may arise through the averaging of both exposures and disease outcomes, which may be heterogeneous and may be distributed differently in different populations; through the unrepresentative selection of the units across which the comparisons are made; through failure or inability to take into account known or unknown confounding variables; or through inconsistencies in the recording of data. Nonetheless, these studies may be useful in providing initial clues for the generation of hypotheses, and may provide additional evidence supporting that arising from studies based on individuals.

The evidence from the covariation studies reviewed does little to clarify the role of alcohol consumption in ischemic heart disease. The extensive temporal analysis by Schmidt and Popham (1981a, 1982), carried out on data from each of 19 countries, to which may be added the analysis of United States data (LaPorte et al., 1980a), indicates that the relationship between *per capita* alcohol consumption and IHD mortality is inconsistent, both in direction and significance. The spatial analyses also have provided mixed results. Some of the analyses indicated negative relationships, which were statistically significant, while others failed to find significant relationships in either direction. Evidence from covariation studies concerning the relationship between IHD and cirrhosis (as an index of heavy drinking) was also mixed. Some studies suggested significant relationships, both negative and positive, and others found no relationships at all.

Schmidt and Popham (1981a), in attempting to explain these inconsistencies, postulated that the association between alcohol consumption and IHD in general populations is dependent upon variation in a third, as yet unidentified, factor. They suggested that a nutritional factor might be involved, one which is reasonably constant through time but variable from country to country. Manku et al. (1979) postulated a dietary explanation involving an interaction of alcohol with intake of essential fatty acids and suggested that beverage differences per se were artifacts. Specific beverages may be indicators of total alcohol consumption or of certain food habits, or both. As Schmidt and Popham (1981a) pointed out, among European countries high wine consumption tends to go with high overall alcohol consumption. It is also joined with a relatively high intake of vegetable oils and cheese and a low intake of milk and butter, these dietary elements being, respectively, negatively and positively correlated with IHD mortality.

Prevalence Studies

Wood et al. (1981) have presented the results of a prevalence study conducted on a random sample of 448 men aged 45–54 drawn from the lists of general practitioners in North Edinburgh and West Fife, Scotland. Reported alcohol consumption was recorded in units per week, where one unit was equal to $^1/_2$ pint of beer. The prevalence of IHD in three diagnostic groups, myocardial infarction (MI), angina pectoris (AP), and "possible" IHD, was determined from medical records and through the use of the Chest Pain Questionnaire of the World Health Organization. Alcohol consumption in units/week was similar in men with and without IHD. The number of drinking days/week was also not significantly different. However, there was a tendency for men with IHD, and particularly, those with MI, to consume their weekly intake of alcohol in a lesser number of drinking days so that units of alcohol consumed per drinking day tended to be greater in the IHD groups than in the non-IHD group ($p = 0.06$). In the MI, AP, and "possible" IHD groups average alcohol consumption per drinking day was 8.6 units, 7.8 units, and 6.7 units, respectively, while in the men without IHD it was 6.2 units.

The results of prevalence studies must also be interpreted with caution. Although the study by Wood et al. (1981) indicated somewhat more intensive drinking (consumption of the same amount of alcohol over a shorter time) in men with IHD than in men free of disease, and especially so in men with MI, one must be careful not to assign an unwarranted time sequence. The more intensive consumption pattern may be a consequence rather than a cause of IHD. Also, patients with IHD identified in a prevalence survey may not be representative, in that this excludes patients dying suddenly and may overrepresent cases of longer duration. Furthermore, this may underrepresent patients that leave the population to enter specialized treatment facilities elsewhere.

Case–Control (Retrospective) Studies

Several case–control studies provide evidence of a negative association between alcohol consumption and the risk of IHD. In a study of myocardial infarction (MI) in women, Petitti et al. (1979) found any drinking (not defined further) to be associated with a decreased risk of MI. Of the controls 66.4% drank alcohol compared with only 39.5% of the cases ($p < 0.01$). The risk of MI in drinkers relative to nondrinkers (not defined) was 0.32 (90% confidence limits 0.17, 0.63). Adjustment for smoking, hypertension, hypercholesterolemia, obesity, and history of gallbladder disease, using a multivariate logistic regression analysis, further enhanced the favorable relative risk experience of the drinkers.

Similarly, Rosenberg et al. (1981), in a study of MI in women less than 50 years of age, found the risk in current drinkers (defined as those drinking within the last year) to be 0.6 relative to that in never-drinkers (95% confidence limits 0.5, 0.8). After adjustment by multiple logistic regression analysis for 13 possible confounding variables (age, location of hospital, religion, years of education, menopausal status, number of visits to a physician or clinic in preceding year, cigarette smoking, hypertension, diabetes, history of abnormal blood lipids, history of obesity, year of admission, and oral contraceptive use within preceding year), the relative risk was 0.7 (95% confidence limits 0.5, 1.0). The estimated risk, relative to never-drinkers, of those drinking on less than 4 days a week was 0.6, while that for those drinking more often was 0.7. For exdrinkers, however, the risk relative to never-drinkers was 0.8 (95% confidence limits 0.4, 1.6). A further analysis by beverage type showed that the point estimate of relative risk was less than unity for each type of beverage (beer, 0.8; spirits, 0.9; wine, 0.4) but only that for wine was statistically significant ($p < 0.001$). Upon allowance for confounding factors, these findings were not materially altered: for drinkers of wine, in particular, the relative risk estimate was 0.5 (95% confidence limits, 0.3, 0.8).

In neither study were data provided on risk according to amount consumed daily. The main findings of three case–control studies providing such information are shown in Table 3. To assist comparison alcohol intake has been expressed in "standard drinks" (SD).* Klatsky et al. (1974, 1979), in an analysis which controlled for a number of coronary risk factors (electrocardiographic abnormality, current cigarette smoking, serum cholesterol, systolic and diastolic blood pressure, serum glucose, and triceps skin folds, as well as for age, sex and "skin color" [as an indication of race]), found the regular use of alcohol to be associated with a decreased risk of both myocardial infarction (MI) and sudden cardiac death (SCD).† For both major coronary events the decreased risk was evident at all levels of daily use studied, although it was most marked in the highest consumption group (those consuming 4.3 + SD per day). Although the trend toward less use of alcohol among SCD patients was not statistically significant, the findings were consistent with those for MI and provided strong evidence that the inverse relation between the risk of MI and daily alcohol use was not an artifact produced by selective survival of drinkers over nondrinkers with regard to hospitalization. Eighty-seven percent of the SCD victims died outside hospital.

Stason et al. (1976), in a study of alcohol consumption and nonfatal myocardial infarction, failed to find any major overall association between daily drinking and risk. The relative risk of daily drinkers compared with never or occasional drinkers was 0.9 (95% confidence limits 0.6, 1.2). They found some evidence, however, that the risk may be less in subjects consuming six or more

---

* A "standard drink" (SD) is defined as 0.6 Imperial oz (13.6 g; 17.1 ml) of absolute alcohol, the amount contained in $1^1/_2$ Imperial oz spirits (40% v/v), 12 Imperial oz beer (5% v/v), or 5 Imperial oz table wine (12% v/v). An Imperial ounce is equal to 0.96 of an American ounce.

† Sudden cardiac death (SCD) was defined as death due primarily to heart disease which occurred within 24 hr of the apparent onset of symptoms in persons who were not so ill that death could have been considered imminent or expected. There were 197 men who met rigid criteria for SCD, most of whom (87%) died outside hospital. Forty-nine (13%) died "instantaneously" or within 1 min of apparent onset of symptoms. More than two-thirds (68%) had clinical evidence of coronary disease before or during the fatal event. Of the 141 autopsied victims, all but one has moderate or severe coronary narrowing. The authors considered this series of deaths to be sudden deaths due to coronary disease.

Table 3. Risk of IHD by Level of Daily Alcohol Consumption Expressed as Standard Drinks (SD): Findings from Selected Case–Control Studies.

Kaiser–Permanente Study[a]				Boston Collaborative Drug Study[b]				Florida Two Counties Study[c]		
Daily consumption[d]		MI	SCD	Daily consumption[f]		MI (nonfatal)		Daily consumption[i] (SD)	CHD (fatal within 24 hr)	
"Drinks"[e]	SD	Relative risk		"Drinks"[g]	SD	Rate ratio	95% C.L.[h]		Rate ratio[j]	95% C.L.[h]
None[e]	None[e]	(1.0)	(1.0)	Never/occasional[g]	Never/occasional[g]	(1.0)	—	None[k]	(1.0)	—
≤2	≤1.4	0.7	0.8	<6	<4.3	1.0	0.7, 1.5	<3.5	0.4	0.3, 0.6
3–5	2.1–3.6	0.7	0.8	6+	4.3+	0.6	0.3, 1.1	3.5+	0.7	0.4, 1.1
6+	4.3+	0.4	0.7							

[a] Klatsky et al. (1974, 1979).
[b] Stason et al. (1976).
[c] Hennekens et al. (1978).
[d] Not further defined in questionnaires; the authors suggest that the amount of absolute alcohol contained in a "drink" of beer, wine, or spirits would be about 12.1 ml on average. If so, a "drink" would be approximately equivalent to 0.71 SD.
[e] Alcohol not consumed in past year. Contains lifetime abstainers and former drinkers.
[f] Each subject used his or her own definition of a "drink." Under the assumptions of Klatsky et al. (1974) a "drink" would contain on average about 12.1 ml of absolute alcohol, the equivalent of about 0.71 SD.
[g] Former drinkers not included.
[h] Confidence limits.
[i] Daily drinkers were divided into those drinking 59.2 ml (2 American oz) of absolute alcohol or less and those drinking more, these amounts, respectively, being approximately equal to <3.5 SD and 3.5+ SD.
[j] Adjusted matched pair ratio.
[k] Lifetime abstainers and former drinkers.

"drinks" daily (approximately 4.3+ SD), although there did not appear to be an effect at lower levels of consumption. Confounding by potential risk factors other than alcohol consumption was controlled by stratification by a multivariate confounder-summarizing score based on a linear discriminant function separating cases and controls. The confounder-summarizing score included smoking, hypertension, diabetes, age, sex, ponderosity, and a number of other factors. Within each quintile of the confounder-summarizing score the relative advantage of those drinking 4.3+ SD per day compared with the never or occasional drinkers was evident.

Hennekens et al. (1978) found a decreased risk of fatal coronary heart disease (CHD) in men who were daily drinkers compared to men who were nondrinkers. The method of analysis permitted control, if indicated, for several possible confounding variables including cigarette consumption, histories of diabetes, hypertension, and elevated cholesterol, and relative weight (Hennekens et al., 1976). The adjusted estimated relative risk for any versus no daily drinking was 0.6 (95% confidence limits 0.5, 0.8). The decreased risk, however, was limited to those drinkers consuming 59.2 ml of alcohol per day or less (<3.5 SD) (Table 3). For those consuming more (3.5 + SD) a statistically significant decrease in risk was not found, the risk ratio being 0.7 (95% confidence limits 0.4, 1.1). In a further analysis (Hennekens et al., 1979) it was shown that the reduced risk for drinkers of <3.5 SD per day was similar for each type of alcohol. The adjusted relative risks for beer, wine, and spirits, respectively, were 0.3, 0.3, 0.2, all of which were statistically significant ($p < 0.001$).

Several potential weaknesses are inherent in the methodology of case–control studies. These involve the selection of representative cases and appropriate controls, and more importantly, because it is harder to identify and overcome, bias in ascertainment of the factor. Hennekens et al. (1978), for example, noted the potential for this ascertainment bias in their study of fatal MI, in which information about alcohol consumption was obtained from the wives of both the cases and the controls. If wives thought daily alcohol consumption was harmful, wives of cases might have overestimated their husbands' consumption, whereas wives of controls might have underestimated. This systematic error would lead to an overestimation of a protective role of alcohol in lighter drinkers and an underestimation of a protective effect in heavier drinkers. A systematic error in the opposite sense could also be supposed. In that study it was possible to check for potential bias in the replies of the wives of controls (by comparing their responses with those obtained directly from husbands), but, of course, it was impossible to check for bias in the responses of the wives of the cases.

In that study the participation rate among the wives of the eligible cases was only 58%. Seventeen percent were nonrespondents and another 19% refused to cooperate in the study procedures. It is possible, therefore, that the cases included in the analysis were not representative. A potential bias also existed in the selection of wives for interview; wives of cases may have been more or less available for interview than wives of controls. Therefore, as far as availability might be associated with husbands' daily alcohol consumption, a systematic error in either direction in the estimate of the effect of alcohol might have occurred.

Furthermore, information on alcohol consumption was obtained only for the 3 months prior to death. It is possible that some patients were experiencing prodromal symptoms during this period and might have reduced their drinking, leading to an underestimation of a protective effect of alcohol consumption at lower levels. Hennekens et al. (1978) were able to examine their data for this particular bias. The analysis was repeated, restricting the sample to those pairs in which there was no history of a prior event. The results of this restricted analysis were identical to those of the principal analysis, indicating that changes in daily alcohol consumption following a coronary event did not account for the findings.

The case–control studies reviewed are in accord in some respects, and discrepant in others. All suggest a protective effect associated with drinking. However, Rosenberg et al. (1981) reported that this protective effect was confined to wine drinkers, while Hennekens et al. (1979) found a protective effect which was similar in drinkers of wine, beer and spirits. Klatsky et al. (1974, 1979) and Stason et al. (1976) found a statistically significant protective effect only in those drinking 4.3 SD or more, whereas Hennekens et al. (1978) found a statistically significant protective effect that was confined to drinkers of <3.5 SD daily. Despite these discrepancies, overall, the results suggest that alcohol consumption, at least at the levels studied, may be protective against IHD.

Cohort (Prospective) Studies

In three prospective studies no significant associations between alcohol consumption and IHD* were reported. Doyle et al. (1957) and Morris et al. (1966) failed to find any relationship between alcohol consumption (not further defined) and the risk of IHD in New York civil servants and London bus drivers and conductors, respectively. More recently, in a study of Belfast men, in which a multivariate analysis to control for major coronary risk factors was used, Greig et al. (1980) found no association between the frequency of alcohol consumption and the risk of IHD. In these studies the data from which the conclusions arose were not reported.

Ten cohort studies provide data on the risk of IHD in relation to alcohol consumption. The results from each are summarized below.

Tecumseh Health Study.   In the Tecumseh Health Study (Ullman et al., 1974; National Institute on Alcohol Abuse and Alcoholism, 1974) the relationship

---

* The reader is referred to the original papers for complete descriptions of the cases.

between alcohol use and CHD (nonfatal MI and CHD death) was studied in an 8–10-year follow-up of men aged 45–59 at entry. In the overall analysis, which did not control for coronary risk factors, the risk of MI was the same (8/1000 person-years) in those who never drank, in those consuming 4 American oz or less of absolute alcohol per week (about 7 SD), and in those consuming more (Table 4). In former drinkers the rates were about three times higher in both those who classified themselves as former lighter drinkers (about 7 SD per week or less) and as former heavier drinkers, 23 and 27/1000 person-years, respectively, suggesting that the increased risk was related to factors other than drinking. In analyses that attempted to control for cigarette smoking, blood pressure, and blood cholesterol, common patterns did not emerge. A multivariate analysis was not carried out.

Los Angeles Heart Study.   In the Los Angeles Heart Study of male civil servants, Chapman et al. (1974) found little overall association between alcohol consumption and the incidence of either total CHD or the combined diagnoses, MI and SCD, in a cohort of men aged 40–64 at entry in 1959 and followed for 12 years (Table 5). However, in a 1967 cohort (containing some of the 1959 cohort), which was followed for $4^1/_2$ years, there were statistically significant differences in the incidence of MI and SCD among the alcohol-consuming groups, with the rates being considerably lower in the two higher consumption groups. A similar, but not statistically significant trend was observed for the incidence

Table 4. Incidence of CHD (Nonfatal MI and CHD Death) in Men Age 45–59 at Entry in Relation to Alcohol Consumption: The Tecumseh Health Study[a]

| Alcohol consumption | All men | Incidence rates per 1000 per year | | | | Cholesterol[c] | |
| | | Blood pressure[b] | | Cigarette smoking | | Blood cholesterol[c] | |
		Low	High	<20/day	≥20/day	Low	High
Never drank	8	4	15	3	26	4	14
Current drinker							
Lighter[d]	8	8	10	7	9	7	13
Heavier	8	8	9	3	11	3	23
Former drinker							
Lighter[d]	23	26	17	25	20	0	78
Heavier	27	12	66	26	29	0	63

[a] Adapted from Ullman et al. (1974), National Institute on Alcohol Abuse and Alcoholism (1974).
[b] The low-blood-pressure group was comprised of the lower two-thirds of the systolic blood pressure distribution (systolic pressure lower than approximately 150 mm Hg).
[c] The low-blood-cholesterol group was comprised of the lower two-thirds of the distribution (approximately 250 mg/100 ml. or lower).
[d] 4 American oz or less of absolute alcohol per week (about 7 SD or less per week or averaged over the week; about 1 SD per day or less). The heavier drinking group reported drinking in excess of this amount.

Table 5. Incidence of Total Coronary Heart Disease and Myocardial Infarction and Sudden Cardiac Death in Men Free of Heart Disease at Entry: The Los Angeles Heart Study[a] (Incidence Rates per 1000[b])

Alcohol consumption		1959 Cohort (12-year follow-up)			1967 Cohort ($4^1/_2$-year follow-up)		
(oz/wk)	SD/day[c]	N	Total CHD	MI and SCD	N	Total CHD	MI and SCD
<0.1	<0.02	130	238	131	232	86	60
0.1–1.9	0.02–0.4	77	273	104	271	66	41
2.0–6.4	0.5–1.6	154	156	71	172	35	12
6.5+	>1.6	95	189	105	183	31	16
X^2 test of rates			$p > 0.10$	$p > 0.10$		$p < 0.10$	$p < 0.025$

[a] Adapted from Chapman et al. (1974).
[b] Not age-adjusted.
[c] Approximate equivalent.

of all CHD. An examination of incidence rates in three age groups (35–49, 50–64, and 65 and over) revealed lower rates in those men drinking 0.5 SD per day or more compared with those drinking less in the two older age groups, but in the 35–49 age groups the rates were virtually identical. A multivariate analysis controlling for coronary risk factors was not carried out. There was some evidence that persons in the 1959 cohort who decreased or stopped drinking during the 12-year follow-up had higher rates of CHD.

Framingham Study. In the 6-year follow-up of the Framingham Study, Dawber et al. (1959) reported that alcohol consumption *per se* did not appear to have any relation to the development of coronary heart disease. However, they remarked further that women who developed coronary heart disease had lower alcohol consumption than expected. Kannel (1976), in reporting on the 16-year follow-up, observed that alcohol intake appeared to have little effect on the incidence of CHD. He noted, however, that the regression coefficients were negative, suggesting the possibility of a protective effect. Stason et al. (1976) reanalyzed these data and, controlling for age and sex only, showed that persons consuming at least 30 oz (885 ml) of ethanol per month (an average of 1.7+ SD daily) had a statistically significant lower risk of CHD (rate ratio point estimate 0.7; 95% confidence limits 0.5, 0.9) than persons consuming lesser amounts.

Recently, Gordon et al. (1981a) reported on the results of a 4-year follow-up of 859 Framingham men aged 45–64 who were free of heart disease when examined between 1966 and 1969 and who completed a baseline 24-hr dietary recall interview. The men who remained free of CHD had a baseline mean daily alcohol intake of 25 g compared with mean intakes of 12, 10 and 16 g, respectively, in men who subsequently developed CHD (all forms), nonfatal MI and CHD death, and other forms of CHD (Table 6). The incidence of CHD death

Table 6. Age-Adjusted Mean Baseline Daily Alcohol Consumption[a] in Men Aged 45–64 Years according to Subsequent Coronary Heart Disease Experience[b]

CHD status study	Framingham 4-year follow-up			Honolulu 6-year follow-up			Puerto Rico 6-year follow-up		
	No.	g	SD	No.	g	SD	No.	g	SD
No CHD	780	25	1.8	7008	14	1.0	7932	12	0.9
All CHD	79	12[c]	0.9[c]	264	8[c]	0.6[c]	286	8	0.6
MI[d] and CHD death	51	10[c]	0.7[c]	164	5[c]	0.4[c]	163	9	0.7
Other CHD	28	16	1.2	100	14	1.0	123	8	0.6

[a] Based on a standardized 24-hr dietary recall obtained by interview.
[b] Data adapted from Gordon et al. (1981a).
[c] $p < 0.01$ versus no CHD.
[d] Nonfatal MI.

and MI declined significantly as daily intake increased up to a level of 40 or more grams (>2.9 SD) (Table 7). A multivariate analysis, which controlled for age, relative weight, systolic blood pressure, serum cholesterol, cigarettes smoked, and diabetes, revealed a statistically significant negative standardized logistic regression coefficient for CHD on alcohol consumption.

Honolulu Heart Study. In the 6-year follow-up of the Honolulu Heart Study, Yano et al. (1977) found a strong negative association between alcohol consumption and the risk of total coronary heart disease and nonfatal myocardial

Table 7. Incidence of Nonfatal Myocardial Infarction and Coronary Heart Disease Death by Level of Daily Alcohol Intake in Men Aged 45–64 Years (Rates per 1000)[a]

Daily alcohol intake[b]		Framingham (4-year follow-up)[c]	Honolulu (6-year follow-up)	Puerto Rico (6-year follow-up)
g	SD			
None	None	91.6	25.8	21.5
1–14	≤1.0	44.8	24.9	11.0
15–39	1.1–2.9	39.4	17.1	13.5
40+	>2.9	20.4	7.2	17.0
Significance of trend analysis		($p < 0.05$)	($p < 0.05$)	($p > 0.05$)
Standardized logistic regression coefficient[d]		-0.99 ($p < 0.01$)	-0.71 ($p < 0.01$)	-0.20 ($p > 0.05$)

[a] Adapted from Gordon et al. (1981a).
[b] Based on a standardized 24-hr dietary recall obtained by interview.
[c] The authors report that recently analyzed 6-year follow-up data yield the same conclusions.
[d] See text for description of multivariate model.

Table 8. Age-Adjusted 6-Year Incidence of Coronary Heart Disease and 9-Year Incidence of Coronary Heart Disease Death by Usual Daily Alcohol Consumption in Men 45–68 Years of Age at Entry. The Honolulu Heart Disease Study[a]

Usual daily alcohol consumption		Men at risk	Age-adjusted incidence rates per 1000					
ml	SD[b]		Total CHD (6-year)	CHD death (6-year)	Myocardial infarction (6-year)	Coronary insufficiency (6-year)	Angina pectoris (6-year)	CHD death (9-year)
None[c]	None[c]	3565	46.0[d]	6.8	21.2	2.8	15.2	24.6
1–6	<0.4	1034	41.2	6.3	22.0	3.2	9.7	16.6
7–15	0.4–0.9	962	30.7	4.2	14.0	3.1	9.4	20.9
16–39	1.0–2.2	1024	26.7	4.0	15.6	3.0	4.1	7.8
40–59	2.3–3.4	1006 }	21.2 }	3.0 }	4.2 }	6.6 }	7.4 }	5.9
60+	3.5+							12.8
Significance test for trend			$p < 0.001$		$p < 0.001$			

[a] Adapted from Yano et al. (1977) and Kagan et al. (1981).
[b] Approximate equivalent.
[c] Includes former drinkers.
[d] A separate analysis for never drinkers and former drinkers was carried out. The age-adjusted 6-year incidence rate in each group, respectively, was 43.6 and 55.7 per 1000.

infarction in Japanese men living in Hawaii (Table 8). A similar, but not statistically significant, trend was found for CHD death. No trend was noted for either acute coronary insufficiency (CI)* or angina pectoris (AP).† The authors noted that the declining incidence of coronary heart disease was not seen beyond a daily level of alcohol consumption exceeding 60 ml (>3.5 SD), although the number of subjects consuming this amount was too small to assure the detection of a statistically significant increase or decrease in incidence. However, data from the 9-year follow-up (Kagan et al., 1981) provided further clarification, at least with regard to CHD death. Again, a decreasing incidence of CHD death with increasing alcohol consumption up to 60 ml/day (3.5 SD) (Table 8) was found. At this level and beyond, however, the rate was double that of the next lowest consumption group.

The authors noted that the higher CHD incidence rate in nondrinkers might be an artifact due to inclusion of men who had stopped drinking for health reasons. Further analysis separating lifetime abstainers from former drinkers showed the highest incidence rate to be in the former drinkers, with lifetime abstainers having an intermediate rate. Multivariate analysis which controlled for cigarette smoking and other coronary risk factors (serum lipids, serum uric acid, systolic blood pressure, subscapular skin fold, relative weight, and age) showed that the statistically significant negative association between alcohol consumption and the risk of death from CHD and nonfatal MI (the so-called "hard" cases) became stronger when other risk factors were taken into account. Similarly, in an analysis of the 8-year follow-up data, Blackwelder et al. (1980) reported that in a univariate analysis the standardized coefficient for the association between alcohol consumption and the risk of CHD death was -0.421 ($p < 0.01$), whereas in a multivariate analysis, which controlled for age, cigarettes smoked per day, relative weight, systolic blood pressure, and serum cholesterol, the standardized coefficient was -0.485 ($p < 0.01$).

Analysis of the incidence rates of total CHD by beverage type revealed a statistically significant negative trend for beer consumption only (Yano et al., 1977). A negative trend was found for spirits, which was not statistically significant. Wine drinkers had lower rates of CHD than non-wine drinkers but the numbers were small and no trend was evident. A further comparison of total

---

* In the Honolulu Heart Study, acute coronary insufficiency (CI) was defined as coronary-type chest pain usually lasting more than 30 min with documented transient specific electrocardiographic change and without a rise in myocardial serum enzymes (Kagan et al., 1975).

† Angina pectoris (413, ICD-8) is defined as an episode of chest pain due to a temporary discrepancy between the supply and demand of oxygen to the heart. This may be due to low oxygen levels in the blood, restricted blood flow to the heart (coronary insufficiency), or an increase in heart work beyond normal levels. Most often, angina pectoris is a chronic condition caused by a restricted blood flow due to hardening and narrowing of the coronary arteries supplying the heart muscle (coronary atherosclerosis). (Public Health Service, 1979a.)

alcohol and specific beverage consumption between men who developed the grouped diagnoses, nonfatal MI and CHD deaths, the so-called "hard" cases, and the no-CHD group showed lower mean consumptions of total alcohol, beer and spirits in the CHD group (Kagan et al., 1981). The differences were statistically significant. The mean consumption of wine was also lower, but the numbers were small and the difference was not statistically significant.

The findings noted above were obtained from baseline alcohol consumption data collected by asking about usual consumption per day or per week (for nondaily drinkers). However, baseline alcohol consumption data also were obtained as part of a 24-hr dietary recall interview. The two methods produced similar results. Men aged 45–68 at entry who developed MI or died of CHD in the 6-year follow-up period had a statistically significant lower mean daily alcohol intake (6 g or 0.4 SD) than did those who remained free of CHD (14 g or 1.0 SD). Men who developed CI and AP had mean daily alcohol intakes of 23 g and 9 g, respectively, which were not statistically different from the intake of the men who remained free of disease (Yano et al., 1978). Corresponding data for men aged 45–64 at entry (Table 6) indicated similar findings. Multivariate analysis using a logistic model, which controlled for age, systolic blood pressure, serum cholesterol, relative weight, cigarettes smoked per day, and carbohydrate intake, showed a highly significant standardized coefficient for the grouped outcomes, nonfatal MI and CHD death, on mean daily alcohol consumption. The results of a similar analysis which employed age, relative weight, systolic blood pressure, serum cholesterol, cigarettes smoked per day, total starch intake per day, ability to read or write Japanese, and diabetes, as well as alcohol, produced a similar highly significant negative standardized logistic coefficient (Table 7).

It is of interest that Yano et al. (1977), in commenting on the significantly lower mean daily alcohol intake of Japanese men living in Hawaii who had nonfatal MI and died of CHD compared with those who remained free of CHD, noted that the reverse was found in Japanese men living in California, namely, a positive association between CHD and mean alcohol intake. It was indicated, however, that the latter data were unpublished.

Puerto Rico Heart Health Program.   In the 6-year follow-up of 8218 urban and rural Puerto Rican men, aged 45–64 years and free of CHD at entry, alcohol intake at baseline was found to be lower in the men who subsequently developed various manifestations of CHD than in the men who remained free of disease (Table 6), although the differences were not statistically significant. The findings published in an earlier report (Garcia-Palmieri et al., 1980), in which this analysis was carried out separately for urban and rural men, revealed differences in the same direction in both groups, those for rural men being statistically significant. The incidence of nonfatal MI and CHD death (Table 7) did not show the striking negative relationship with daily alcohol intake evident in the Framingham and Honolulu studies. Nevertheless, a multivariate logistic regression analysis employing nine variables (alcohol, total starch, area of residence, blood glucose, age, relative weight, systolic blood pressure, serum cholesterol, and cigarettes smoked) revealed standardized coefficients of total coronary heart disease, nonfatal MI and CHD death, and other CHD on daily alcohol consumption of -0.23 ($p < 0.05$), -0.20, and -0.26, respectively, yielding the conclusion that alcohol is negatively associated with CHD when major coronary risk factors are taken into account. This was further confirmed in univariate analysis and in multivariate analysis employing six variables (alcohol, carbohydrate, systolic blood pressure, serum cholesterol, cigarettes smoked, and blood glucose), which were undertaken in each of two age groups (45–54 and 55–64) within each residence area (urban and rural) (Garcia-Palmieri et al., 1980). The logistic function coefficients regressing nonfatal MI and CHD death on daily alcohol consumption were consistently negative in all age–area subgroups, the rural coefficients being larger. None, however, was statistically significant.

Chicago Western Electric Company Study. In an early report based on a $4^1/_2$ year follow-up of men employed in the Chicago Western Electric Company, Paul et al. (1963) reported that "no positive association was found between the development of coronary disease and the intake of alcoholic beverages." No data were supplied. However, a 17-year follow-up of 1,832 white men aged 40–55

Table 9. Seventeen-Year Age-Adjusted Coronary Heart Disease Mortality Rates in 1832 White Men Aged 40–55 and Free of Disease at Entry, and Mortality Rates in Two Periods of Follow-up, by Alcohol Consumption: The Chicago Western Electric Company Study[a]

Maximum number of "drinks"[b] per month	Approximate average SD per day	Age-adjusted CHD mortality rates per 1000		
		17 Years	≤10 Years	>10 Years
Former drinker	Former drinker	201	[c]	[c]
None	None	49 ⎫	⎫	⎫
		⎬ 80	⎬ 31	⎬ 52
1–14	<0.4	84 ⎭	⎭	⎭
15–44	0.4–1.1	77	28	51
45–104	1.2–2.8	73	30	45
105–164	2.9–4.4	55	17	42
165+	4.5+	155	48	123
X^2 test of rates[d]		$p < 0.10$	[c]	[c]
<165	<4.5	75	29	49
165+	4.5+	155	48	123
t test of rates[d]		$p < 0.10$	[c]	[c]

[a] Adapted from Dyer et al. (1980, 1981). All figures are from the 1980 paper except that for former drinkers.
[b] A "drink" contained the amount of alcohol in 12 American oz of 4% v/v beer, (about 0.83 SD).
[c] Not available.
[d] Adjusted for age, diastolic blood pressure, cigarettes/day, and serum cholesterol.

Table 10. Rates of Mortality and Hospitalization for Selected Cardiovascular Diagnoses (ICD Rubrics) according to Usual Daily Alcohol Consumption: Kaiser–Permanente Study[a]

Usual daily alcohol consumption		Percent mortality (10-year follow-up)[c]			Percent hospitalization (6-year follow-up)[d]				
"Drinks"[b]	SD	All CHD (410–414)	Acute MI (410)	Other CHD (411–414)	All CHD (410–414)	Acute MI (410)	Cardiomyopathy (425)	Congestive heart failure (427.0, 427.1)	Arrhythmias (427.2, 427.6, 427.9)
None[e]	None[e]	3.3	1.8	1.5	8.0	3.8	0.1	2.1	2.6
≤2	≤1.4	2.0	1.1	0.9	5.9	3.2	0.0	1.3	2.7
3–5	2.1–3.6	2.3	1.4	0.9	5.8	2.6	0.1	1.6	3.1
6+	4.3+	2.7	1.1	1.6	5.6	2.5	0.4	2.3	3.6

[a] Adapted from Klatsky et al. (1981a,b).

[b] The authors suggest that the amount of absolute alcohol contained in a "drink" would be about 12.1 ml on average. If so, a drink would be approximately equivalent to 0.71 SD.

[c] Each drinking group contained 2015 persons. Each person drinking 6+ "drinks" daily was carefully matched with a nondrinker, a drinker of <2 drinks daily, and a drinker of 3–5 drinks daily for age, sex, presence or absence of established cigarette smoking, and examination date. Of the nondrinkers who were established smokers 50.5% smoked 1 pack or more daily compared with 53.7% of each of the other drinking groups, and 11.8% smoked two packs or more daily compared with 11.0, 15.7, and 23.4% in each of the other drinking groups in order.

[d] The same groups as described in the preceding footnote were studied. Hospitalization data by diagnosis were available only for 6 years. The hospitalization incidence rates were based on the mean number of persons in each alcohol consumption group on June 30th of each year from 1971 to 1976. The persons at risk in each drinking group from none to 6+, respectively, were 1406, 1435, 1374, and 1320.

[e] Alcohol not consumed in past year. This group contains lifetime abstainers and former drinkers.

and free of coronary heart disease at entry has been published recently (Dyer et al., 1980, 1981). Former drinkers had the highest risk of CHD death (Table 9). Nondrinkers had the lowest risk of CHD. Among current drinkers the rates of CHD death appeared to decline as drinking increased across the four drinking groups, from occasional drinkers (1–14 "drinks" per month for an average of <0.4 SD per day) to the group drinking an average of 2.9–4.4 SD per day. However, those in the fifth drinking group (165 "drinks" per month for an average of 4.5+ SD per day) had higher rates of CHD: about double the rates observed in those drinking less than that amount. An analysis of the age-adjusted mortality rates in the five groups showed that the rates of CHD death were different among the five drinking groups ($p < 0.05$). With adjustment of the CHD rates for age, cigarettes/day, diastolic blood pressure, and serum cholesterol, the differences ceased to be statistically significant ($p < 0.10$). When alcohol intake was dichotomized at 4.5+ and <4.5 SD per day, results of t tests for differences in the rates of CHD death were significant after adjustment for age and cigarettes/day. However, the differences between these two drinking groups did not remain statistically significant when other factors, particularly diastolic blood pressure, were taken into account.

A further analysis comparing the 10-year with the 10–17-year mortality rates revealed weak associations between alcohol consumption and CHD death in the earlier period. Much stronger associations were found for the later period and, in particular, those consuming 4.5+ SD per day had rates two and one-half times higher than those consuming lesser amounts.

Kaiser–Permanente Study.   An 11-year follow-up of some 4000 men and women aged 35–45 who were enrolled in the Kaiser–Permanente Medical Care Program, with regard to the mortality risk associated with smoking, multivariate analysis of alcohol consumption, smoking, other coronary risk factors, and several other variables, failed to show a statistically significant relationship between alcohol consumption and the risk of CHD mortality (Friedman et al., 1979). The adjusted risk of those consuming three or more "drinks" per day (2.1+ SD) relative to those consuming lesser amounts, including none at all, was 0.9 ($p = 0.74$). However, recently published follow-up data on CHD mortality in this population over a range of daily alcohol consumption (Table 10), make it clear that the earlier statistically insignificant finding was an artifact arising from the imposed dichotomous definition of the risk variable. This resulted in the amalgamation of alcohol consumption groups with different cardiovascular risks, which obscured significant intergroup differences.

Klatsky et al. (1981a,b), in follow-up studies in which the drinking groups were carefully matched for age, sex, race, smoking, and examination date, provide data on the risk of both mortality and hospitalization due to CHD (Table 10). The highest mortality risks for coronary heart disease were found in the nondrinking group, followed by the highest drinking group (4.3+ SD daily), suggesting a "U-shaped" mortality curve. However, in the subgroups of CHD, "acute" and "other", this "U-shaped" pattern was found only in the "other" group, which overall accounted for just under half the CHD deaths and consisted almost completely of chronic ischemic heart disease (412.9, ICD-8).

The incidence of hospitalization for total CHD and for acute MI was highest in the nondrinking group and it declined as daily drinking increased up to and including the six or more "drinks" (4.3 + SD) daily group. Cardiomyopathy was an infrequent cause of hospitalization, but the rate was highest in the heaviest drinking group. The incidence of hospitalization for arrhythmias was lowest in the nondrinking group and tended to increase as drinking increased. On the other hand, rates of hospitalization for congestive heart failure suggested a "U-shaped" pattern, with the highest rates being in the nondrinking and the highest drinking (4.3 + SD daily) groups.

Yugoslavia Cardiovascular Disease Study.   In the 7-year follow-up of 11,121 Yugoslav men aged 35–62 at entry the incidence of CHD death and nonfatal MI was negatively related to alcohol consumption (Kozarevic et al., 1980) (Table 11). In a multivariate logistic regression analysis which controlled for age, systolic blood pressure, serum cholesterol, and cigarettes smoked per day, the standardized logistic coefficient for CHD (CHD death and nonfatal MI)

Table 11. Seven-Year Incidence of Coronary Heart Disease by Baseline Alcohol Consumption in Men Aged 35–62 at Entry: The Yugoslavia Cardiovascular Disease Study[a]

	7-Year incidence rates	
Consumption (summary score)[b]	CHD death and non-fatal MI	CHD death
<4	13.7	9.5
4–7	16.8	9.2
30–36	14.8	11.1
60+	8.1	5.6
Standardized logistic coefficient[c]	−0.312	−0.215
	($p < 0.01$)	($p > 0.05$)

[a] Adapted from Kozarevic et al. (1980).
[b] Not a continuous variable; cannot be converted to SD. A score of 4–7 indicates the drinking of one alcoholic beverage type at least once a week and other beverages less often. A score of 30–36 indicates daily consumption of one beverage and others less often. A score of 60+ indicates daily consumption of at least two beverage types. Those with lower scores were presumably lighter drinkers than those with higher scores.
[c] See text for description of multivariate model.

on alcohol consumption was negative and statistically significant. A lower incidence of CHD deaths and CHD (CHD deaths and nonfatal MI) was evident, however, only in the group with the highest alcohol consumption summary scores (60+) who were "presumably the heaviest drinkers" (p. 615). A further comparison of "daily" (score 30+) versus other drinkers by beverage type (wine, beer, and other alcohol beverages) revealed a consistently lower incidence of CHD (CHD death and nonfatal MI) and CHD death among daily drinkers of each beverage type, although none of the differences were statistically significant.

Some interesting differences emerged in further analyses of these data (Kozarevic et al., 1982), in which the relationships between alcohol consumption and various manifestations of CHD were examined in more detail, and according to the urban or rural residence of the subjects. In a multivariate analysis which controlled for cigarette smoking, age, years of schooling, serum cholesterol, hematocrit, systolic blood pressure, and adiposity (Quetelet index), the standardized logistic regression coefficient of CHD incidence on alcohol consumption was negative in both urban and rural dwellers, but only the former was statistically significant. For the entire group of subjects alcohol consumption was inversely related to the incidence of MI and of nonsudden CHD death, but not to sudden CHD death. In urban residents the multivariate analysis showed that alcohol consumption was inversely related to nonfatal MI ($p < 0.01$) and nonsudden CHD death ($p < 0.05$). However, for sudden CHD death the standardized logistic regression coefficient, although negative, was not statistically significant. In rural residents the standardized regression coefficients were negative, but not statistically significant, for nonfatal MI and nonsudden CHD death. For sudden CHD death the coefficient was positive, but not statistically significant. Thus, the findings in both urban and rural residents suggested that alcohol consumption was related differently to sudden CHD death than to the other two manifestations, although these differences could have arisen by chance. An additional analysis of drunkenness in relation to CHD death indicated that the more recently drunkenness had occurred, the more likely was the CHD death to be sudden. This relationship of inebriation to sudden CHD death was confirmed in multivariate regression analyses by area of residence which controlled for age, systolic blood pressure, cigarette smoking, serum cholesterol level, frequency of drinking, and the Quetelet index. In both urban and rural areas recent drunkenness was related positively to sudden CHD death, but the relationship was statistically significant ($p < 0.05$) only in urban residents. These investigators postulated that the apparent absence of protection against sudden death might reflect the deleterious effects of heavy alcohol consumption, including drunkenness, on the myocardium with an increased vulnerability to lethal arrhythmias.

Whitehall Study.   In the 10-year follow-up of 1422 male civil servants aged 40–64 at entry the age-adjusted cardiovascular death rate was found to be higher in the "nondrinkers" (defined on the basis of a 3-day dietary record completed at entry) than in those consuming 0.1–9.0, 9.1–34.0, or more than

34 g alcohol (>2.5 SD) per day (Marmot et al., 1981). Among the drinking groups there was little difference in mortality. Restricting the analysis to CHD deaths did not change this pattern. However, a multivariate analysis adjusting for coronary risk factors was not reported.

Lipid Research Clinics Program Study.   Preliminary data involving an average follow-up of 5.6 years in 4073 men and 3519 women aged 30 and older have been reported (Criqui et al., 1982). In a multivariate analysis which controlled for age, systolic blood pressure, obesity, triglycerides, and estrogen use in women, cigarettes per day were positively associated with CHD mortality in both men and women ($p < 0.02$), while alcohol consumption (ml/day) was only weakly associated negatively with CHD mortality ($p > 0.05$). The addition of HDL and LDL cholesterol to the model as covariates produced little change in these results, suggesting that the reported mortality associations were largely independent of the previously reported relationships of both smoking and alcohol consumption to HDL and LDL cholesterol.

Interpretation of Findings.   Cohort studies are also subject to certain systematic errors. For example, ascertainment of the disease may be biased by knowledge about the risk factor status of the individual. Loss of subjects either at the beginning or during the follow-up period may introduce systematic error in determining the associations between the risk factor and the disease. Nonetheless, of the various types of observational studies these are thought to provide the most definitive information about the causes of disease.

The results of 13 cohort studies have been reviewed. In several, no significant association between alcohol consumption and IHD was reported, and in others, interpretation was difficult because multivariate analyses controlling for coronary risk factors were not carried out. However, from the findings of seven studies certain common patterns emerged, of which some were statistically significant. In the Yugoslavia Heart Disease Study (Kozarevic et al., 1980) daily drinkers appeared to have lower incidence and death rates from CHD than other drinkers, which was evident for all beverage types. CHD incidence and death rates were lowest in the group with the highest consumption scores, who were "presumably the heaviest drinkers" (p. 615). Unfortunately, it is not possible to define further this level of consumption. Multivariate analysis which controlled for several coronary risk factors revealed a statistically significant negative association between alcohol consumption and the occurrence of nonfatal MI and CHD death. A multivariate analysis of preliminary data from the Lipid Research Clinics Program Study (Criqui et al., 1982) showed a weak negative relationship between alcohol consumption and CHD mortality. This relationship, however, was not statistically significant, and from the data available it is not possible to comment further on the risk relationship over a defined range of consumption.

Both the Framingham and the Puerto Rico studies (Gordon et al., 1981a) provide additional evidence that drinkers have lower rates of non-fatal MI and CHD death than nondrinkers (Table 7). Furthermore, in the Framingham Study these rates decreased as drinking increased and were lowest in the highest drinking group (>2.9 SD per day). Multivariate analysis controlling for CHD risk factors revealed a statistically significant negative relationship between rates of nonfatal MI and CHD death and alcohol consumption. In a similar analysis undertaken in the Puerto Rico Study, a negative relationship, which was not statistically significant, was found. However, that for total CHD and alcohol consumption was statistically significant. These studies yield the conclusion that increasing daily alcohol consumption up to and including a level of more than 2.9 SD per day is associated with a decreasing risk of CHD. The data from the Honolulu Study, when analyzed in a similar fashion (Table 7), support this conclusion.

The findings of three studies, the Chicago Western Electric Company Study, the Honolulu Heart Disease Study, and the Kaiser–Permanente Study, are more informative in that the range of alcohol consumption over which risk was determined was more extensive. A somewhat consistent pattern emerged with regard to the risk of CHD death in relation to estimated daily alcohol consumption (Table 12), that of a so-called "U-shaped" mortality curve. The results of these studies were similar in suggesting that the risk of CHD death was higher in nondrinkers than in drinkers and that the risk decreased as alcohol consumption increased up to levels of about 3.5–4.5 SD per day. However, when consumption was at this level and beyond, rates of CHD death increased.

The interpretation of these findings and those of several other cohort studies is complicated by the composition of the baseline comparison group. In the Honolulu Study the nondrinking group against which the other consumption groups were compared contained former drinkers as well as lifetime abstainers. In several studies (e.g., the Tecumseh Study and the Chicago Western Electric Company Study) former drinkers were shown to have the highest rates of CHD. These persons may quit drinking because of prodromal indications of CHD or because of other health conditions which increase the risk of CHD. In the Honolulu Study a separate analysis was done (Yano et al., 1977) which showed that former drinkers had the highest rates of total CHD incidence whereas lifetime abstainers had intermediate rates (Table 8). The published rate for the nondrinking group for CHD death (Table 12), however, is a composite of the two risks. In these studies and in the other cohort studies, in which the nondrinker group could and would contain former drinkers (e.g., those based on dietary recall during an immediately preceding period), an apparent negative or "U-shaped" dose–response relationship could be accentuated by an artifactually high rate in the nondrinking group due to the inclusion of former drinkers.

In the Chicago Western Electric Company Study, in which former drinkers were excluded from the baseline nondrinking group, the nondrinking and the other drinking groups were not statistically different in risk experience when other CHD risk factors were controlled. For statistical analysis the baseline group consisted of abstainers and occasional drinkers. The rate for never drinkers, however, was calculated separately, and was found to be the lowest of all groups (Table 9), providing no evidence to support a "U-shaped" mortality curve.

In evaluating the finding of a "U-shaped" mortality curve in the Kaiser-

Table 12. Mortality from CHD by Daily Alcohol Consumption in Three Prospective Studies[a]

Chicago Western Electric Study		Honolulu Heart Study		Kaiser–Permanente Study	
Average daily consumption (SD)	Age-adjusted CHD mortality per 1000 men (17-year follow-up)	Usual daily consumption (SD)[c]	Age-adjusted CHD mortality per 1000 men (9-year follow-up)	Usual daily consumption[b] (SD)	CHD mortality per 100 men and women (10-year follow-up)
Former drinker	201	None[c]	24.6	None[d]	3.3
0–<0.4	80	<0.4	16.6	<1.4	2.0
0.4–1.1	77	0.4–0.9	20.9	2.1–3.6	2.3
1.2–2.8	73	1.0–2.2	7.8	4.3+	2.7
2.9–4.4	55	2.3–3.4	5.9		
4.5+	155	3.5+	12.8		

[a] Adapted from Dyer et al. (1980, 1981), Kagan et al. (1981), and Klatsky et al. (1981a).
[b] Consumption groups were matched on age, sex, and cigarette smoking habit.
[c] Includes former drinkers.
[d] Alcohol not consumed in last year. This group contains lifetime abstainers and former drinkers.

Permanente Study, again it is essential to consider the composition of the non-drinking group, defined in this study as those not consuming alcohol in the past year. Several possible sources of bias related to this group have been suggested (Craven, 1981; Meltzer, 1981; Rosman, 1982; Miller, 1982). Especially note-worthy are concerns about possible contamination of this group by current heavy drinkers or former heavy drinkers who, for whatever reason, understate their consumption of alcohol. The finding of higher rates of cirrhosis and accidents in the nondrinking group compared with the two or fewer "drinks" ($<$1.4 SD) per day group raises serious concern about the credibility of the self-reported nondrinking status. Certainly, the inclusion in the nondrinking group of heavy drinkers, either current or former, could explain the higher mortality experience by this group compared to the two or fewer "drinks" per day group, giving an erroneous impression of a "U-shaped" alcohol-mortality curve. In view of the strong association between heavy drinking and heavy smoking (Ashley and Rankin, 1980) it is of interest that in an analysis which controlled for smoking and drinking (Klatsky et al., 1981a) the "U-shaped" mortality patterns was most apparent in the heaviest smokers, but was not found at all in nonsmokers. In reply to these concerns (Klatsky et al., 1981c; Klatsky, 1982) the investigators conceded that part of the observed "U-shaped" alcohol-mortality curve could be related to unstable drinking patterns in the nondrinking group. Therefore, the conclusion that a negative or a "U-shaped" relationship exists between alcohol consumption and CHD mortality must be viewed as highly tentative.

Furthermore, data from three studies, the Honolulu Heart Disease Study, the Kaiser–Permanente Study, and the Yugoslavia Heart Disease Study, in which the risk relationships of alcohol consumption and various clinical presentations of CHD were examined separately, suggest that the dose–response relationships may not be the same for the different clinical categories of CHD. In the Kai-ser–Permanente Study the pattern for acute MI mortality was different from that of other CHD (mostly chronic IHD) mortality (Table 10). In the Honolulu study (Table 8), the dose–response relationship for the incidence of acute MI was dissimilar to that for CI and AP. In the Yugoslavia Study alcohol consumption was inversely related to the incidence of MI and nonsudden CHD death, but not to sudden CHD death.

From the cohort study data, therefore, it is not possible to draw definitive conclusions about the relationship of alcohol consumption to ischemic heart disease.

3. STUDIES OF ALCOHOLICS AND PROBLEM DRINKERS

Excess mortality from IHD has been found in a number of follow-up studies of clinically treated alcoholics and problem drinkers. Data from seven studies, summarized in Table 13, all indicate an increased ratio of observed to expected

Table 13. Ischemic Heart Disease Mortality in Clinically Treated Alcoholics and in Problem Drinkers, and Percentage of Total Excess Mortality in Groups Due to IHD and to Cirrhosis

Reference	Locale and period of study	Source of sample	Sex	Ratio of observed/ expected deaths	Percentage of excess mortality	
					IHD	Cirrhosis
Sundby, 1967	Norway, 1925–1962	Admissions for alcoholism to inpatient psychiatric department	M	1.3[a] (420, 422, ICD-6,7)	6	4
Schmidt and Popham, 1980	Canada, 1951–1970	Alcoholism clinic admissions	M F	1.4 2.4 (420–422, ICD-6,7)	20 18	16 15
Thorarinsson, 1979	Iceland, 1951–1974	First admissions to alcoholism treatment facilities	M	1.9 (420,ICD-7)	19	4
Robinette et al., 1979	U.S., 1946–1974	U.S. army servicemen hospitalized	M	1.2[b] (410–413,ICD-8)	8	7

Davies, 1965	U.S., 1940–1962	Insurance policy holders substandard due to alcohol habits	M and F	2.6 (arteriosclerotic and degenerative heart disease)	36	[d]
Pell and D'Alonzo, 1973	U.S., 1965–1969	Problem drinkers employed in industry	M and F	1.9 (coronary heart disease)	26	16
Dyer et al., 1977	U.S., 1958–1973	Problem drinkers employed in industry	M	2.2[c] (coronary heart disease)	68	[d]

[a] Ratios are based on Oslo's mortality for the period 1925–1962. Sundby also provided data for the period 1951–1962. For this period the ratio of observed to expected deaths was 1.2. The percentages of excess mortality due to IHD and cirrhosis, respectively, were 12% and 3%.
[b] Ratio not corrected for competing risks.
[c] Adjusted for age, diastolic blood pressure, heart rate, serum cholesterol, relative weight, and cigarettes/day.
[d] Not available from data provided.

deaths, varying from 1.2 to 2.6. In several studies the excess of IHD deaths contributed substantially to the overall excess of observed deaths from all causes, often more so than deaths from cirrhosis of the liver.

However, one exception to these findings is noteworthy. In an analysis that separated mortality experiences of vagrant and nonvagrant alcoholics, Sundby (1967) found that the excess mortality from IHD was confined to the nonvagrant group, the ratio of observed to expected deaths being 1.35 for the observation period 1950–1962. In contrast, among vagrants there was actually a deficit of observed to expected deaths from this cause, the ratio being 0.86. Sundby noted that nutritional differences could be assumed between the two groups and he speculated that the vagrants might have been less at risk from the nutritional influences leading to atherosclerosis. It is possible, also, that some kinds of deaths might have been ascribed differently in the two groups. It was found that deaths ascribed to symptoms referable to the cardiovascular system [acute heart failure, undefined (782.4, ICD-6)] and sudden deaths [cause unknown (795.2, ICD-6)] occurred more often than expected in both groups of alcoholics, but particularly in the vagrant group. When these latter categories of deaths were considered together with IHD deaths, the ratio in the vagrant group exceeded unity (1.10), although it was still less than the comparable ratio in the nonvagrant group (1.37).

The mortality experiences of problem drinkers in general population studies generally confirm these findings. In the Göteborg prospective study (Wilhelmsen et al., 1973) a positive association between problem drinking (as indicated by registration with the Temperance Board) and the incidence of CHD (nonfatal acute myocardial infarction and sudden deaths due to CHD) was observed (relative risk, 2.4). A multivariate analysis with eight other risk variables (high systolic blood pressure, smoking, high blood cholesterol, dyspnea, physical activity during work, hematocrit, triglycerides, and place of birth), confirmed a significant positive association between registration and the risk of CHD. The investigators raised the possibility that this association might be due to a complex variable (perhaps a psychosocial one), which was correlated with intemperance.

In an 11-year follow-up study of mortality in twins in relation to smoking habits and alcohol problems (as evidenced by registration in a nationwide registry for misconduct while under the influence of alcoholic beverages), Friberg et al. (1973) further confirmed the excess mortality in problem drinkers from CHD (410–414, ICD-8) and sudden death (795.99, ICD-8),* an excess found among

---

* The rubric 795.99, ICD-8, refers to sudden death (cause unknown). In the study by Friberg et al. (1973), 88 of the 223 cases considered were ascribed to this rubric. Of these cases 42 died within 2 hr and had a history of CHD and/or autopsy findings of CHD, and 23 died within 24 hr and had similar signs of CHD. In another 14 cases death occurred within 2 hr, but without other evidence of CHD. In the remaining nine cases, death occurred within 24 hr without objective signs of CHD.

nonsmokers and smokers alike. The authors pointed out, however, that it was not known to what extent registration reflected personality factors rather than drinking habits. Indeed, in an extension of this study involving death-discordant twin pairs, it was shown that registration was somewhat more common among deceased twins, irrespective of cause (deFaire, 1974). Also, in a comparison of twin pairs discordant for heavy drinking in which smoking was controlled, Hrubec et al. (1976) found heavy drinking to be associated with angina pectoris in monozygotic twin pairs ($p < 0.05$). In dizygotic twin pairs there was a weaker association, which was not statistically significant.

In a 12-year follow-up study of Finnish male drinkers in two social classes, Poikolainen (1983) found statistically significant ($p < 0.05$) positive linear associations between CHD mortality and episodes of intoxication, hangover, and hangover drinking, identified in the baseline year, which were confined to the upper class. The point estimate of risk of death in men intoxicated once a week relative to men never intoxicated during the baseline year was 2.5 ($p < 0.05$). The respective figures for hangover and hangover drinking were 2.0 ($p < 0.05$) and 2.1 ($p \geq 0.05$).

Elmfeldt et al. (1976) found alcohol intemperance (registration by a Temperance Board) to be more common among myocardial infarction patients dying outside hospital than in the general population. However, for those patients who survived to reach hospital the registration rate was similar to that found in the general population. These investigators wondered whether registered persons might be more at risk of early fatality because of a predisposition to malignant arrhythmias, which might be further accentuated by myocardial ischemia, or because of psychosocial factors. Fraser and Upsdell (1981) compared patients experiencing sudden cardiac death with those experiencing acute myocardial infarction who did not die suddenly, selected randomly from a community register of acute coronary events. In a multivariate logistic analysis which included smoking, alcohol consumption over the previous year was one of several factors that discriminated between the two groups. Heavy alcohol consumers had a higher proportion of coronary events in the form of sudden deaths. This proportionate increase was probably due, at least in part, to an absolute increase in the frequency of sudden death, supporting the contention that persons with high alcohol consumption, who also have myocardial infarction, are more likely to die suddenly of the infarction.

The findings of studies of alcoholics and problem drinkers are consistent and provide evidence that the risk of IHD in very heavy drinkers is increased. In so doing, they provide a further dimension to the evidence from the three cohort studies summarized in Table 12, which indicated that the risk of CHD death is increased at levels of consumption exceeding 3.5–4.5 SD per day.

The problems inherent in studies of alcoholic, problem, and heavy drinkers with regard to establishing the risk of IHD, however, have been recognized (e.g.,

Dyer et al., 1977). The potential for error in ascribing cause of death in heavy drinkers must be considered. Deaths (particularly, sudden deaths) due to other causes may be attributed to IHD either because of similarities in the clinical presentations, or for social or other reasons. For example, deaths due to alcoholic cardiomyopathy may be erroneously coded IHD, as may deaths due to liver disease (Randall, 1980a,b; May et al., 1980). A propensity for the unwarranted ascription of deaths to vascular disease and, in particular, to IHD has been noted in general populations (Lundberg and Voigt, 1979; Engel et al., 1980). Whether and to what extent this tendency pertains in alcoholic and heavy drinking populations is not known. It is possible that systematic errors in the opposite direction might occur as well. Sudden deaths due to IHD might be ascribed to unknown causes, especially in certain subgroups of heavy drinkers, such as vagrants.

The importance of controlling for cardiovascular risk factors in assessing the role of heavy drinking in ischemic heart disease must not be forgotten. This is particularly true with regard to smoking, since heavy smoking and heavy drinking are frequently associated (Ashley and Rankin, 1980). In only one of the studies cited in Table 13, by Dyer et al. (1977), were other cardiovascular risk factors taken into account. In that study of problem drinkers employed in industry the ratio of observed to expected CHD deaths, adjusted for age, diastolic blood pressure, heart rate, serum cholesterol, relative weight, and cigarettes smoked per day, was found to be 2.2. Schmidt and Popham (1981b) analyzed further their data on alcoholics reported in Table 13. A strategy for the control of cigarette smoking was employed and they found that the excess mortality from total heart and circulatory disease (400–468, ICD-7) was reduced quite substantially, although it was still present and statistically significant ($p < 0.01$).

4. ETIOLOGIC SIGNIFICANCE OF THE EPIDEMIOLOGIC FINDINGS

The findings from a number of epidemiologic studies involving a variety of populations and methods have been reviewed and summarized. From these findings a tentative picture of the possible relationship between alcohol consumption and IHD has begun to emerge. The results of cohort studies of general populations, in particular, are helpful in characterizing the relationship between IHD and alcohol consumption. It appears that the risk of some manifestations of IHD may be inversely related to alcohol consumption up to a level of about 3.5–4.5 SD per day. In multivariate analyses, which have controlled for major coronary risk factors, statistically significant negative associations have been observed. On the other hand, evidence from both cohort studies of general populations and studies of alcoholics and problem drinking populations suggest that at higher levels of consumption statistically significant positive associations exist.

Alternative Explanations

In evaluating the possible causal significance of these associations several alternative explanations must be considered. Could the findings be due to chance, bias, or confounding factors? The statistically significant associations which have emerged have come from a number of studies employing different methodologies, and it is therefore unlikely that chance is the explanation. Some potential sources of bias, which may have affected the validity of the findings of the various types of observational studies, already have been considered. Of particular general concern is the potential bias in ascertainment of the risk factor. The difficulties inherent in measuring alcohol consumption in individuals are well-known, and the potential for systematic error is recognized (Popham and Schmidt, 1981a,b; Cahalan, 1981). There appears to be a tendency for heavier drinkers to underestimate their drinking. Such a tendency might lead to an invalid attentuation of a negative relationship as consumption increases and, consequently, the possible protective effect of alcohol might be underestimated. Conversely, it might lead to the identification of an invalidly low level of consumption at which increased risk of CHD first appears.

The problem of confounding is also a major concern. In many of the case–control and cohort studies, and in some of the studies of alcoholics, problem, and heavy drinkers, controls for other coronary risk factors, which could confound the alcohol–IHD relationship, have been incorporated into the analyses. But certainly, control for potential confounders cannot be considered to be complete. For example, personality type, other psychosocial variables, diet, and other factors of lifestyle and the environment also may be risk or protective factors for IHD and may be associated with various alcohol consumption patterns. If, for example, persons who drink in the lower range of alcohol consumption also tend to have Type B personalities, a protective factor against IHD (Rosenman et al., 1966; Haynes et al., 1980), then the protective effect seen in such persons may reflect the effect of personality factors rather than the effect of alcohol. Similarly, it is possible that anxiety-prone individuals who report distressing life events may be more likely to have relatively high alcohol intakes (Spicer et al., 1981). An increased risk of IHD in such persons may reflect the effect of anxiety-proneness rather than high alcohol intake. Another possibility for confounding relates to dietary differences associated with drinking patterns. In a population study, Jones et al. (1982) found that men drinking moderately (1.8–3.6 SD per day) consumed significantly less fat, cholesterol, protein, and carbohydrates, than did other drinking groups. They suggested that such a dietary difference might account for the apparent protective effect of alcohol. Nonetheless, some of the associations, both negative and positive, which have been observed, have persisted in multivariate analyses controlling for at least some major known coronary risk factors, suggesting that the associations may be causal.

Criteria for Establishing Causal Associations

Several criteria have been identified against which observed statistically significant associations may be evaluated in determining their causal significance. The evidence will be considered against three of these criteria.

Strength of the Association.   This criterion refers to the ratio of disease rates in those with and without the presumed factor. The greater the deviation from unity in either direction the greater is the likelihood that the factor is causally related to the outcome. The demonstration of a dose–response gradient further increases the likelihood that the association is causal.

In these studies an assessment of the strength of association is difficult because of the heterogeneous nature of the nonexposed group. As has been noted earlier, in some studies nondrinkers included lifetime abstainers (never drinkers) and former drinkers; these groups have widely differing risks. The risk estimates of drinkers versus nondrinkers may be reduced too much by the inclusion in the nonexposed group of former drinkers having a high risk. The strength of association was found to vary considerably among the studies. In some it was relatively weak (e.g., Dyer et al., 1980), while in others it was relatively strong (e.g., Pettiti et al., 1979). As has been discussed, a statistically significant negative dose–response gradient at lower levels of consumption was found in several cohort studies. However, in others the dose–response gradient was not statistically significant when other risk variables were controlled.

Consistency of the Association.   This criterion requires that the association be found in other studies involving different populations and different methods. The more often an association is found under diverse circumstances the more likely it is causal.

The evidence concerning heavy drinking and the risk of IHD is reasonably consistent. The results of several cohort studies of different general populations and studies of alcoholics, problem, and heavy drinkers living in different places, support the contention that the risk of IHD is increased. The evidence concerning the risk of IHD at lower levels of consumption is not so consistent. Although on balance the evidence from both case–control and cohort studies conducted in a variety of settings seems to favor a negative association, some results have been obtained that do not support this conclusion. In several cohort studies no statistically significant associations have been observed. Furthermore, in covariation studies based on total populations, the results have been mixed both in direction and significance.

Biological Plausibility. Additional support for the causal nature of an association exists if a causal interpretation is biologically plausible in terms of what is known about the factor and the disease. This would include knowledge of the mechanism or mechanisms by which the factor might exert its effect.

Atherosclerosis and HDL-Cholesterol. The relationship of alcohol to atherosclerosis, a major underlying component of the pathogenesis of IHD, has been the subject of much controversy. In autopsy studies spanning many years, comparisons of the degree of atherosclerosis and/or the frequencies of coronary occlusion and myocardial infarction in patients with and without cirrhosis of the liver have been attempted. Some of these studies have suffered from serious methodologic deficiencies. The findings of most (e.g., Berrios and Rodriguez, 1959; Creed et al., 1955; Grant et al., 1959; Hall et al., 1953; Hirst et al., 1965; Howell and Manion, 1960; Kane and Aronson, 1971; MacDonald and Mallory, 1958; Nicolas et al., 1970; Ruebner et al., 1961; Vaněček, 1976) suggest that cirrhosis is associated with lesser degrees of atherosclerosis and lower frequencies of coronary occlusion and myocardial infarction. Nonetheless, in several studies (Ishitoya et al., 1967; Jaworski, 1981; Parrish and Eberly, 1961; Restrepo et al., 1968; Wilens, 1947) no differences in atherosclerosis between patients with and without cirrhosis were found at autopsy. Several investigators have postulated mechanisms by which liver cirrhosis might be protective against atherosclerosis (e.g., Creed et al., 1955; Grant et al., 1959; Howell and Manion, 1960), while others have suggested that the chronic heavy drinking which leads to cirrhosis is, itself, the protective factor (e.g., Schmidt and Popham, 1981a). Still others have stressed the importance of selection biases that must be recognized in assessing the validity of the purported differences (Cornfield, 1959; Mainland, 1953; Parrish and Eberly, 1961; Ruebner et al., 1961).

The relationship between chronic heavy alcohol consumption and atherosclerosis in the absence of cirrhosis of the liver also has been studied at autopsy. Wilens (1947), in a study of 423 male alcoholics and 434 male controls of comparable age, blood pressure, and nutritional status, showed that the degree of atherosclerosis was almost identical in the two groups. Hirst et al. (1965) found that alcoholics without cirrhosis had coronary narrowing and occlusion of a degree comparable to that found in controls and that the frequencies of myocardial infarction in the two groups were equal. Viel et al. (1966, 1970) found no statistically significant differences between alcoholics, heavy drinkers, and others in the frequency of coronary atherosclerosis or fibrous plaques in the aorta. In a medical–legal autopsy series of 39 male chronic alcoholics (of whom only five had cirrhosis) and 145 men dying violent deaths in whom a history of alcoholism was excluded, Rissanen (1974) found no differences between the groups in the severity of coronary stenosis or in the extent of coronary and aortic raised lesions. Coronary fatty streaks, however, were less extensive in the chronic alcoholic group.

In autopsy investigations of atherosclerosis, which were not confined to chronic alcoholics, mixed findings have been obtained. Sackett et al. (1968), in a well-designed study that controlled for the influence of smoking, failed to demonstrate any relationship of significance between aortic atherosclerosis and the use of alcohol by "social" drinkers. In a multivariate analysis of the autopsy findings in men enrolled in the Honolulu Heart Disease Study, Rhoads et al. (1978) found alcohol consumption to be negatively correlated with the severity of both aortic and coronary atherosclerosis, but the correlations were low and not statistically significant. These investigators postulated that the protective effect of alcohol consumption against MI found in the Honolulu Study might be due to other incidental characteristics of persons who drink or to mechanisms that inhibit the precipitation of MI, possibly involving platelet aggregation. On the other hand, Moore et al. (1975) found alcohol consumption to be a significant discriminator between the quartiles of autopsied white men with the highest and lowest raised coronary lesion scores; the highest quartile consumed less alcohol. Lifšic (1976), in a univariate analysis involving 737 autopsies of men aged 30–69 years who died from a variety of causes, found that men consuming alcohol on 16 or more days per month had less extensive coronary atherosclerosis than those drinking less frequently or not at all. Fresh myocardial lesions and large myocardial scars were also somewhat less common among those in the higher consumption group.

The findings of these two latter studies are in keeping with the results of clinical studies of atherosclerosis conducted by Barboriak and colleagues (Barboriak et al., 1977, 1979a,b; Barboriak, 1977; Anderson et al., 1978) on more than 2000 men and women. A statistically significant inverse relationship between the level of alcohol consumption and the degree of coronary artery occlusion, as determined by arteriography, has been demonstrated. This was accompanied by a corresponding trend in the prevalence of previous myocardial infarction.

It should be noted, however, that Barboriak (1977) found that some patients with higher alcohol intakes developed symptoms of heart disease at an earlier age than abstainers despite less extensive coronary artery disease. He suggested the possibility that the chronic intake of excessive amounts of alcohol may produce more severe symptoms than expected on the basis of coronary artery disease, possibly through direct cardiotoxic effects. It must be stressed that the patients in these studies were self-selected on the basis of symptoms.

More recently, Barboriak and colleagues (Gruchow et al., 1982) have reported that the pattern of drinking is related to the degree of coronary occlusion. Regular drinkers who had, on average, a weekly consumption of about 17.5 SD (about 2.5 SD daily) were further characterized as relatively consistent or as variable drinkers. Higher levels of occlusion were found among nondrinkers, occasional drinkers, and regular drinkers with a variable intake, while significantly lower levels of occlusion were found in regular drinkers with a consistent pattern of intake ($p < 0.02$). Furthermore, while occlusion scores were inversely correlated with consumption by regular drinkers who drank relatively consistent amounts ($p < 0.02$), drinkers with variable drinking patterns had higher occlusion scores regardless of amounts consumed. These findings suggest that the attenuating effect of alcohol consumption on coronary occlusion is reversed by a variable or sporadic pattern of alcohol intake.

The biological plausibility of a negative relationship between alcohol consumption and IHD by means of an antiatherogenic effect has been greatly strengthened by the rapidly accumulating body of evidence that alcohol consumption is positively associated with plasma levels of high-density lipoprotein cholesterol (HDL-C), this factor, in turn being negatively associated with atherogenesis and the risk of CHD (e.g., Heiss et al., 1980).* In their clinical studies, Barboriak et al. (1979b) demonstrated that the degree of coronary occlusion was inversely related to levels of HDL-C. They suggested that the direct relationship between alcohol and this antiatherogenic cholesterol fraction provides a plausible underlying mechanism for their findings of an inverse relationship between alcohol consumption and coronary artery occlusion. Furthermore, in their analysis of coronary occlusion in relation to pattern of intake, cited above, there were significant associations between the total cholesterol to HDL-C ratio and the pattern of drinking which were in keeping with the findings concerning coronary artery occlusion.

Very substantial epidemiologic evidence supports the inference that alcohol consumption is a determinant of HDL-C. For example, prevalence data from the Lipid Research Clinics Program Study indicate a consistent, strong, independent, positive gradient between HDL-C levels and reported alcohol intake in both men and women, regardless of type of beverage consumed (Ernst et al., 1980; Gordon et al., 1981b). Also, persons who never drank had lower levels of HDL-C than drinkers who had not drunk in the preceding week, suggesting that even occasional drinking may raise the HDL-C level and that this elevation may persist, for at least a while, even though the person has not been drinking. Prevalence data from the Cooperative Lipoprotein Phenotyping Study, prevalence and longitudinal data from the Multiple Risk Factor Intervention Trial, and experimental data (Hulley and Gordon, 1981), as well as studies of alcoholics (e.g., LaPorte et al., 1981), and of other diversely characterized populations (e.g., Willett et al., 1980), all provide consistent evidence that alcohol is a determinant of HDL-C.

While this relationship could explain the observed negative relationship

---

* For recent comprehensive reviews of the epidemiology of plasma HDL-C and the relationship to alcohol consumption the reader is referred to two supplements to *Circulation* [See Tyroler (1980) and Kaelber and Barboriak (1981).]

between alcohol intake and CHD over the lower part of the consumption range, it is not clear how it relates to a positive association in the heavy drinking range. In this respect, the recent observations of Gruchow et al. (1982) on pattern of drinking, described above, are of interest, as are reports that alcoholics with liver disease do not have the high levels of HDL-C found in alcoholics without liver disease (Devenyi et al., 1981; Sabesin, 1981), and that the proportions of the subfractions of HDL-C may be altered in the presence of cirrhosis (Okazaki et al., 1981), alcoholic liver disease (Sabesin, 1981), and alcoholism (Danielsson et al., 1978; Cushman et al., 1981). Much remains to be learned about the functions of HDL and its subfractions (Levy and Rifkind, 1980). No doubt, the relationships between HDL, alcohol consumption, and the risk of CHD will be further clarified, as the process of defining these functions continues.

Blood Pressure.   The evidence linking alcohol consumption and blood pressure has increased substantially in the last several years and this relationship must be considered as a possible mechanism underlying an association between alcohol consumption and IHD. The evidence from several types of studies involving a variety of populations has been reviewed and evaluated recently (Wallace et al., 1981; Celentano et al., 1981). It is clear that the interpretation of the available evidence is complicated by differences among these various studies which make comparisons difficult, and by problems inherent in the studies themselves. Although both cross-sectional and prospective epidemiologic evidence (e.g., Dyer et al., 1981) suggests that alcohol consumption at higher levels is associated with increased blood pressure levels, not all the evidence supports this conclusion, including experimental studies of acute administration of alcohol in men and acute and chronic administration studies in animals. The epidemiologic findings also suggest age and sex differences in the alcohol–blood-pressure relationships, some of the age–sex patterns being compatible with the postulated "U-shaped" relationship between alcohol consumption and the risk of IHD. For example, Klatsky et al. (1977) showed that systolic blood pressure increased linearly in men with increasing alcohol consumption, but not in women. Women who drank two or fewer drinks daily had lower systolic blood pressures than nondrinkers or women drinking three or more drinks daily. Harburg et al. (1980), using data from the Tecumseh Community Health Study, showed that age–weight-adjusted systolic blood pressure varied with alcohol use linearly in men with a slight dip at one to two drinks per week, whereas in women the relationship was curvilinear with a low point at about four drinks per week. Similarly, in the Lipid Research Clinics Study, Wallace et al. (1981) demonstrated a linear relationship in men aged 35–49 and 50–64, but in men aged 20–34 and in women in all these age groups the relationship was curvilinear. In these studies somewhat similar sex differences in diastolic blood pressure were also found. The explanation for these potentially important age and sex differences is not readily evident. However, further analysis of the Lipid Research Clinics Program Study data suggests that the "U-shaped" alcohol–blood-pressure relationship in women was confounded by the effects of obesity and lack of cigarette smoking in nondrinking women and by very low alcohol consumption in hypertensive women using medication (Criqui et al., 1981).

The findings of Dyer et al. (1981) in "problem" drinkers employed by the Chicago Peoples Gas Company and in "heavy" drinkers (4.5+ SD daily) employed by the Chicago Western Electric Company illustrate the difficulty in determining from the available data the possible role of alcohol-related blood pressure changes with regard to the risk of IHD in heavy drinkers. In both studies there were significant cross-sectional associations between heavy alcohol use and the level of blood pressure and the prevalence of hypertension. Furthermore, significant prospective relationships were shown in both studies between heavy alcohol use and the incidence of high blood pressure in men who were normotensive at entry. However, the "problem" drinkers had a higher risk of IHD, which persisted after adjustment of blood pressure and other risk variables, whereas in the heavy drinkers the excess risk was no longer significant when blood pressure was entered into the analysis.

Platelet Aggregation and Fibrinolytic Activity. Alcohol consumption might exert a protective or a risk effect through an influence on platelet aggregation. Alterations in platelet count are well-recognized in association with heavy drinking (Lindenbaum, 1974; Cowan, 1975). Thrombocytopenia is a common finding. It is reversible upon cessation of drinking and a rebound thrombocytosis may follow (Haut and Cowan, 1974; Haselager and Vreeken, 1977; Cowan, 1980). While alcohol-related thrombocytopenia may be associated with decreased platelet aggregation, rebound thrombocytosis has been implicated as a possible factor in thromboembolic disease. Manku et al. (1979) have suggested a possible mechanism whereby alcohol in moderate amounts may decrease platelet aggregation by enhancing antiaggregatory prostaglandin formation. Deykin et al. (1982) have reported a potentiation of aspirin-induced prolongation of the bleeding time by the consumption of a moderate amount of ethanol (50 g or 3.7 SD), although this amount of ethanol by itself had no effect on bleeding time. They suggested that this might provide an explanation for the protective effect of alcohol against IHD that has been observed in population studies, since both ethanol and aspirin ingestion are widespread in general populations. Elmèr et al. (1981) have shown that acute intoxication in nonalcoholics inhibits platelet aggregation and prolongs bleeding time.

Meade et al. (1979), in a study of employed men in whom the average daily consumption was 1.7 SD, found an increase in fibrinolytic activity and a decrease in fibrinogen concentration in drinkers compared with nondrinkers. They suggested that these effects would be protective as far as the risk of IHD was concerned. On the other hand, Lee et al. (1980), in a study of male alcoholics and controls under 50 years of age found fibrinolytic activity to be reduced in the alcoholics, while there were no differences in platelet aggregation. Again, further study appears warranted.

Cardiac Effects.   The adverse effects of heavy alcohol consumption on cardiac morphology, metabolism, and function are well-documented (e.g., Rubin, 1980, 1982; Regan and Haider, 1981; Khetarpal and Volicer, 1981). They suggest several mechanisms by which alcohol might increase the risk of IHD. Alcohol-induced cardiac arrhythmias, in particular ventricular fibrillation, with or without alcohol-related derangements in magnesium and/or other electrolytes, have been postulated as one mechanism underlying the increased risk of SCD in alcoholics and heavy drinkers. There is some evidence that persons with underlying ischemia (Kentala et al., 1976) or with other IHD risk factors (Waern et al., 1982) may be more susceptible to such arrhythmias.

Alcoholic cardiomyopathy and its consequence, congestive heart failure, are now well-recognized in association with chronic heavy alcohol use. Subclinical manifestations include decreased cardiac contractibility and myocardial depression. It is possible that ischemic effects may be enhanced in the presence of alcohol-induced myocardial toxicity. For example, Orlando et al. (1976), in a study of patients with angina pectoris and proven coronary artery disease, showed that alcohol decreased the time required to precipitate exercise-induced angina and significantly increased the ischemic changes in the electrocardiogram. These changes were attributed to a myocardial effect of alcohol, involving an increase in oxygen demand along with myocardial depression in subjects who were unable to increase their coronary blood flow because of underlying coronary artery disease. Fraser and Upsdell (1981) have postulated that the so-called "U-shaped" mortality curve for IHD in relation to alcohol consumption may represent the interaction of the effects of alcohol-related protection against atheroma and alcohol-induced cardiomyopathy.

The evidence concerning the effects of alcohol consumption on coronary blood flow is controversial (Bing and Tillmanns, 1977). Some investigators have reported increases in blood flow with consumption, while others have reported decreases or no changes at all. It is possible that the effects vary according to the amount consumed and the presence or absence of underlying coronary artery disease. Recent experimental work by Talesnik et al. (1980) suggests that ethanol in concentrations found in the blood *in vivo* may be beneficial in facilitating coronary reaction during cardiac exertion. On the other hand, cardiodepressant effects also were noted, particularly at higher levels. In another study in which coronary blood flow was measured in patients with cardiac disease, Gould et al. (1972) found that alcohol consumption produced a significant increase in coronary blood flow. However, this was associated with increased myocardial oxygen requirements, suggesting a detrimental effect in patients with coronary artery disase.

Atypical myocardial infarction in the absence of minimal narrowing of the coronary arteries has been documented in chronic alcoholics (Regan et al., 1975, 1977; Regan and Ettinger, 1979). Morphologic examination of the myocardium revealed concentric periarterial fibrosis which was postulated to restrict coronary flow increments during periods of high flow requirements. This phenomenon was considered analogous to the cardiac muscle necrosis associated with the periarterial lesions of constrictive pericarditis, perhaps conditioned by the abnormal metabolism of cardiac cells in chronic alcoholism. Factor (1976) has described intramyocardial small-vessel disease in chronic alcoholism. Magnesium deficiency is also a well-known concomitant of heavy alcohol consumption (Flink, 1971), and it recently has been shown that magnesium deficiency can produce spasm of the coronary arteries which may result in sudden death due to ischemic heart disease (Turlapaty and Altura, 1980). Moreyra et al. (1982) have reported acute myocardial infarction in nonalcoholic patients with normal coronary arteries after acute ethanol intoxication. They speculated that alcohol-induced coronary spasm might be the underlying mechanism.

It must be concluded that alcohol consumption, at least heavy consumption, has adverse effects on cardiac function which could increase the risk of IHD.

5. CONCLUSION

The relationship between alcohol consumption and the risk of IHD remains incompletely understood. The evidence that heavy drinking is associated with an increased risk of IHD is reasonably consistent and several underlying mechanisms have been postulated. While there seems little doubt that heavy alcohol consumption increases the risk of myocardial disease and other adverse cardiac effects, it still remains to be proved that the IHD risk association *per se* is causal. It could be artifactual, arising from misclassification, or indirect and explainable through personality, lifestyle, or other psychosocial characteristics of heavy drinkers. Evidence supporting a negative relationship between alcohol consumption at lower levels and IHD risk is not so consistent, although it is growing. Several underlying mechanisms can be postulated, but it is far too early to claim causal association. Again, much of the apparent effect may be explainable by diet, personality, or other attributes of persons who consume alcohol at low levels.

Much more clarification is required. The recent report by Gruchow et al. (1982), in which "binge" drinkers did not show the relative protection from coronary artery occlusion that was found in moderate regular drinkers raises questions about the influence of pattern of drinking. The report of Stason et al. (1976), suggesting that the protective effects of alcohol consumption were more marked in smokers than in nonsmokers, raises issues concerning the interactions of risk and protective factors. Most of the studies to date have dealt with men. More data are needed on women, in whom the risk of IHD is lower.

Clearly, preventive or therapeutic advice concerning the possible benefits of light to moderate drinking with regard to IHD would be premature and ill-founded. At the individual level it is important to note that the risks of certain diseases and conditions may be increased at levels of alcohol consumption associated with decreased risks of IHD. This may be particularly true for women. For example, Péquignot and Tuyns (1975) showed that women consuming as little as 20–40 g of alcohol daily (about $1^1/_2$–3 SD) may be at increased risk of liver cirrhosis, and several studies suggest that drinking as little as 2 SD daily may increase the risk of adverse fetal outcomes (e.g., Little and Streissguth, 1981). At the population level the strong relationship between overall consumption and heavy consumption (e.g., Schmidt and Popham, 1978) must be borne in mind, and any measure that might tend to increase overall consumption must be carefully evaluated for its risks as well as for its benefits. At the present time, emphasis must remain on the many adverse consequences which accompany excessive alcohol use (Eckardt et al., 1981), and on defining more precisely what constitutes hazardous drinking.

REFERENCES

Anderson, A. J., Barboriak, J. J., and Rimm, A. A., 1978, Risk factors and angiographically determined coronary occlusion, *Am. J. Epidemiol.* **107**:8.

Ashley, M. J., and Rankin, J. G., 1980, Hazardous alcohol consumption and diseases of the circulatory system, *J. Stud. Alcohol* **41**:1040.

Barboriak, J. J., 1977, Alcohol and coronary-artery disease, *Lancet* **1**:1212.

Barboriak, J. J., Rimm, A. A., Anderson, A. J., Schmidhoffer, M., and Tristani, F. E., 1977, Coronary artery occlusion and alcohol intake, *Br. Heart J.* **39**:289.

Barboriak, J. J., Anderson, A. J., Rimm, A. A., and Tristani, F. E., 1979a, Alcohol and coronary arteries, *Alcohol Clin. Exp. Res.* **3**:29.

Barboriak, J. J., Anderson, A. J., and Hoffmann, R. G., 1979b, Interrelationship between coronary artery occlusion, high-density lipoprotein cholesterol, and alcohol intake, *J. Lab. Clin. Med.* **94**:348.

Berrios, J. R., and Rodriguez, R. C., 1959, The incidence of atherosclerotic heart disease among cirrhotics, *Bol. Asoc. Med. P. R.* **51**:205.

Bing, R. J., and Tillmanns, H., 1977, The effect of alcohol on the heart, in: *Metabolic Aspects of Alcoholism* (C. S. Lieber, ed.), pp. 117–134, University Park Press, Baltimore.

Blackwelder, W. C., Yano, K., Rhoads, G. G., Kagan, A., Gordon, T., and Palesch, Y., 1980, Alcohol and mortality: The Honolulu Heart Study, *Am. J. Med.* **68**:164.

Brummer, P., 1969, Coronary mortality and living standard, II. Coffee, tea, cocoa, alcohol and tobacco, *Acta Med. Scand.* **186**:61.

Buck, C., Donner, A. P., and Simpson, H., 1982, Garlic oil and ischemic heart disease, *Int. J. Epidemiol.* **11**:294.

Cahalan, D., 1981, Quantifying alcohol consumption: patterns and problems, *Circulation* **64**(Supplement III):7.

Celentano, D. D., Martinez, R. M., and McQueen, D. V., 1981, The association of alcohol consumption and hypertension, *Prev. Med.* **10**:590.

Chapman, J. M., Massey, F. J., Coulson, A., and Sayre, J., 1974, A Study of Relationships Between Alcohol Consumption and Heart Disease. Final Report to the National Institute on Alcohol Abuse and Alcoholism (Contract HSM-42-73-113).

Cornfield, J., 1959, Principles of research, *Am. J. Ment. Defic.* **64**:240.

Cowan, D. H., 1975, The platelet defect in alcoholism, *Ann. N.Y. Acad. Sci.* **252**:328.

Cowan, D. H., 1980, Effect of alcoholism on hemostasis, *Semin. Hematol.* **17**:137.

Craven, P. C., 1981, Alcohol and mortality, *Ann. Intern. Med.* **95**:780.

Creed, D. L., Baird, W. F., and Fisher, E. R., 1955, The severity of aortic arteriosclerosis in certain diseases: A necropsy study, *Am. J. Med. Sci.* **230**:385.

Criqui, M. H., Wallace, R. B., Miskel, M., Barrett-Connor, E., and Heiss, G., 1981, Alcohol consumption and blood pressure: The Lipid Research Clinics Prevalence Study, *Hypertension* **3**:557.

Criqui, M. H., Cowan, L. D., Tyroler, H. A., Bangdiwala, S., Heiss, G., Wallace, R. B., and Davis, C. E., 1982, Cigarette smoking, alcohol consumption, and coronary, cardiovascular, and all cause mortality in men and women. Preliminary results from the Lipid Research Clinics follow-up study. *Am. J. Epidemiol.* **116**:560.

Cushman, P., Alaupovic, P., and Barboriak, J., 1981, High density lipoproteins in alcoholics: Changes with abstinence, *Alcohol. Clin. Exp. Res.* **5**:146.

Danielsson, B, Ekman, R., Fex, G., Johansson, B. G., Kristensson, H., Nilsson-Ehle, P., and Wadstein, J., 1978, Changes in plasma high density lipoproteins in chronic male alcoholics during and after abuse, *Scand. J. Clin. Lab. Invest.* **38**:113.

Davies, K. M., 1965, The influence of alcohol on mortality. *Proc. Home Office Life Underwriters Assoc.* **46**:159.

Dawber, T. R., 1980, *The Framingham Study: The Epidemiology of Atherosclerotic Disease*, pp. 91–120, Harvard University Press, Cambridge.

Dawber, T. R., Kannel, W. B., Revotskie, N., Stokes, J., Kagan, A., and Gordon, T., 1959, Some factors associated with the development of coronary heart disease: six years' follow-up experience in the Framingham Study, *Am. J. Public Health* **49**:1349.

deFaire, U., 1974, Ischaemic heart disease in death discordant twins: A study on 205 male and female pairs, *Acta Med. Scand. [Suppl]* **568**:1.

Devenyi, P., Robinson, G. M., Kapur, B. M., and Roncari, D. A. K., 1981, High-density lipoprotein cholesterol in male alcoholics with and without severe liver disease, *Am. J. Med.* **71**:589.

Deykin, D., Janson, P., and McMahon, L., 1982, Ethanol potentiation of aspirin-induced prolongation of the bleeding time, *New Engl. J. Med.* **306**:852.

Doyle, J. T., Heslin, A. S., Hilleboe, H. E., Formel, P. F., and Korns, R. F., 1957, III. A prospective study of degenerative cardiovascular disease in Albany: Report of three years' experience—1. Ischemic heart disease, *Am. J. Public Health* **47**(Suppl.):25.

Dyer, A. R., Stamler, J., Paul, O., Berkson, D. M., Lepper, M. H., McKean, H., Shekelle, R. B., Lindberg, H. A., and Garside, D., 1977, Alcohol consumption, cardiovascular risk factors, and mortality in two Chicago epidemiologic studies, *Circulation* **56**:1067.

Dyer, A. R., Stamler, J., Paul, O., Lepper, M., Shekelle, R. B., McKean, H., and Garside, D., 1980, Alcohol consumption and 17-year mortality in the Chicago Western Electric Company Study, *Prev. Med.* **9**:78.

Dyer, A. R., Stamler, J., Paul, O., Berkson, D. M., Shekelle, R. B., Lepper, M. H., McKean, H., Lindberg, H. A., Garside, D., and Tokich, T., 1981, Alcohol, cardiovascular risk factors and mortality: The Chicago experience, *Circulation* **64**(Suppl. III):20.

Eckardt, M. J., Harford, T. C., Kaelber, C. T., Parker, E. S., Rosenthal, L. S., Ryback, R. S., Salmoiraghi, G. C., Vanderveen, E., and Warren, K. R., 1981, Health hazards associated with alcohol consumption, *JAMA* **246**:648.

Elmėr, O., Göransson, G., and Zoucas, E., 1981, Alcohol and blood loss, *Lancet* **2**:639.

Elmfeldt, D., Wilhelmsson, C., Vedin, A., Tibblin, G., and Wilhelmsen, L., 1976, Characteristics of representative male survivors of myocardial infarction compared with representative population samples, *Acta Med. Scand.* **199**:387.

Engel, L. W., Strauchen, J. A., Chiazze, L., and Heid, M., 1980, Accuracy of death certification in an autopsied population with specific attention to malignant neoplasms and vascular diseases, *Am. J. Epidemiol.* **111**:99.

Ernst, N., Fisher, M., Smith, W., Gordon, T., Rifkind, B. M., Little, A., Mishkel, M. A., and Williams, O. D., 1980, The association of plasma high-density lipoprotein cholesterol with dietary intake and alcohol consumption. The Lipid Research Clinics Program Prevalence Study, *Circulation*, **62**(Suppl. IV):41.

Factor, S. M., 1976, Intramyocardial small-vessel disease in chronic alcoholism, *Am. Heart J.* **92**:561.

Flink, E. B., 1971, Mineral metabolism in alcoholism, in: *The Biology of Alcoholism*, Volume 1 (B. Kissin and H. Begleiter, eds.), pp. 377–395, Plenum Press, New York.

Fraser, G. E., and Upsdell, M., 1981, Alcohol and other discriminants between cases of death and myocardial infarction, *Am. J. Epidemiol.* **114**:462.

Friberg, L., Cederlöf, R., Lorich, U., Lundman, T., and deFaire, U., 1973, Mortality in twins in relation to smoking habits and alcohol problems, *Arch. Environ. Health* **27**:294.

Friedman, G. D., Dales, L. F., and Ury, H. K., 1979, Mortality in middle-aged smokers and non-smokers, *N. Engl. J. Med.* **300**:213.

Garcia-Palmieri, M. R., Sorlie, P., Tillotson, J., Costas, R., Cordero, E., and Rodriguez, M., 1980, Relationship of dietary intake to subsequent coronary heart disease incidence: The Puerto Rico Heart Health Program, *Am. J. Clin. Nutr.* **33**:1818.

Gordon, T., Kagan, A., Garcia-Palmieri, M., Kannel, W. B., Zukel, W. J., Tillotson, J., Sorlie, P., and Hjortland, M., 1981a, Diet and its relation to coronary heart disease and death in three populations, *Circulation* **63**:500.

Gordon, T., Ernst, N., Fisher, M., and Rifkind, B. M., 1981b, Alcohol and high-density lipoprotein cholesterol, *Circulation* **64**(Suppl. III):63.

Gould, L., Collica, C., Zahir, M., and Gomprecht, R. F., 1972, Ethyl alcohol: Effects on coronary blood flow in man, *Br. Heart J.* **34**:815.

Grant, W. C., Wasserman, F., Rodensky, P. L., and Thomson, R. V., 1959, The incidence of myocardial infarction in portal cirrhosis, *Ann. Intern. Med.* **51**:774.

Greig, M., Pemberton, J., Hay, I., and MacKenzie, G., 1980, A prospective study of the development of coronary heart disease in a group of 1202 middle-aged men, *J. Epidemiol. Community Health* **34**:23.

Gruchow, H. W., Hoffman, R. G., Anderson, A. J., and Barboriak, J. J., 1982, Effects of drinking patterns on the relationship between alcohol and coronary occlusion, *Atherosclerosis* **43**:393.

Hall, E. M., Olsen, A. Y., and Davis, F. E., 1953, Portal cirrhosis: clinical and pathologic review of 782 cases from 16,600 necropsies, *Am. J. Pathol.* **24**:993.

Harburg, E., Ozgoren, F., Hawthorne, V. M., and Schork, M. A., 1980, Community norms of alcohol usage and blood pressure: Tecumseh, Michigan, *Am. J. Public Health* **70**:813.

Haselager, E. M. and Vreeken, J., 1977, Rebound thrombocytosis after alcohol abuse: A possible factor in the pathogenesis of thromboembolic disease, *Lancet* **1**:774.

Haut, M. J., and Cowan, D. H., 1974, The effect of ethanol on hemostatic properties of human blood platelets, *Am. J. Med.* **56**:22.

Haynes, S. G., Feinleib, M., and Kannel, W. B., 1980, The relationship of psychosocial factors to coronary heart disease in the Framingham study. III. Eight-year incidence of coronary heart disease, 1980, *Am. J. Epidemiol.* **111**:37.

Heiss, G., Johnson, N. J., Reiland, S., Davis, C. E., and Tyroler, H. A., 1980, The epidemiology of plasma high-density lipoprotein cholesterol levels. The Lipid Research Clinics Program Prevalence Study, Summary, *Circulation* **62**(Suppl. IV):116.

Hennekens, C. H., Drolette, M. E., Jesse, M. J., Davies, J. E., and Hutchison, G. B., 1976, Coffee drinking and death due to coronary heart disease, *N. Engl. J. Med.* **294**:633.

Hennekens, C. H., Rosner, B., and Cole, D. S., 1978, Daily alcohol consumption and fatal coronary heart disease, *Am. J. Epidemiol.* **107**:196.

Hennekens, C. H., Willett, W., Rosner, B., Cole, D. S., and Mayrent, S. L., 1979, Effects of beer, wine, and liquor in coronary deaths, *JAMA* **242**:1973.

Hirst, A. E., Hadley, G. G., and Gore, I., 1965, The effect of chronic alcoholism and cirrhosis of the liver on atherosclerosis, *Am. J. Med. Sci.* **249:**143.

Howell, W. L., and Manion, W. C., 1960, The low incidence of myocardial infarction in patients with portal cirrhosis of the liver: A review of 639 cases of cirrhosis of the liver from 17,731 autopsies, *Am. Heart J.* **60:**341.

Hrubec, Z., Cederlöf, R., and Friberg, L., 1976, Background of angina pectoris: Social and environmental factors in relation to smoking, *Am. J. Epidemiol.* **103:**16.

Hulley, S. B., and Gordon, S., 1981, Alcohol and high-density lipoprotein cholesterol. Causal inference from diverse study designs, *Circulation* **64**(Suppl. III):57.

Ishitoya, Y., Namiki, T., and Itoh, C., 1967, Quantitative survey on lesions of aortic atherosclerosis in Sendai area, *Tohoku J. Exp. Med.* **92:**379.

Jaworski, R. C., 1981, Alcohol and atherosclerosis, *Med. J. of Aust.* **2:**456.

Jones, B. R., Barrett-Connor, E., Criqui, M. H., and Holdbrook, M. J., 1982, A community study of calorie and nutrient intake in drinkers and nondrinkers of alcohol, *Am. J. Clin. Nutr.* **35:**135.

Kaelber, C. T., and Barboriak, J., (eds.), 1981, Symposium on Alcohol and Cardiovascular Diseases, American Heart Association Monograph Number 81, *Circulation* **64**(II):Suppl. III.

Kagan, A., Gordon, T., Rhoads, G. G., and Schiffman, J. C., 1975, Some factors related to coronary heart disease incidence in Honolulu Japanese men: The Honolulu Heart Study, *Int. J. Epidemiol.* **4:**271.

Kagan, A., Yano, K., Rhoads, G. G., and McGee, D. L., 1981, Alcohol and cardiovascular disease: The Hawaiian experience, *Circulation* **64**(Suppl. III):27.

Kane, W. C., and Aronson, S. M., 1971, Cerebrovascular disease in an autopsy population. 4. Reduced frequency of stroke in patients with liver cirrhosis, *Trans. Am. Neurol. Assoc.* **96:**259.

Kannel, W. B., 1976, Some lessons in cardiovascular epidemiology from Framingham, *Am. J. Cardiol.* **37:**269.

Kentala, E., Luurila, O., and Salaspuro, M. P., 1976, Effects of alcohol ingestion on cardiac rhythm in patients with ischemic heart disease, *Ann. Clin. Res.* **8:**408.

Khetarpal, V. K., and Volicer, L., 1981, Alcohol and cardiovascular disorders, *Drug Alcohol Depend.* **7:**1.

Klatsky, A. L., 1982, Alcohol and mortality, *Ann. Intern. Med.* **96:**120.

Klatsky, A. L., Friedman, G. D., and Siegelaub, A. B., 1974, Alcohol consumption before myocardial infarction: Results from the Kaiser–Permanente epidemiologic study of myocardial infarction, *Ann. Intern. Med.* **81:**294.

Klatsky, A. L., Friedman, G. D., Siegelaub, A. B., and Gérard, M. J., 1977, Alcohol consumption and blood pressure: Kaiser–Permanente multiphasic health examination data, *N. Engl. J. Med.* **296:**1194.

Klatsky, A. L., Friedman, G. D., and Siegelaub, A. B., 1979, Alcohol use, myocardial infarction, sudden cardiac death, and hypertension, *Alcohol. Clin. Exp. Res.* **3:**33.

Klatsky, A. L., Friedman, G. D., and Siegelaub, A. B., 1981a, Alcohol and mortality: A ten-year Kaiser–Permanente experience, *Ann. Intern. Med.* **95:**139.

Klatsky, A. L., Friedman, G. D., and Siegelaub, A. B., 1981b, Alcohol use and cardiovascular disease: The Kaiser–Permanente experience, *Circulation* **64**(Suppl. III):32.

Klatsky, A. L., Friedman, G. D., and Siegelaub, A. B., 1981c, Alcohol and mortality, *Ann. Intern. Med.* **95:**780.

Kozarevic, Dj., McGee, D., Vojvodic, N., Racic, Z., Dawber, T., Gordon, T., and Zukel, W., 1980, Fequency of alcohol consumption and morbidity and mortality: The Yugoslavia Cardiovascular Disease Study, *Lancet* **1:**613.

Kozarevic, Dj., Demirovic, J., Gordon, T., Kaelber, C. T., McGee, D., and Zukel, W. J., 1982, Drinking habits and coronary heart disease: The Yugoslavia Cardiovascular Disease Study, *Am. J. Epidemiol.* **116:**748.

Kuller, L. H., LaPorte, R. E., and Weinberg, G. B., 1979, The decline in ischemic heart disease mortality: Environmental and social variables, in: *Proceedings of the Conference on the Decline in Coronary Heart Disease Mortality* (R. J. Havlik and M. Feinleib, eds.), pp. 312–339, U.S. Department of Health, Education, and Welfare, Public Health Service, National Institutes of Health, NIH Publication No. 79-1610.

LaPorte, R. E., and Cauley, J. A., 1981, Wine, age, and coronary heart disease, *Lancet* 1:105.

LaPorte, R. E., Cresanta, J. L., and Kuller, L. H., 1980a, The relationship of alcohol consumption to atherosclerotic heart disease, *Prev. Med.* 9:22.

LaPorte, R. E., Cresanta, J. L., and Kuller, L. H., 1980b, The relation of alcohol to coronary heart disease and mortality: Implications for public health policy, *J. Public Health Policy* 1:198.

LaPorte, R., Valvo-Gerard, L., Kuller, L., Dai, W., Bates, M., Cresanta, J., Williams, K., and Palkin, D., 1981, The relationship between alcohol consumption, liver enzymes and high-density lipoprotein cholesterol, *Circulation* 64(Suppl. III):67.

Lee, K., Nielsen, D., Zeeberg, I., and Gormsen, J., 1980, Platelet aggregation and fibrinolytic activity in young alcoholics, *Acta Neurol. Scand.* 62:287.

Levy, R. I., and Rifkind, B. M., 1980, The structure, function and metabolism of high-density lipoproteins: A status report, *Circulation* 62(Suppl. IV):4.

Lifšic, A. M., 1976, Alcohol consumption and atherosclerosis, *Bull. WHO* 53:623.

Lindenbaum, J., 1974, Hematologic effects of alcohol, in: *The Biology of Alcoholism,* Volume 3 (B. Kissin and H. Begleiter, eds.), pp. 461–480, Plenum Press, New York.

Little, R. E., and Streissguth, A. P., 1981, Effects of alcohol on the fetus: Impact and prevention, *Can. Med. Assoc. J.* 125:159.

Lundberg, G. D., and Voigt, G. E., 1979, Reliability of a presumptive diagnosis in sudden unexpected death in adults, *JAMA* 242:2328.

MacDonald, R. A., and Mallory, G. K., 1958, The natural history of postnecrotic cirrhosis, *Am. J. Med.* 24:334.

Mainland, D., 1953, The risk of fallacious conclusions from autopsy data on the incidence of diseases with applications to heart disease, *Am. Heart J.* 45:644.

Manku, M. S., Oka, M., and Horrobin, D. F., 1979, Alcohol consumption and coronary heart-disease, *Lancet* 1:1404.

Marmot, M. G., Rose, G., Shipley, M. J., and Thomas, B. J., 1981, Alcohol and mortality: A U-shaped curve, *Lancet* 1:580.

Mathews, J. D., 1976, Alcohol usage as a possible explanation for socio-economic and occupational differentials in mortality from hypertension and coronary heart disease in England and Wales, *Aust. NZ J. Med.* 6:393.

May, S. J., Kuller, L. H., and Perper, J. A., 1980, The relationship of alcohol to sudden natural death, *J. Stud. Alcohol* 41:693.

Meade, T. W., Chakrabarti, R., Haines, A. P., North, W. R. S., and Stirling, Y., 1979, Characteristics affecting fibrinolytic activity and plasma fibrinogen concentrations, *Br. Med. J.* 1:153.

Meltzer, S. J., 1981, Alcohol and mortality, *Ann. Intern. Med.* 95:780.

Miller, G. H., 1982, Alcohol and mortality, *Ann. Intern. Med.* 96:120.

Moore, M. C., Guzman, M. A., Schilling, P. E., and Strong, J. P., 1975, Dietary–atherosclerosis study on deceased persons, *J. Am. Diet. Assoc.* 67:22.

Moreyra, A. E., Kostis, J. B., Passannante, A. J., and Kuo, P. T., 1982, Acute myocardial infarction in patients with normal coronary arteries after acute ethanol intoxication, *Clin. Cardiol.* 5:425.

Moriyama, I. M., Krueger, D. E., and Stamler, J., 1971, *Cardiovascular Diseases in the United States,* Harvard University Press, Cambridge.

Morris, J. N., Kagan, A., Pattison, D. C., Gardner, M. J., and Raffle, P. A. B., 1966, Incidence and prediction of ischemic heart-disease in London busmen, *Lancet* 2:553.

National Institute on Alcohol Abuse and Alcoholism, Public Health Service, U.S. Department of Health, Education, and Welfare 1974, Second Special Report to the U.S. Congress on Alcohol and Health, Rockville, Md., June, pp. 68–72.

Nicolas, G., Gaillard, A., Bouhour, J. B., and Horeau, J., 1970, L'athérosclerose coronarienne chez les cirrhotiques. (Étude anatomique de 50 cas), *Rev. Alcool.* **16**:274.

Okazaki, M., Hara, I., Tanaka, A., Kodama, T., and Yokoyama, S., 1981, Decreased serum HDL_3 cholesterol levels in cirrhosis of the liver, *N. Engl. J. Med.* **304**:1608.

Orlando, J., Aronow, W. S., Cassidy, J., and Prakash, R., 1976, Effect of ethanol on angina pectoris, *Ann. Intern. Med.* **84**:652.

Parrish, H. M., and Eberly, A. L., 1961, Negative association of coronary atherosclerosis with liver cirrhosis and chronic alcoholism—a statistical fallacy, *J. Indiana State Med. Assoc.* **54**:341.

Paul, O., Lepper, M. H., Phelan, W. H., Dupertuis, G. W., MacMillan, A., McKean, H., and Park, H., 1963, A longitudinal study of coronary heart disease, *Circulation* **28**:20.

Pell, S., and D'Alonzo, C. A., 1973, A five-year mortality study of alcoholics, *J. Occup. Med.* **15**:120.

Péquignot, G., and Tuyns, A., 1975, "Self-reported" alcohol consumption and pathological risks (translation from the French of an article entitled "Rations d'alcool consommées "declarées" et risques pathologiques", in: *INSERM, Anglo-French Symposium on Alcoholism*, pp. 1–15, Paris, Selected Translations of International Alcoholism Research (STIAR 6); Available from the National Clearinghouse for Alcohol Information, Department STIAR, P.O. Box 2345, Rockville, Maryland 20852.

Petitti, D. B., Wingerd, J., Pellegrin, F., and Ramcharan, S., 1979, Risk of vascular disease in women: Smoking, oral contraceptives, noncontraceptive estrogens, and other factors, *JAMA* **242**:1150.

Poikolainen, K., 1983, Inebriation and mortality, *Int. J. Epidemiol.* **12**:151.

Popham, R. E., and Schmidt, W., 1981*a*, Words and deeds: The validity of self-report data on alcohol consumption. *J. Stud. Alcohol* **42**:355.

Popham, R. E., and Schmidt, W., 1981*b*, "Words and Deeds": A rejoinder, *J. Stud. Alcohol* **42**:533.

Public Health Service, U.S. Department of Health, Education, and Welfare, 1975, Comparability of Mortality Statistics for the Seventh and Eighth Revisions of the International Classification of Diseases, United States, DHEW Publication No. (HRA) 76-1340, National Center for Health Statistics, Rockville, October. pp. 5–16.

Public Health Service, U.S. Department of Health, Education, and Welfare, 1979*a*, Report: National Heart, Lung, and Blood Institute, Working Group on Heart Disease Epidemiology, NIH Publication No. 79-1667, June.

Public Health Service, U.S. Department of Health, Education, and Welfare, 1979*b*, Smoking and Health: A Report of the Surgeon General, DHEW Publication No. (PHS) 79-50066, Rockville, Md.

Randall, B., 1980*a*, Fatty liver and sudden death: A review, *Hum. Pathol.* **11**:147.

Randall, B., 1980*b*, Sudden death and hepatic fatty metamorphosis: A North Carolina survey, *JAMA* **243**:1723.

Regan, T. J., and Ettinger, P. O., 1979, Varied cardiac abnormalities in alcoholics, *Alcohol. Clin. Exp. Res.* **3**:40.

Regan, T. J., and Haider, B., 1981, Ethanol abuse and heart disease, *Circulation* **64**(Suppl. III):14.

Regan, T. J., Wu, C. F., Weisse, A. B., Moschos, C. B., Ahmed, S. S., Lyons, M. M., and Haider, B., 1975, Acute myocardial infarction in toxic cardiomyopathy without coronary obstruction, *Circulation* **51**:453.

Regan, T. J., Ettinger, P. O., Lyons, M. M., Moschos, C. B., and Weisse, A. B., 1977, Ethyl alcohol as a cardiac risk factor, in: *Current Problems in Cardiology* (W. P. Harvey, ed.), pp. 7–35, Year Book Medical Publishers, Chicago.

Restrepo, C., Montenegro, M. R., and Solberg, L. A., 1968, Atherosclerosis in persons with selected diseases, *Lab. Invest.* **18**:92.

Rhoads, G. G., Blackwelder, W. C., Stemmermann, G. N., Hayashi, T., and Kagan, A., 1978, Coronary risk factors and autopsy findings in Japanese-American men. *Lab. Invest.* **38**:304.

Rissanen, V., 1974, Coronary and aortic atherosclerosis in chronic alcoholics, *Z. Rechtsmed.* **75**:183.

Robinette, C. D., Hrubec, Z., and Fraumeni, J. F., Chronic alcoholism and subsequent mortality in World War II veterans, 1979, *Am. J. Epidemiol.* **109**:687.

Rosenberg, L., Slone, D., Shapiro, S., Kaufman, D. W., Miettinen, O. S., and Stolley, P. D., 1981, Alcoholic beverages and myocardial infarction in young women, *Am. J. Public Health* **71**:82.

Rosenman, R. H., Friedman, M., Jenkins, D., Straus, R., Wurm, M., and Kositchek, R., 1966, The prediction of immunity to coronary heart disease, *JAMA* **198**:1159.

Rosman, J. K., 1982, Alcohol and mortality, *Ann. Intern. Med.* **96**:120.

Rubin, E., 1980, Cardiovascular effects of alcohol, *Pharmacol. Biochem. Behav.* **13**(Suppl. 1):37.

Rubin, E., 1982, Alcohol and the heart: Theoretical considerations. *Fed. Proc.* **41**:2460.

Ruebner, B. H., Miyai, K., and Abbey, H., 1961, The low incidence of myocardial infarction in hepatic cirrhosis. A statistical artefact? *Lancet* **2**:1435.

Sabesin, S. M., 1981, Lipid and lipoprotein abnormalities in alcoholic liver disease, *Circulation* **64**(Suppl. III):72.

Sackett, D. L., Gibson, R. W., Bross, I. D. J., and Pickren, J. W., 1968, Relation between aortic atherosclerosis and the use of cigarettes and alcohol: An autopsy study, 1968, *N. Engl. J. Med.* **279**:1413.

St. Leger, A. S., Cochrane, A. L., and Moore, F., 1979, Factors associated with cardiac mortality in developed countries with particular reference to the consumption of wine, *Lancet* **1**:1017.

Schmidt, W., and Popham, R. E., 1978, The single distribution theory of alcohol consumption: A rejoinder to the critique of Parker and Harman. *J. Stud. Alcohol* **39**:400.

Schmidt, W., and Popham, R. E., 1980, Sex differences in mortality: A comparison of male and female alcoholics, in: *Alcohol and Drug Problems in Women: Research Advances in Alcohol and Drug Problems*, Volume 5 (O. J. Kalant, ed.), pp. 365–384, Plenum Press, New York.

Schmidt, W., and Popham, R. E., 1981*a*, Alcohol consumption and ischemic heart disease: Some evidence from population studies, *Br. J. Addict.* **76**:407.

Schmidt, W., and Popham, R. E., 1981*b*, The role of drinking and smoking in mortality from cancer and other causes in male alcoholics, *Cancer* **47**:1031.

Schmidt, W., and Popham, R. E., 1982, Alcohol consumption and ischemic heart disease: An addendum, *Br. J. Addict.* **77**:91.

Spicer, J., McLeod, W. R., O'Brien, K. P., and Scott, P. J., 1981, Psychosomatic patterns of coronary risk in a community sample of New Zealand men, *J. Chron. Dis.* **34**:271.

Stason, W. B., Neff, R. K., Miettinen, O. S., and Jick, H., 1976, Alcohol consumption and nonfatal myocardial infarction, *Am. J. Epidemiol.* **104**:603.

Statistics Canada, 1982, Causes of Death: Provinces by Sex and Canada by Age and Sex, Detailed Categories of the "International Classification of Diseases"—ICD, 1981, Health Division, Vital Statistics and Diseases Registries Section, Ministry of Supply and Services Canada, Ottawa, December 1982, (Catalogue 84-203 Annual).

Sundby, P., 1967, *Alcoholism and Mortality*, The National Institute for Alcohol Research, Publication Number 6, Universitetsforlaget, Oslo.

Talesnik, J., Belo, S., and Israel, Y., 1980, Enhancement of noradreneline-induced metabolic coronary dilation by ethanol, *Eur. J. Pharmacol.* **61**:279.

Thorarinsson, A. A., 1979, Mortality among men alcoholics in Iceland, 1951–1974, *J. Stud. Alcohol* **40**:704.

Turlapaty, P. D. M. V., and Altura, B. M., 1980, Magnesium deficiency produces spasms of coronary arteries: Relationship to etiology of sudden death ischemic heart disease, *Science* **208**:198.

Tyroler, H. A. (ed.), 1980, Epidemiology of plasma high-density lipoprotein cholesterol levels, The Lipid Research Clinics Program Prevalence Study, Lipid Research Clinics Program, American Heart Association Monograph Number 73, *Circulation* **62**:Suppl. IV.

Ullman, B. M., Lamphiear, D. E., Luton, J. R., Ross, H. W., and Wheeler, N. C., 1974, Alcohol Consumption and Coronary Heart Disease. Final Report to the National Institute on Alcohol Abuse and Alcoholism [Contract LHSM 42-73-72 (NIA)].

Vaněček, R., 1976, Atherosclerosis and cirrhosis of the liver, *Bull. WHO* **53:**567.

Viel, B., Donoso, S., Salcedo, D., Rojas, P., Varela, A., and Alessandri, R., 1966, Alcoholism and socioeconomic status, hepatic damage, and arteriosclerosis, *Arch. Intern. Med.* **117:**84.

Viel, B., Salcedo, D., Donoso, S., and Varela, A., 1970, Alcoholism, accidents, atherosclerosis, and hepatic damage, in: *Alcohol and Alcoholism* (R. E. Popham, ed.), pp. 319–337, University of Toronto Press, Toronto.

Waern, A. U., Lidell, C., and Hellsing, K., 1982, Alcohol intake, serum β_2-microglobulin and ventricular extrasystoles: Factors related to death in five-year follow-up of middle-aged men, *Ups. J. Med. Sci.* **87:**119.

Wallace, R. B., Lynch, C. F., Pomrehn, P. R., Criqui, M. H., and Heiss, G., 1981, Alcohol and hypertension: Epidemiologic and experimental considerations, The Lipid Research Clinics Program, *Circulation* **64**(Suppl. III):41.

Werth, J., 1980, A little wine for thy heart's sake, *Lancet* **2:**1141.

Wilens, S. L., 1947, The relationship of chronic alcoholism to atherosclerosis, *JAMA* **135:**1136.

Wilhelmsen, L., Wedel, H., and Tibblin, G., 1973, Multivariate analysis of risk factors for coronary heart disease, *Circulation* **48:**950.

Willett, W., Hennekens, C. H., Siegel, A. J., Adner, M. M., and Castelli, W. P., 1980, Alcohol consumption and high-density lipoprotein cholesterol in marathon runners, *N. Engl. J. Med.* **303:**1159.

Wood, D. A., Birtwhistle, A., Riemersma, R. A., and Fulton, M., 1981, Alcohol consumption and risk factors for ischemic heart disease, Paper presented at the 9th International Scientific Meeting of the International Epidemiological Association, August 22–28, Edinburgh.

Yano, K., Rhoads, G. G., and Kagan, A., 1977, Coffee, alcohol and risk of coronary heart disease among Japanese men living in Hawaii, *N. Engl. J. Med.* **297:**405.

Yano, K., Rhoads, G. G., Kagan, A., and Tillotson, J., 1978, Dietary intake and the risk of coronary heart disease in Japanese men living in Hawaii, *Am. J. Clin. Nutr.* **31:**1270.

Heavy Alcohol Consumption and Physical Health Problems

A Review of the Epidemiologic Evidence

ROBERT E. POPHAM, WOLFGANG SCHMIDT, and
STEPHEN ISRAELSTAM

1. INTRODUCTION

The epidemiologic evidence for physical health problems associated with chronic heavy alcohol use derives mainly from studies of (1) the drinking histories of persons suffering from a particular disease or trauma; (2) the prevalence of physical health problems in samples of heavy drinkers; (3) the mortality experience of such drinkers; (4) the drinking histories of persons who have died from a particular cause; (5) regional and temporal variation in cause-specific death rates in relation to variation in indices of the prevalence of heavy drinking; and (6), more recently, the morbidity and mortality experience of general population groups evaluated for a number of years and characterized at entry according to alcohol consumption. There have also been many experimental studies but, with human subjects, these have necessarily been short-term and confined to acute effects presumed to be relevant to the disease conditions of interest.

Until recently, morbidity studies (especially those in the first two categories above) have rarely included control samples, and as a result, seldom provide the data required to calculate risk factors. However, such studies taken as a whole present a reasonably consistent picture of the types of physical health problems likely to be encountered among heavy drinkers. Typically found are

ROBERT E. POPHAM, WOLFGANG SCHMIDT, AND STEPHEN ISRAELSTAM ● Addiction Research Foundation, Toronto, Ontario, Canada.

(1) certain diseases of the nervous system such as peripheral neuropathy, and various indications of brain damage; (2) certain diseases of the digestive system, especially those of the liver, acute and chronic gastritis, peptic ulcers, and pancreatitis; (3) certain respiratory diseases, especially chronic bronchitis, pneumonia, and tuberculosis; (4) certain heart and vascular diseases, especially cardiomyopathy and hypertension; (5) certain cancers, especially those of the upper respiratory and upper digestive tracts; and (6) injuries resulting from motor vehicle and other accidents. For recent reviews of the literature on morbidity, see Kissin and Begleiter (1974), Lelbach (1974), Seixas et al. (1975), Fallon and Lesesne (1978) and Eckardt et al. (1981). The morbidity picture for clinical alcoholics is well-illustrated by the data of Ashley et al. (1977) for a sample of admissions to a voluntary inpatient treatment unit in Toronto, Canada (Table 1). It is noteworthy that, contrary to a currently popular notion, the female alcoholics in this population apparently did not exhibit greater morbidity than the male.

The incidence of the various diagnoses listed may be influenced by patient selection and thoroughness of physical examination, among other factors. However, the kinds of illness and trauma encountered would seem to be typical.

Table 1. Prevalence at Admission of Various Disease Entities, Complications, and Trauma in a Clinical Sample of Alcoholics[a]

	Men (N = 736)		Women (N = 135)	
	No	Percent	No	Percent
Fatty liver	351	47.7	37	27.4
Chronic obstructive lung disease	89	12.1	8	5.9
Trauma (all)	88	11.4	10	7.4
Hypertension	64	8.7	9	6.7
Malnutrition	57	7.7	12	8.9
Gastritis	45	6.1	4	3.0
Fractures	42	5.7	5	3.7
Peripheral neuritis	34	4.6	3	2.2
Hiatus hernia	33	4.5	8	5.9
Cirrhosis	32	4.4	4	3.0
Anemia (all)	31	4.2	18	13.3
Peptic ulcer	30	4.1	5	3.7
Chronic brain damage	27	3.7	4	3.0
Obesity	23	3.1	8	5.9
Ischemic heart disease	23	3.1	0	0.0
Cardiomyopathy	20	2.7	6	4.4
Pneumonia	19	2.6	3	2.2
Epileptic disorders	19	2.6	1	0.7
Diabetes	18	2.4	1	0.7
Gastrointestinal hemorrhage	17	2.3	3	2.2
Urinary tract infection	12	1.6	4	3.0
Acute brain syndromes	12	1.6	1	0.7
Pancreatitis	6	0.8	1	0.7

[a] From Ashley et al. (1977).

2.   GENERAL MORTALITY

As might be expected in view of the morbidity data, mortality studies have consistently revealed significantly elevated death rates in samples of alcoholics and other heavy drinkers. In all but one case, the studies listed in Table 2 found a ratio of observed over expected mortality approaching two or more. The variations in the ratio reflect mainly differences in type and size of sample, length of follow-up, and source of expected value. There is also a consistent difference in the size of the ratios for men and women. This difference will be discussed in Section 4.9. A satisfactory explanation cannot be offered for the departure of Dahlgren's ratio from that obtained in all other studies. This is particularly puzzling since his compatriot, Dahlberg—whose sample likewise included Swedish Temperance Board patients—found considerable excess mortality.

An important question about the generality of these findings is the possibility that excess mortality is greater among heavy drinkers who seek treatment than among those who do not. Schmidt and de Lint (1972) found a relatively high death rate in the first year after clinic admission. This would suggest that some of their patients sought treatment as a consequence of seriously deteriorated physical health. Although omitting first-year deaths reduced the ratio by less than 5%, this does not allow for the possibility that some persons who sought help were suffering from chronic diseases not immediately fatal. On the other hand, the last nine studies in Table 2 employed essentially nonclinical samples and show—with the exception noted—significantly high ratios. A mean could not be calculated for Dahlberg's sample, but the ratios for most of his age groups ranged from 1.4 to 1.9. As in other studies, only those patients aged 65 and older failed to exhibit a notable excess mortality.

The ratio of 1.8 based on the study of Newman (1965) is of special interest since his largely male sample was secured through a case-finding survey in Ontario employing many different sources in addition to hospitals and physicians. The follow-up period overlaps that of Schmidt and Popham (1980) who found a male ratio of 1.9 for their sample of clinic alcoholics drawn from the same province.

The ratios for samples of industrial workers (Pell and D'Alonzo, 1973; Dyer et al., 1981) and for insurance policy holders (Menge, 1950; Davies, 1965) would also seem to indicate that elevated death rates are not confined to clinically identified heavy drinkers. The samples of policy holders included not only persons with a history of alcoholism treatment or frequent episodes of intoxication, but also "steady free users" without a history of drunkenness and persons with only occasional episodes of intoxication. When the several categories of heavy alcohol users were examined separately, both authors found mortality ratios of 2 or more in every case. Moreover, care was taken to assure that the excess mortality was not attributable to selection factors resulting from the insurance risk categories used for comparison. Indeed, Davies' study was restricted to a comparison of groups differing only with respect to alcohol habits.

Table 2. The Ratio of Alcohol to Expected Mortality in Samples of Alcoholics and Other Heavy Drinkers

Reference	Country	Source of sample	Sex	Size of sample	No. of deaths	Follow-up period	Exposure (years)	Source of expected values	Ratio
Gabriel (1935)[a]	Austria	Admissions to a mental hospital alcoholism treatment facility	M	1,107	148	1922–1935	?	Male mortality for the age group 21–70 in Austria 1932	2.5
Kendall and Staton (1966)	U.K.	Untreated referrals to an alcoholism treatment facility	M and F	59	11	1950–1963	?	Age–sex-specific mortality in the U.K.	5.2
Gorwitz et al. (1966)	U.S.A.	Admissions for alcoholism to mental hospitals	M F	1,355 298	70 15	1961–1964 1961–1964	n.a. n.a.	Age–sex-specific mortality in Maryland	6.0 9.0
Brenner (1967)	U.S.A.	Admissions to alcoholism treatment facilities	M and F	1,343	217	1954–1961	7,289	Age–sex-specific mortality in California	3.0
Sundby (1967)	Norway	Admissions for alcoholism to a psychiatric facility	M	1,722	1,061	1925–1962	34,951	Age-specific male mortality in Oslo	1.7
Babigian and Odoroff (1969)	U.S.A.	Admissions for alcoholism to psychiatric facilities	M F	? ?	362 78	1960–1966 1960–1966	? ?	Age–sex-specific mortality in Monroe County, N.Y.	2.8 4.9

Gillis (1969)	South Africa	Admissions to an alcoholism treatment facility	M F	707 95	81 9	1959–1964 1959–1964	3,080 399	Age–sex-specific mortality of whites in South Africa	3.9 4.5
Lindelius and Salum (1972)	Sweden	Admissions to a hospital ward for the treatment of acute alcoholic psychosis	M	1,021	275	1956–1966	?	Age-specific male mortality in Stockholm	3.6
Lindelius et al. (1974)	Sweden	Admissions for alcohol abuse to a psychiatric facility	M F	139 118	25 19	1962–1971 1962–1971	910 761	Age–sex-specific mortality in Stockholm	2.3 3.9
Nicholls et al. (1974)	U.K.	Admissions for alcoholism to mental hospitals	M F	678 257	233 76	1953–1967 1953–1967	? ?	Age–sex-specific mortality in the U.K.	2.7 3.1
de Lint and Levinson (1975)	Canada	Admissions to an alcoholism treatment facility	M F	154	20 2	1969–1974 1969–1974	584 167	Age–sex-specific mortality in Ontario	3.6 3.9
Ciompi and Medvecka (1976)	Switzerland	Admissions for alcoholism to a psychiatric facility	M F	1,229 155	968 110	1915–1962 1915–1962	? ?	Age–sex-specific mortality in Switzerland	1.7 2.2
Dahlgren and Myrhed (1977)	Sweden	Admissions for alcoholism to a hospital department of alcoholic diseases	M F	100 100	16 18	1963–1975 1963–1975	713 791	Age–sex-specific mortality in Stockholm	3.0 5.6

(continued)

Table 2. (*Continued*)

Reference	Country	Source of sample	Sex	Size of sample	No. of deaths	Follow-up period	Exposure (years)	Source of expected values	Ratio
Robinette et al. (1979)	U.S.A.	Admissions for alcoholism to V.A. hospitals	M	4,401	1,438	1946–1974	112,244	Age-specific mortality of male patient control group	1.9
Thorarinsson (1979)	Iceland	Treated alcoholics listed in the Psychiatric Register	M	2,863	573	1951–1974	37,200	Age-specific male mortality in Iceland	2.2
Schmidt and Popham (1980)	Canada	Admissions to an alcoholism treatment facility	M F	9,889 2,158	1,823 296	1951–1971 1951–1971	85,641 17,805	Age–sex-specific mortality in Ontario	1.9 3.0
Menge (1950)	U.S.A.	Insurance policy holders classed as substandard due to alcohol habits	Mostly M	?	758	1931–1948	72,658	Standard male mortality risk for insurance policy holders	3.1
Dahlgren (1951)	Sweden	Cases recorded by Malmö and Malmöhus Temperance Boards	M	10,616	564	1939–1947	?	Age-, marital status-, and residence-specific male mortality in Sweden	1.1
Dahlberg (1952)[b]	Sweden	Cases recorded by Stockholm Temperance Board	M	2,365	?	?	?	Age-specific male mortality in Stockholm	1.0–1.9
Davies (1965)	U.S.A.	Insurance policy holders classed	Mostly M	2,582	140	1940–1962	?	Standard male mortality risk	2.7

		as substandard due to alcohol habits						for insurance policy holders
Newman (1965)[c]	Canada	Case-finding survey of one county	Mostly M	339	84	1953–1964	3,594	Age-specific average male mortality in Ontario, 1953–1964	1.8
Giffen et al. (1971)	Canada	Persons arrested for drunkenness 3 or more times within a year	M	343	191	1940–1961	4,214	Age-specific male mortality in Ontario	2.3
Pell and D'Alonzo (1973)[d]	U.S.A.	Alcoholic employees of one industry	M / F	842 / 57	96 / 6	1965–1969 / 1965–1969	n.a. / n.a.	Age-sex-specific mortality of fellow-worker control group	3.1 / 6.7
Dyer et al. (1981)	U.S.A.	Problem drinker employees of one industry	M	38	20	1958–1973	n.a.	Age-specific male mortality of fellow-worker cohort	2.6
Dyer et al. (1981)	U.S.A.	Consumers of 6 or more drinks per day among the employees of one industry	M	78	32	1957–1974	n.a.	Age-specific male mortality of fellow-worker cohort	2.7

[a] In the calculation of the mortality rate, Gabriel did not take into account the number of years of exposure to the risk of death. As a consequence, his ratio of actual to expected mortality was erroneously inflated. On the assumption that the admissions and deaths were evenly distributed over Gabriel's observation period, an estimate of exposure years could be obtained. The ratio shown in the table was calculated on the basis of this estimate.

[b] A risk ratio for all age groups combined was not reported. For those up to 65 years of age, the reported ratios ranged from 1.4 to 1.9; for the older groups, from 1.0 to 1.2.

[c] The ratio of 1.0 reported in this study resulted from a misapplication of life tables and a faulty comparison of the sample with the general population. Thus, number of survivors rather than number of deaths was employed in the comparison. This is particularly unjustified when the number of exposure years is small. However, the primary data provided in the report permitted calculation of the ratio shown in the table in accord with accepted methods.

[d] The alcoholic sample comprised persons identified by the company physician as "recovered," "suspected," or "known" alcoholics.

Although there are many earlier mortality studies in which insurance data were utilized (see Pearl, 1926 for a review of these), the typical approach was to contrast abstainers with nonabstainers. The abstaining group always lived longer, but it would be erroneous to conclude that drinking in any amount shortens life expectancy. Among nonabstainers there is bound to be a proportion of habitually heavy drinkers whose unfavorable mortality experience adversely affects the picture for the group as a whole. The studies of Menge and Davies, on the other hand, employed standard risks for comparison, and these reflected the mortality experience of both abstainers and moderate drinkers.

We must emphasize our agreement with Room and Day (1974) that samples of the type described cannot be assumed *a priori* to be representative of all heavy drinkers. Certainly, in the case of those identified as "alcoholic" the mortality ratio might be expected to exceed that found in a suitable population sample of heavy drinkers. However, that a significantly elevated death rate probably does characterize the latter, is indicated by the work of Pearl (1926) and Room and Day (1974) who determined the death rates of drinkers identified through studies of general population samples in the United States. In both studies, heavy users were found to have higher death rates than moderate users and abstainers. The findings differed in that the group for which Room and Day found elevated mortality constituted only 2–3% of the total sample (probably 4–5% of the drinkers). In contrast, the proportion of all drinkers found by Pearl to have an elevated risk seems to have been at least three times as great. The two studies were conducted at quite different times, appear to have used somewhat different definitions of "heavy drinking," and employed different techniques to assure the accuracy and comprehensiveness of the data on drinking habits. Furthermore, Pearl's sample was, in effect, a random sample of the white working-class population of a single city (Pearl, 1923); that of Room and Day was more generally representative of the national population.

A prospective study of the employees of an industrial company in Chicago also found elevated death rates among the heavy drinkers (five or more drinks/day) compared with the moderate drinkers and abstainers (Dyer et al., 1977). In addition, the abstainers were found to have a slightly higher mortality rate than the moderate drinkers. Such a difference was also reported by Pearl (1926) and by Room and Day (1974), but only Pearl attempted to control for the possibility that the difference might be due to factors not connected with alcohol use. He selected from his original records pairs of brothers, one of whom had been a moderate drinker throughout his life, and the other, an abstainer. Under these conditions, the difference in mortality rates disappeared.*

---

* Another possible explanation for such a difference is that, even when an effort is made to avoid the difficulty, some persons for whom drinking has been contraindicated on health grounds (including former heavy drinkers) are recorded as abstainers. Certainly where a distinction has been made (e.g., Dyer et al., 1977, 1981), former drinkers are found to have a substantially higher mortality rate than abstainers and moderate drinkers.

Our conclusions are that (1) chronic heavy users of alcohol, whether identified in a clinical situation, through a case-finding survey, on the basis of drunkenness arrest record, or through reported drinking habits, have a substantially elevated risk of premature death; and (2) there is no convincing evidence that moderate consumption has an adverse effect or that a lifetime of abstinence has a beneficial effect on life expectancy.

3. HAZARDOUS CONSUMPTION LEVELS*

Clearly these conclusions beg a question of very great practical importance: at what point is consumption "heavy" in that there is a significant reduction in life expectancy? In the studies reviewed, heavy consumption was either implied *a priori* (e.g., in studies of clinic alcoholic samples), or defined without reference to a minimum volume of intake. A valid quantitative estimate of the upper limit of safe consumption (from a public health rather than an individual standpoint) has yet to be made. Of several early attempts perhaps the best known is that of Anstie. He concluded (1870, p.222), from observations of essentially acute effects, that a daily intake of 1.5 oz (4.4 cl) of absolute alcohol was "about the limit of what can be habitually taken . . . without provoking symptoms of chronic malaise indicative of actual alcohol poisoning." Later, Becker (1908) postulated the safe level to be 15 g (1.9 cl). This represented the daily equivalent of the average lifetime intake of a sample of men over 90 years of age.

More recently, Schmidt and de Lint (1970) approached the problem from the standpoint of the lower limit of consumption of clinic alcoholics. They found that a daily intake at or above 15 cl of absolute alcohol described the reported consumption of 96% of the alcoholics in their sample. Moreover, the mean for the sample was in remarkably close agreement with that for samples in several other countries (Schmidt, 1973). Given the mortality data reviewed in the previous section, and the reasonable assumption that clinical alcoholics are among the heaviest of heavy drinkers, there can be little doubt that, as a group, drinkers of such quantities have a markedly elevated risk of premature death. However, it is evident from the results of retrospective studies that this is very unlikely to be the lowest point at which a significant increase in the risk of alcohol-related disease occurs. Thus, Péquignot (1974) showed that an average daily intake of 40–60 g (5–7.5 cl) was potentially cirrhogenic, and similar levels have been

---

* Quantities of alcohol consumed have been expressed in a variety of ways by the authors whose work is cited in this review. However, in the interest of comparability we have translated all quantities reported into centiliters (cl) of absolute alcohol. For the benefit of those who prefer to think in terms of 'drinks', a 1.5 ounce drink of distilled spirits at 40% alcohol by volume contains 1.7 cl of absolute alcohol. This amount is also contained in a 12 ounce bottle of 5% beer or a 5 ounce glass of 12% table wine.

implicated in the development of certain cancers. For example, Tuyns et al. (1977) reported that the same consumption level carries a significant risk of cancer of the esophagus. Keller and Terris (1965) found a strong positive association between the consumption of 35 g (4.4 cl) of absolute alcohol per day, when combined with cigarette smoking, and the occurrence of carcinomata of the mouth and pharynx.* Wynder and Mabuchi (1972) found that the risk of carcinomata of the larynx and mouth rose significantly among drinkers of 7 or more oz of whiskey per day (approximately 8 cl of absolute alcohol).

It should be noted that the daily averages to which we have been referring simply represented the total volume of alcohol consumed over a given period of time divided by the number of days involved. Clearly, such a measure does not take into account differences in pattern, for example, whether the drinking is steady or intermittent. However, with respect to most alcohol-related organic complications, the pattern appears to be irrelevant except in so far as it affects the total amount of alcohol consumed. The crucial variables appear to be the duration of drinking and the volume of alcohol consumed (Lelbach, 1974).

On the other hand, it is evident that the pattern of drinking is highly relevant to the acute problems of alcohol use. Here the issue is the concentration of alcohol in the body at a given time rather than the amount consumed over a long period. Only in the field of vehicular accidents have attempts been made to establish risk levels. It is generally agreed that blood alcohol concentrations in drivers below 0.05% are safe. Above this level, risk rates increase exponentially with blood alcohol concentration.

In summary, the lowest level of alcohol consumption that constitutes a significant hazard to longevity has yet to be determined. It is almost certainly substantially less than a daily equivalent of 15 cl and quite possibly below 7.5 cl. The derivation of a valid estimate of this point is a critical problem for future research. The task is unusually difficult, among other reasons, because of the apparent inability of current survey techniques to elicit accurate quantitative data on volume of consumption. Accordingly, as we have noted elsewhere (Popham and Schmidt, 1978), the most promising route would seem to be through a combination of animal mortality studies and controlled retrospective studies of persons with or who have died from alcohol-related trauma or disease.

4. MORTALITY BY CAUSE

Another question posed by the occurrence of a higher death rate among heavy drinkers is the relative etiologic significance of the alcohol consumption and frequently associated factors such as heavy smoking, emotional problems,

---

* Risk ratios calculated from the Keller and Terris data by Rothman (1980) suggest an elevated risk among nonsmokers at even lower alcohol consumption levels.

and personal neglect. Clearly the issue is of considerable importance in planning preventive action. It may be approached through examination of the specific causes of death which have a higher than expected frequency among such drinkers. These are shown for various samples in Table 3. Although "alcoholism" is a commonly reported cause of death, especially in clinical samples of heavy drinkers, it has not been included because the role of alcohol is self-evident, and because coincident pathologies are typically present; its selection as the primary cause is therefore usually an arbitrary decision. An indication that the causes listed are those mainly responsible for the overall excess mortality is provided by the relatively small size of the ratio for deaths from all other causes (except alcoholism) found by Schmidt and de Lint (1972). These residual deaths made up 11% of the total mortality. The ratio, calculated from the data reported by the authors, was 1.3. That it exceeded one can be attributed to the inclusion of deaths from several rare causes such as homicide, nutritional diseases, and epilepsy which also occur more often than expected among heavy drinkers (Riddick and Luke, 1978; Schmidt and Popham, 1981).

Tuberculosis

The excess mortality from tuberculosis, observed in several samples, is in accord with morbidity data indicating an apparently higher than expected rate of alcoholism among tubercular patients (Feingold, 1976; Bailey et al., 1978), and of tuberculosis among heavy drinkers (Lyons and Saltzman, 1974). However, with one notable exception (Robinette et al., 1979), for which we cannot offer an explanation, there were few deaths from this cause, especially in the recent samples. For example, Schmidt and de Lint (1972) found no deaths from tuberculosis, and Schmidt and Popham (1981) found only five deaths from this cause in their much larger sample of the same population. Doubtless this reflects both the general decline in the incidence of tuberculosis and the very marked decline in the likelihood of death from the disease. The excess mortality found seems to be at least partly attributable to factors often associated with chronic heavy drinking, such as poor food habits, living in close contact with high-risk groups, heavy smoking, and failure to seek or remain in treatment (Brown and Campbell, 1961; Doll and Hill, 1964; Sundby, 1967). Whether or not chronic alcohol use exerts an additional independent effect remains open to question.

Cancer

Excess mortality from cancer involves only a limited number of sites: the lung and upper digestive and upper respiratory tracts. Hepatic carinoma has frequently been found at autopsy in association with cirrhosis (Parker, 1957; Leevy et al., 1964; Lee, 1966), and Tuyns (1979) suggests, in his recent review, that between 5 and 30% of all cirrhosis patients also show liver cancer. However, hepatic carinoma has rarely been reported as a cause of death in mortality studies of heavy drinkers. It may be that it is overlooked in the absence of an autopsy or that the associated cirrhosis tends to be favored as the primary cause for purposes of death certification. Another possibility is suggested by Purtilo and Gottlieb's (1973) observation that hepatic carcinoma typically appears about 8 years after cirrhotic damage is first discerned. Given the extremely heavy alcohol consumption characteristic of the clinical samples in Table 3, and the notoriously low rate of success of efforts to induce the alcoholic to remain abstinent, it is conceivable that many do not live long enough after the onset of cirrhosis to develop cancer of the liver. Some support for this explanation was recently advanced by Schmidt and Popham (1981). Cirrhosis deaths among their alcoholic patients occurred some 5 years, on the average, after the disease was first diagnosed.

Carcinomata at the other sites mentioned clearly contribute to the elevated death rate. In the case of lung cancer, it would now seem that the excess mortality

Table 3. Ratio of Actual to Expected Mortality by Cause in Male

| | | Cancer | | |
	Tuberculosis	All	Upper digestive/ respiratory	Cardiovascular disease
Clinical samples				
Gabriel (1935)[a]	4.1	2.5		2.2
Sundby (1967)[c]	2.8	1.7	4.2	1.1
Lindelius and Salum (1972)[d]	2.2[b]	1.8		
Nicholls et al. (1974)[e]	3.0[b]	1.7		
Robinette et al. (1979)[f]	10.2	1.2		1.3
Thorarinsson (1979)[g]		1.5		1.8
Schmidt and Popham (1981)[h]		1.3	3.9	1.4
Nonclinical samples				
Menge (1950)[i]	1.7	2.7		4.1
Dahlgren (1951)[j]	0.8	1.2		0.9
Davies (1965)[k]		2.3		2.6
Giffen et al. (1971)[l]		1.4		
Dyer et al. (1981)[m]		3.7		1.9

[a] See Note [a], Table 2.
[b] Actual frequency less than six deaths.
[c] The ratios based on Oslo mortality were taken in all instances. Those for tuberculosis and all cancers were for the 1951–1962 subsample. All other ratios were for the total (1925–1962) sample. The ratio for pneumonia was obtained from Table 42, for general mortality from Table 5, and for all other causes from Table 17 of Sundby (1967).
[d] Based on the data provided in Table 35 of Lindelius and Salum (1972).
[e] Taken from Table 1 of Nicholls et al. (1974).
[f] Taken from Table 3 of Robinette et al. (1979).

is entirely attributable to heavy smoking (Schmidt and Popham, 1981). On the other hand, in the development of carcinomata of the oral cavity, pharynx, larynx, and esophagus, most workers contend that alcohol consumption and smoking each have an independent effect, and a synergistic effect when combined (Keller and Terris, 1965; Rothman and Keller, 1972; Kissin and Kaley, 1974; Tuyns et al., 1977, 1978; Rothman, 1978; McMichael, 1979). However, Wynder (1978), who pioneered research on this question, maintains that heavy alcohol use in the absence of smoking does not increase the risk of these cancers.

Mortality studies of alcoholics are probably not capable of resolving this issue. Our impression, which is supported by at least one systematic study (Dreher and Fraser, 1967), is that nearly all alcoholics are more or less heavy smokers. The consequence is that the effects of the heavy drinking alone cannot be isolated. Schmidt and Popham (1981) compared the mortality experience of their alcoholics with that of nonalcoholic veterans who were equally heavy smokers. Although much reduced, there remained a significant excess mortality from cancers of the head and neck with smoking thus controlled. While this result

Samples of Alcoholics and Other Heavy Drinkers

Cerebrovascular lesions	Pneumonia	Peptic ulcer	Liver cirrhosis	Suicide	All accidents	All causes
	1.9		4.0[b]	9.0		2.5
1.6	1.9		3.6	5.8	2.8	1.7
			14.8	9.4	8.2	3.6
						2.7
1.3	2.3		2.1	2.0	3.4	1.9
1.1	3.3		11.3	4.4	4.0	2.2
1.2	2.8	2.8	11.1	5.8	3.6	1.9
			14.0	4.0	3.2	3.1
0.5			3.0	3.2	2.5	1.1
				2.7[b]	4.7	2.8
						2.3
						2.7

[g] Taken from Table 2 of Thorarinsson (1979).

[h] Taken from Table 2 of Schmidt and Popham (1981).

[i] Calculated from the percentage in Table VI of Menge (1950) using the general mortality ratio provided by the author.

[j] Taken from Table 4 of Dahlgren (1951) except that for cerebrovascular lesions which was calculated on the basis of data for cerebral hemorrhage given in the text.

[k] Taken from Table 3 of Davies (1965).

[l] Taken from Table XIII-16 of Giffen et al. (1971).

[m] Taken from Table 10 of Dyer et al. (1981).

confirms that heavy alcohol use elevates the smoking-related risk, the question remains as to whether the effect is additive or synergistic and, for that matter, whether or not there is an independent alcohol effect. An unexpected outcome of the comparison was the finding that the two study populations did not differ in their overall death rates from cancer. In other words, while heavy alcohol use enhanced the smoking-related risk of cancer at certain sites, it apparently did not increase the risk of neoplastic disease in general. This finding is somewhat easier to reconcile with Wynder's view that heavy alcohol use alone may not be carcinogenic than with the contention that there is an independent alcohol effect.

Cardiovascular Diseases

Deaths due to cardiovascular diseases account for a large proportion of the total excess mortality in most samples in Table 3. For example, of the 862 extra deaths in the Schmidt and Popham (1981) sample, 175 were ascribed to this causal category. An excess of deaths from cerebrovascular lesions (apoplexy) was also found in two of the samples (Sundby, 1967; Robinette et al., 1979). However, with the exception of alcoholic cardiomyopathy (Burch and Giles, 1974)—a comparatively rare cause of death—there is as yet no clear evidence that heavy alcohol consumption *per se* is or is not of etiologic importance in the development of these conditions (Sundby, 1967; Knott and Beard, 1972). Heavy smoking has been clearly implicated in the development of arteriosclerotic heart disease (Weir and Dunn, 1970), the most common cause of death in the category. However, it has been shown in at least three studies that, although much reduced, there remains a significant excess mortality from cardiovascular disease in heavy drinkers when smoking is controlled (Tibblin et al., 1975; Dyer et al., 1981; Schmidt and Popham, 1981).

It is not unlikely that lack of exercise, other neglect of personal health care, and dietary factors also contribute to the excess cardiovascular mortality. In this regard, it is noteworthy that an excess was not found in Sundby's (1967) sub-sample of homeless alcoholics. Possibly, a relatively lower intake of food by this type of heavy drinker—who is likely to be notably underweight (Olin, 1966)—reduces the risk of developing certain cardiovascular diseases. Another possibility, currently under debate, is that at least moderate dosages of alcohol have a protective function in regard to ischemic heart disease (see, for example, Klatsky et al., 1979). This has been rendered plausible by the results of experimental and clinical research indicating an enhancing effect of alcohol on coronary blood flow (Bing and Tillmanns, 1977) and on high-density lipoprotein levels (Williams et al., 1979), and supported by the findings of several prospective, case–control, and other epidemiologic studies. The relevant literature has been thoroughly reviewed in Chapter 4.

Pneumonia

While pneumonia is no longer one of the most common causes of death among heavy drinkers, owing to the introduction of antibiotics, it is evident from the ratios in Table 3 that a substantial excess of observed over expected deaths continues to be found. Moreover, studies of pneumonia victims indicate a more prolonged course of illness, a greater frequency of complications, and a higher death rate from the disease when there is a history of heavy alcohol consumption (Lyons and Saltzman, 1974). Both acute and chronic effects of alcohol are probably involved. Acute effects include the fact that heavy drinkers often display a characteristic form of pneumonia resulting from frequent aspiration of foreign matter into the lungs, and which may be especially severe if the foreign matter contains alcohol (Lyons and Saltzman, 1974). Chronic heavy alcohol consumption appears to lower resistance to infections, for example, through reduction in the number and impairment of the mobilization of leukocytes (Lindenbaum, 1974). Specific toxic effects of alcohol on respiratory symptoms and pulmonary function have also been suggested (Emirgil et al., 1974; Emirgil and Sobol, 1977), and recently shown to obtain even when smoking and other risk factors were taken into account (Lebowitz, 1981). In addition, it is clear that associated heavy smoking (Rankin et al., 1969; Weir and Dunn, 1970), delays in seeking treatment, and neglect of personal hygiene are also implicated (Chomet and Gach, 1967).

Peptic Ulcer

Comparatively few deaths from peptic ulcers have been reported in any mortality studies, but in one an excess mortality was clearly substantiated and shown to remain significant when the effect of smoking was controlled (Schmidt and Popham, 1981). The evidence from morbidity data as to whether or not the frequency of ulcers exceeds that in the general population is inconclusive (Bingham, 1960; Engeset et al., 1963). It is probably safe to assume that chronic heavy alcohol consumption and commonly associated behaviors such as smoking, poor eating habits, and other neglect of personal care will increase the likelihood of a fatal outcome once an ulcer has occurred. However, whether or not heavy drinking can be the cause of such an ulcer in the first instance has yet to be demonstrated (Lorber et al., 1974).

Cirrhosis

It is evident from Table 3 that a very substantial excess mortality from liver cirrhosis has been consistently found. Indeed, the disease is one of the leading causes of death among heavy drinkers. For example, it ranked second only to cardiovascular diseases in contribution to total excess mortality in the Schmidt and Popham (1981) sample. The etiologic importance of long-term heavy alcohol intake *per se* would seem to have been established beyond doubt. In addition to a large body of epidemiologic and clinical evidence (Lelbach, 1976; Schmidt, 1977; Thaler, 1977), recent experimental work has convincingly shown that a direct hepatotoxic effect of alcohol is mainly responsible rather than a nutritional deficiency (Rubin and Lieber, 1974; Israel et al., 1975). However, malnutrition when present may well heighten susceptibility to the alcohol effect.

Suicide

Generally, suicide is also found to be one of the more frequent causes of death among chronic heavy drinkers, and, conversely, a high proportion of alcoholics among suicide cases has been reported (Robins et al., 1959; Dorpat and Ripley, 1960; Riddick and Luke, 1978). The studies in Table 3 consistently reported a high ratio of actual to expected deaths from this cause. This excess mortality has usually been attributed to predisposing personality factors and to the emotional consequences of progressive dissolution of primary social ties associated with a history of heavy drinking (Rushing, 1969; Lundquist, 1970). On the other hand, Schmidt and de Lint (1972) considered that insufficient attention had been given to the possibility of a direct effect of alcohol. They noted the high rates of suicide among bartenders, restaurateurs, and hoteliers who have in common a very high exposure to alcoholic beverages. There would seem to be little reason to suppose that these occupations have more than their share of persons with personality traits predisposing to suicide. However, as Ignev's (1979) study suggests, this does not rule out the effect of social deterioration resulting from heavy drinking.

Accidents

The data in Table 3 clearly indicate an excess mortality from accidents. Brenner (1967) and Schmidt and de Lint (1972) found the principal types to be vehicular accidents, poisonings, fires, and falls. Traumatic injuries, as shown in Table 1, were among the more common diagnoses in the sample of alcoholic patients studied by Ashley et al. (1977). Various investigators—for example, Goldberg (1955) and Schmidt et al. (1962)—have shown that drinking drivers involved in accidents frequently have a history of excessive drinking, and that clinical alcoholics constitute a high-risk group for such accidents. Perhaps an elevated death rate from accidental injury is hardly surprising given the well-known impairing effects of large dosages of alcohol on judgment and motor functions. Schmidt et al. (1962) argued that the high frequency of such acute impairment was a sufficient explanation for the elevated vehicular accident rate of alcoholics, but Selzer (1975) and McLean and Campbell (1979) considered

that personality traits were also of etiologic significance. With respect to accidental poisoning, Schmidt and de Lint (1972) found that acute alcohol intoxication was mentioned on the death certificate in half their cases, and Schmidt
and Popham (1980) noted that such deaths commonly involved alcohol and
another psychoactive drug, usually a barbiturate. Careless smoking was the
commonly recorded contributing cause in deaths due to fires (Schmidt and de
Lint, 1972). Other factors probably include the greater likelihood of fires and
accidental falls in the deteriorated rooming-houses where heavy drinkers may
often live, and the greater likelihood of death due to a generally poorer state of
health (Brenner, 1967; Hudson, 1978).

Table 4 summarizes the foregoing discussion of the principal factors which
may account for the mortality data on heavy drinkers. Clearly the tabulation
oversimplifies what is probably often a complex picture of interrelated causes
in the individual case. Therefore, it is important to emphasize that the table
refers to the *population* of heavy drinkers, not to the individual drinker, and to
what is known about the causes of *elevated death rates* in this population, rather
than about the causes of the diseases and behaviors which may lead to death.

Sex Differences in Cause of Death

Schmidt and Popham (1980) found that, generally speaking, the same causes
in nearly the same proportions contributed to the excess mortality in male and
female alcoholics. This is illustrated in Fig. 1.

In Table 2, the ratios for women are consistently larger than those for men.
However, this difference is due to the fact that mortality rates for women in
general populations (on which expected values were based) are uniformly much
lower than for men. When Schmidt and Popham compared the mortality of the
female with that of the male alcoholic, the female alcoholic was found to have
a significantly *lower* mortality. The rates did not differ for causes with a clear
alcoholic etiology such as trauma, liver cirrhosis, and cancer of the esophagus.
On the other hand, despite apparently little difference between the two sex groups
in duration and amount of smoking (Dreher and Fraser, 1968), the women had
substantially lower death rates from lung cancer and heart disease. Schmidt and
Popham concluded that these lower death rates reflected a constitutional difference in susceptibility. However, possible contributing factors such as differential
exposure to industrial pollutants, stress, and other occupational hazards cannot
be ruled out.

The absence of a sex difference in cirrhosis mortality in the Schmidt and
Popham samples would seem to be at variance with the observation in several
morbidity studies (e.g., Wilkinson et al., 1969; Péquignot et al., 1974; Krasner
et al., 1977) that women are more susceptible to the disease. However, the lack
of agreement is probably attributable to the broader age range of the mortality
sample. Thus, Schmidt and Popham found a higher rate among women below

Table 4. Etiologic Significance of Alcohol and Associated Variables in the Excess Mortality of Chronic Heavy Drinkers

Cause of death	Effects of alcohol	Heavy tobacco smoking	Emotional problems	Poor food habits	Other personal neglect	Increased environmental hazards
Tuberculosis		x		x	x	x
Carcinoma						
Mouth	xx	xx				
Larynx	xx	xx				
Pharynx	xx	xx				
Esophagus	xx	xx				
Lung		xx				
Alcoholic cardiomyopathy	xx					
Other cardiovascular diseases		xx		x	x	
Pneumonia	xx	xx			xx	
Peptic ulcers	xx	x		x	x	
Liver cirrhosis	xx			x		
Suicide	x		xx			
Accidents	xx	xx	x		x	x

x = probably indicated; xx = clearly indicated. Where a space is left blank, the factor is probably either of no significance or its role, if any, is unknown.

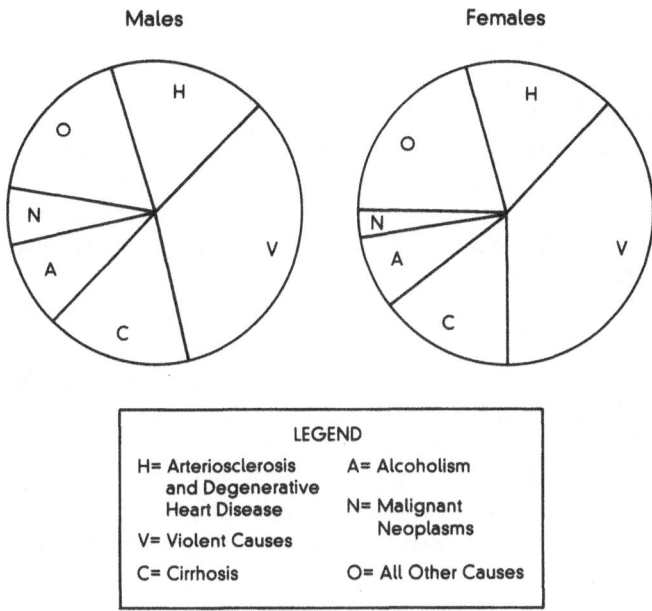

Figure 1. Proportional contribution of major causes of death to the excess mortality of alcoholics. (From Schmidt and Popham, 1980.)

50 years of age (the age group emphasized in clinical samples), but also that this was offset by a lower rate among those aged 50 and older.

Beverage Differences in Cause of Death

In the discussion to this point, there has been an implicit assumption that the important variable in alcohol consumption is the intake of ethyl alcohol. Since alcoholic beverages differ substantially in ethanol concentration and in the types and amounts of other substances present, the possible etiologic relevance of the class of beverage consumed cannot be ignored. For example, heavy beer consumption has been seen to have special significance in alcoholic cardiomyopathy (McDermott et al., 1966; Anonymous, 1968), wine consumption in liver cirrhosis (Whitlock, 1974), and distilled spirits in certain cancers (Tuyns, 1970). For diseases of the heart and liver, the case for a specific beverage effect rests mainly on uncontrolled clinical studies, and epidemiologic analyses in which total alcohol intake was not adequately taken into account. Following a careful assessment of the experimental, clinical, and epidemiologic literature, Lelbach (1974, p. 157) concluded that: "No clear-cut evidence has been presented to prove that, if comparable quantities of ethyl alcohol are absorbed, any of the commonly consumed alcoholic beverages exert a more pronounced influence on the production of organic lesions than others."

On the other hand, Lelbach's conclusion may not apply to the development of alcohol-related carcinomata. Here the evidence for a specific beverage effect derives from retrospective clinical studies in which appropriate controls were employed (Wynder et al., 1957; Tuyns, 1970). The findings suggested that whisky drinking predisposed to the development of carcinomata of the head, neck, and esophagus more than beer or wine. Tuyns et al. (1979) argued that the ethanol concentration in the beverage consumed is the relevant factor, although they noted that apple-based alcoholic beverages may carry a risk independent of concentration.

It has also been argued that since beer has a lower intoxication potential than other beverages, it carries a lower risk of causing alcohol-related accidents (Brewers Assoc. of Canada, 1973). The contention rests on the finding of Dussault et al. (1972)—which is consistent with a large body of experimental evidence (see Takala et al., 1957 for a review)—that beer produced less impairment than the same quantity of alcohol consumed in more concentrated form. The issue, dealt with at length elsewhere (Popham et al., 1976), reduces essentially to two questions: given the experimental conditions imposed by Dussault et al., is their result likely to be applicable to actual drinking situations? and, is there any *direct* evidence of a lower risk of accidents when drinking involves beer?

As to the first question, the experiment involved prolonged fasting of the subjects followed by consumption of a single large dose of alcohol in one or another beverage. Kalant et al. (1975) showed that under more realistic conditions of progressive drinking, interbeverage differences disappeared altogether. With respect to the second question, little evidence has been gathered to date. Borkenstein et al. (1974) found that drivers in alcohol-related accidents reported beer to be their beverage of choice somewhat more often than drivers not involved in accidents, and that the proportion of accident drivers reporting beer increased with increasing blood alcohol level. However, to determine with certainty the role of beer would require knowledge of the type of beverage actually consumed prior to the accident. In any event, it would seem that the experimental basis of the argument favoring beer is doubtfully relevant—at least in a North American context—and such direct evidence as exists does not support the inference that beer drinking is less likely to cause traffic accidents.

Conclusions

Our conclusions include the following. The acute or chronic effects of alcohol explain a considerable part of the excess mortality of heavy drinkers. The etiologic importance of alcohol is clear with respect to deaths from liver cirrhosis, accidents, and carcinomata of the upper digestive and upper respiratory tracts, and unclear with respect to most cardiovascular diseases. Other factors often associated with heavy drinking, such as heavy smoking, emotional problems, poor food habits, personal neglect, and increased environmental hazards are probably largely or entirely responsible for the elevated death rates from tuberculosis, lung cancer, and suicide; certain of these factors also contribute to the excess deaths from cardiovascular diseases. The excess mortality of male and female alcoholics is similarly distributed by cause; however, despite similarity in smoking history, women have a lower overall mortality rate than men owing mainly to lower death rates from heart disease and lung cancer. Finally, the class of alcoholic beverage consumed is probably of little etiologic significance in the mortality of heavy drinkers although the ethanol concentration in the beverage consumed may be related to the risk of certain cancers.

5. MORTALITY RATE VARIATION IN GENERAL POPULATIONS

The question now arises as to whether or not the relationship between heavy drinking and elevated death rates is reflected in mortality and consumption statistics for general populations. It has been shown that the rate of heavy alcohol use (however defined in terms of consumption levels) is likely to rise and fall with the average per drinker consumption in a population (Ledermann, 1956, 1964; de Lint and Schmidt, 1968; Mäkelä, 1971; Skog, 1971). Theoretically, therefore, variation in *per capita* consumption—either from one region to another or through time in the same region—should be accompanied by similar variation in relevant mortality rates. The extent to which this is actually the case is most likely to be revealed by trends in general mortality, and in death rates for the three causes in which direct effects of alcohol are most clearly implicated: carcinomata of the upper digestive and upper respiratory tracts, cirrhosis, and accidents.

It is evident that any dependence of the general mortality level on the amount of alcohol consumption in a population is not likely to be detected through examination of trends in the crude rate of death from all causes. Thus, excess deaths resulting from heavy drinking would not be expected to constitute more than a small percentage of all deaths in any given year. For example, Schmidt and de Lint (1973) estimated the proportion for Ontario in 1969 to be about 5.5% of total deaths in the age range 20–69 years. This means that any alcohol-related variation in the general mortality rate could be masked by comparatively small fluctuations due to other factors. To overcome the difficulty, Bandel (1930) maximized the variation attributable to alcohol use by focusing only on the mortality of men aged between 40 and 69 years of age. He reasoned that alcohol-related deaths would be at a peak in middle age, and that the frequency of heavy drinking would be much greater among men than among women. Therefore, the excess mortality of men over women in this age range seemed to be the measure most likely to reflect variation in alcohol consumption.

To test the hypothesis, Bandel examined temporal series for Bavaria, Prussia, and Italy. Later, Ledermann (1964) extended the Italian series, added one for France, and undertook a more limited analysis of data for Sweden and Denmark. The relationship between the two variables proved remarkably close in nearly all cases, and was especially pronounced in the Bavarian series, as illustrated in Fig. 2. Ledermann noted that many differences between men and women—both constitutional and in life style—might account for excess mortality among men. However, he concluded that none of these, other than differential drinking habits, offered a reasonable explanation for the covariation of the excess through time with rate of alcohol consumption.

Schoenberg et al. (1971) studied the variation in mortality from esophageal carcinoma in 41 American states and found significant correlations with differences in *per capita* sales of both distilled spirits and cigarettes. Schmidt (1974) examined almost the same series, but included total alcohol sales as an independent variable. The latter, however, did not give a better correlation, and the coefficient was reduced to borderline significance when the distilled spirits component was held constant (see Table 5). On the other hand, the correlation remained significant with the influence of cigarette sales removed. In both studies degree of urbanization was found to be a powerful explanatory variable, although its effect was less marked in Schmidt's data than in the series of Schoenberg et al.

An analysis of variation in mortality rates for 88 départements of France utilized the death rate from alcoholism and cirrhosis as the index of the prevalence of heavy drinking (Lasserre et al., 1966). A correlation of $+0.70$ was found between this variable and rates of death from esophageal carcinoma. In the case

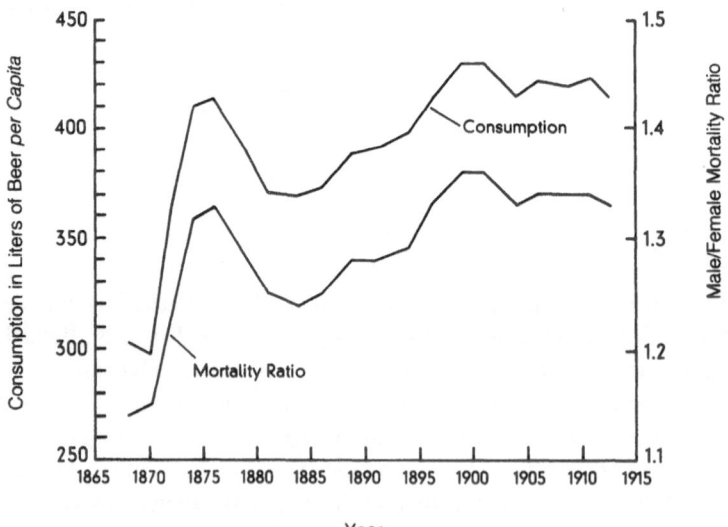

Figure 2. Apparent consumption of beer in liters *per capita* aged 20 and older and excess mortality of men over women: Bavaria, 1867–1913. (From Ledermann, 1964.)

Table 5. Correlation between Various Cancers and Total
Alcohol Consumption: 46 American States in 1960[a]

Type of coefficient	Site of cancer		
	Larynx	Oral cavity	Esophagus
Simple	0.50[b]	0.53[b]	0.61[b]
Partial			
Controlling for the effect of			
Urbanization	0.42[b]	0.41[b]	0.38
Cigarette sales	0.02	0.23	0.47[b]
Distilled spirits sales	0.12	0.22	0.36

[a] From Schmidt et al. (1974).
[b] Significant at the 0.01 level.

of carcinomata of the oral cavity and larynx, the coefficients were 0.44 and 0.40, whereas an analysis of the same variables for 21 countries gave coefficients of 0.68 and 0.80, respectively (Schwartz et al., 1966).

Tuyns and Audigier (1976), in a cohort study of French men born between 1902 and 1916, found that death rates from cancers of the esophagus and larynx, similarly to those from alcoholism and cirrhosis, were responsive to the reduction in availability of alcoholic beverages during World War II. In contrast, mortality from cancers of the lung and pancreas did not show this effect. McMichael (1979) also found death rates from cancers of the esophagus and larynx to be correlated in time series for Australia and Great Britain, and showed that the trends in deaths from these causes differed notably from that in mortality from lung cancer. On the other hand, in Schmidt's series for the United States (Table 5), differences in cigarette sales appeared to explain virtually all of the variation in deaths from carcinoma of the larynx, and most of that in deaths from carcinoma of the oral cavity. Evidently, therefore, the matter cannot be considered entirely settled. There are indications of the expected relationship with regard to esophageal carcinoma, but in the case of cancers of the oral cavity and larynx the findings of correlational studies are not consistent.

The expected correlation with *per capita* alcohol consumption emerges far more clearly in the case of cirrhosis mortality. This is illustrated by the generally high coefficients for various temporal and spatial series shown in Table 6. Moreover, Wallgren et al. (1970) and Skog (1980) demonstrated that even higher coefficients could be obtained when the consumption data were taken for a period preceding that of the mortality data. This procedure allows for the latent period between exposure and death from the disease.

An objection to the inference that these correlations reflect a cause–effect relationship is that several of the time series involve trends that are mainly unidirectional. However, the case is strengthened by the high correlations in the regional series, and by the effect on the mortality trend of a decline in the

Table 6. Temporal and Regional Correlations between Rate of Liver
Cirrhosis Mortality and *Per Capita* Alcohol Consumption[a]

Area	Series	Correlation coefficient	Probability less than
Australia	1938–1959	0.65	0.005
Belgium	1929–1959 (less 1940–1945)	0.75	0.001
Canada	1927–1960	0.88	0.001
Alberta	1929–1960	0.85	0.001
Manitoba	1935–1960	0.86	0.001
Nova Scotia	1932–1960	0.60	0.001
Ontario	1930–1960	0.89	0.001
Quebec	1929–1960	0.43	0.05
Saskatchewan	1943–1960	0.77	0.001
Canada	9 provinces 1955	0.81	0.01
Finland	1933–1957	0.78	0.001
France	1925–1958	0.62	0.001
France	23 départements 1950	0.76	0.001
Holland	1927–1958	0.57	0.001
Sweden	1926–1956	0.45	0.05
United Kingdom	1931–1958	−0.68	0.001
United States	1934–1958	0.60	0.005
United States	45 states, 1939	0.61	0.001
United States	48 states, 1944	0.78	0.001
United States	46 states, 1950	0.76	0.001
United States	46 states, 1957	0.86	0.001
International	11 countries, 1956	0.78	0.005

[a] From Popham (1970). In all instances, liver cirrhosis mortality was expressed as an unstandardized rate: deaths per 100,000 population aged 20 and older; and the measure of alcohol consumption comprised sales of alcoholic beverages expressed as imperial gallons of absolute alcohol *per capita* of population aged 15 and older. The International series comprised Australia, Belgium, Canada, Denmark, France, Finland, the Netherlands, Sweden, Switzerland, the United Kingdom, and the United States.

availability of beverage alcohol. Notable instances of the latter occurred during, and for a period after, World War I when diversion of alcohol to military purposes combined with (in the United States) Prohibition resulted in a severe reduction in supply. In Canada (Popham, 1956), Finland (Bruun et al., 1960), and the United States (Jolliffe and Jellinek, 1941) this was accompanied by a remarkably rapid and very substantial drop in the cirrhosis death rate. In France, extreme shortages occurred during both world wars with particularly dramatic effects on the mortality trend, as shown in Fig. 3. For the country as a whole, the decline in the cirrhosis death rate among middle-aged men was approximately 50%, and for Paris—where there was less possibility of circumventing the rationing system—the decline was more than 80% (Ledermann, 1964).

One probable reason for a more consistent picture with this disease is that, as Jolliffe and Jellinek (1941) argued many years ago, the rate of cirrhosis mortality *not* attributable to heavy drinking probably varies little in many popula-

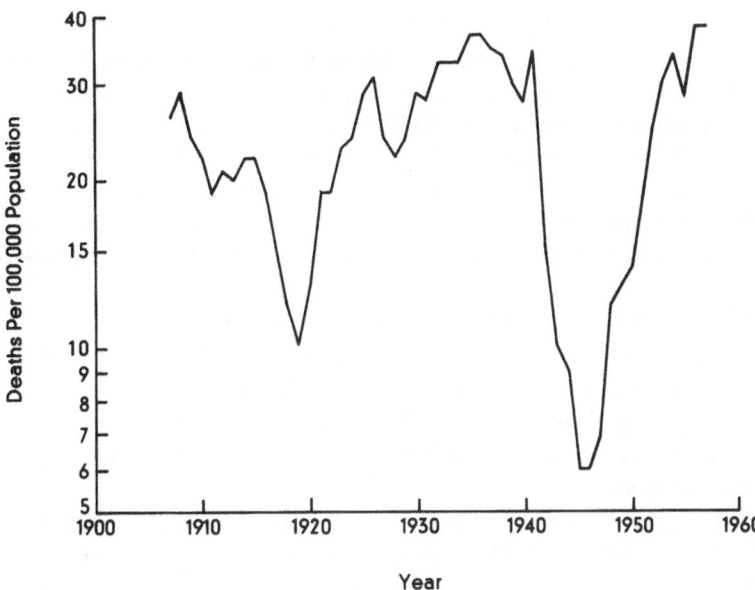

Figure 3. Liver cirrhosis mortality rates: Paris 1907–1956. (Based on data in Ledermann, 1964.)

tions. However, transitory factors unrelated to alcohol use such as the aftermath of an outbreak of epidemic hepatitis, can sometimes distort the trend (Ipsen, 1950). There are also instances where the correlation is weak, and at least one— the United Kingdom—for which a negative coefficient is obtained. The latter may be due to the fact that cirrhosis mortality rates were very low during the period and probably largely attributable to factors other than alcohol. Consequently, there was little more than chance fluctuation in the rate. Under such circumstances, as Popham (1970) suggested, a spurious result might be obtained with Pearsonian correlation analysis. In an examination of British data for a later period, when both cirrhosis and alcohol consumption rates were much higher and changing rapidly, Schmidt (1977) found a highly significant positive correlation. In any case, overall, the results of correlation analyses may be said to confirm the expectation based on morbidity and mortality studies: The rate of death from cirrhosis usually rises and falls with the apparent level of alcohol consumption in general populations. Furthermore, the relationship is evidently curvilinear as shown in Fig. 4, and Schmidt (1975) has demonstrated that an increase in the cirrhosis death rate is likely to be proportional to the square of the corresponding increase in *per capita* consumption.

With the exception of studies of automobile accidents, no attention seems to have been given as yet to the possibility of regional or temporal correlations between accident death rates and apparent alcohol consumption. In a recent study of the effects of a reduction in the legal drinking age in Ontario, Schmidt and Kornaczewski (1975) reported that an increase in *per capita* alcohol sales to the age group affected was paralleled by an increase in the alcohol-related traffic fatalities of the group. On the other hand, the variation among the American states in rate of motor vehicle accidents reported to have involved drinking could not be statistically explained by differences in either alcohol sales or cirrhosis mortality (Schmidt and Smart, 1963). Similarly, the data in Fig. 5 give no indication of a relationship on an international basis between apparent consumption and traffic fatalities. This is particularly noteworthy since the importance of alcohol as a factor in this type of death has been established beyond doubt.

Many sources of considerable variation other than the prevalence of heavy drinking affect accident mortality rates. For example, there are vast differences in traffic density, driver and vehicle safety regulations, and highway engineering from one country to another. Accordingly, it would probably be more surprising to find a significant association between reported accident mortality and *per capita* consumption than not.

The theoretical expectation that the relationship between heavy drinking and excess mortality would be manifested in general populations by covariation of relevant mortality and apparent alcohol consumption rates has been confirmed for cirrhosis mortality. There is highly suggestive evidence that the relationship holds for cancer of the esophagus, and that it is reflected in general mortality rates when the excess mortality of middle-aged men over women is taken as the dependent variable. The correlation data for cancers of the larynx and oral cavity are inconclusive. With respect to accidental death, the question has only been

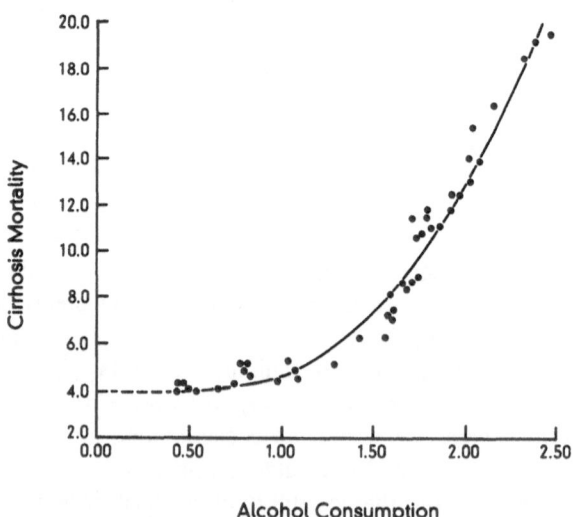

Alcohol Consumption

Figure 4. The regression of cirrhosis mortality (deaths per 100,000 aged 20 and older) on consumption (imperial gallons of alcohol *per capita* aged 15 and older), Ontario data 1932–1974. (From Schmidt, 1977)

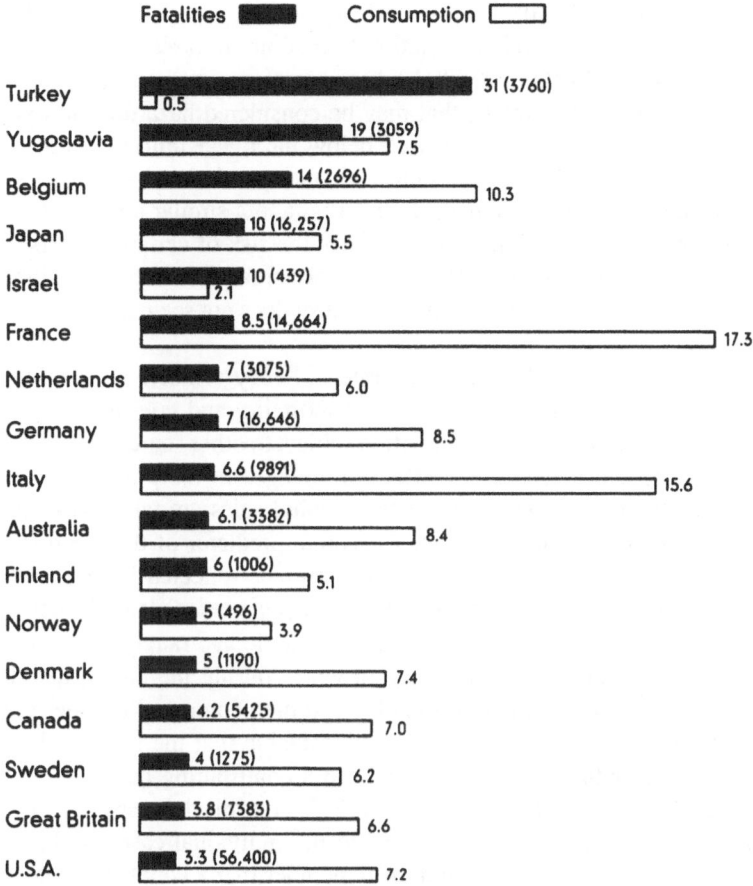

Figure 5. Traffic fatalities per 100 million vehicle kilometers and apparent alcohol consumption in liters *per capita*. All fatality rates are for 1969 except Israel (1968) and Sweden (1966). Total deaths are shown in parentheses. Data were obtained from Borkenstein (1975). Consumption figures are for 1970 or 1971, and were obtained from Sulkunen (1973).

investigated for traffic fatalities. Here the primarily negative results may be attributable to the many sources of variation other than alcohol consumption which affect the mortality trend.

6.   GENERAL CONCLUSION

Our primary purpose in this chapter has been to determine what conclusions about the hazards to physical health of chronic heavy alcohol use may be drawn with confidence from morbidity and mortality studies. The evidence would seem compelling that such use carries a risk of premature death greatly exceeding normal expectancy. While the life style typical of many heavy drinkers un-doubtedly contributes to this elevated risk in some diseases, it is clear that the chronic and acute effects of alcohol *per se* are of central importance in others.

The lowest level of intake that may be considered hazardous has yet to be determined. However, it is surely well below the lower limit (15 cl/day) of the range of consumption of most clinical alcoholics. A level of 5–7.5 cl daily appears to increase the likelihood of cirrhosis, and a similar consumption level, when combined with smoking, may enhance the risk of certain cancers. On the other hand, the consumption of comparatively small amounts (quantities com-monly identified as "moderate" in the literature) does not seem adversely to affect life expectancy.

The impact of the elevated death rate of heavy drinkers on the population at large is not always easy to detect. The mortality and consumption statistics available for general populations are often subject to other sources of considerable variation which can obscure any effect of changes in the prevalence of heavy drinking. Nevertheless, when *per capita* alcohol sales are taken as an index of the latter, it is in most cases a highly accurate predictor of the death rate from cirrhosis. The impact on general mortality may be reflected in the covariation, so far observed in several temporal series, of *per capita* alcohol sales and the excess of male over female mortality in the middle age range. In this regard, a characteristic of mortality in the western world during the past generation has been the lack of improvement in middle-aged men compared to other segments of the population. It is by no means inconceivable that the increase in alcohol consumption over the same period has been a contributing factor.

Finally, we must echo Pearl (1926, p.222) and "disclaim any responsibility" for the application of these conclusions to the individual case. As he so aptly put it "it will be the part of wisdom to remember that a conclusion which is on the average true for a large statistical aggregate may not be so for a particular individual in that aggregate". There will always be the octogenarian who has consumed heavily most of his life; there will be those who seem unable to tolerate the smallest quantities. But such variability in individual susceptibility need not concern us here. The relevant public health objectives are to delineate for the drinking population as a whole the risks of disease and premature death associated with different levels of alcohol consumption, and to seek means to minimize the number of drinkers in the hazardous range.

ACKNOWLEDGMENT

This chapter is a revision and expansion of that published in *Drug and Alcohol Dependence* 1:27–50 (1976). The authors thank their colleague, Dr. Mary Jane Ashley, for a critical reading of the manuscript and many useful suggestions.

REFERENCES

Anonymous, 1968, Heart disease and beer drinking, *Public Health Rep.* **83:**998.

Anstie, F. E., 1870, On the dietetic and medicinal uses of wine, *Practitioner* **4:**219.

Ashley, M. J., Olin, J. S., le Riche, W. H., Kornaczewski, A., Schmidt, W., and Rankin, J. G., 1977, Morbidity in alcoholics. Evidence for accelerated development of physical disease in women, *Arch. Intern. Med.* **137:**883.

Babigian, H. M., and Odoroff, C. L., 1969, The mortality experience of a population with psychiatric illness, *Am. J. Psychiatry* **126:**470.

Bailey, W. C., Sellers, C. A., Sutton, F. D., Sheehy, T. W., and Maetz, H. M., 1978, Tuberculosis and alcoholism: A partial solution through detection, *Chest* **73:**183.

Bandel, R., 1930, Die spezifische Männersterblichkeit als Masstab der Alkoholsterblichkeit, in: *Ergebnisse der Sozialen Hygiene und Gesundheitsfürsorge*, Bd. II, pp. 424–492, Georg Thieme Verlag, Leipzig.

Becker, W. H., 1908, Eine Maximaldosis des Alkohols, *Ther. Monatsch.* **22:**444.

Bing, R. J., and Tillmans, H., 1977, The effect of alcohol on the heart, in: *Metabolic Aspects of Alcoholism* (C. S. Lieber, ed.), pp. 117–134, University Park Press, Baltimore.

Bingham, J. R., 1960, Precipitating factors in peptic ulcer, *Can. Med. Assoc. J.* **83:**205.

Borkenstein, R. F., 1975, Problems of enforcement, adjudication and sanctioning, in: *Proceedings of the 6th International Conference on Alcohol, Drugs, and Traffic Safety* (S. Israelstam, and S. Lambert, eds.), pp. 655–662, Addiction Research Foundation, Toronto.

Borkenstein, R. F., Crowther, R. F., Shumata, R. P., Ziel, W. B., and Zylman, R., 1974, The role of the drinking driver in traffic accidents (The Grand Rapids Study), in *Blutalkohol*, Volume II, Supplement 1 (R. F. Borkenstein, ed.), pp. 1–132, Indiana University, Bloomington.

Brenner, B., 1967, Alcoholism and fatal accidents, *Q. J. Stud. Alcohol* **28:**517.

Brewers Association of Canada, 1973, *Beer, Wine and Spirits: Beverage Differences and Public Policy in Canada*, Alcoholic Beverage Study Committee, Ottawa.

Brown, K. E., and Campbell, A. H., 1961, Tobacco, alcohol and tuberculosis, *Br. J. Dis. Chest* **55:**150.

Bruun, K., Koura, E., Popham, R. E., and Seeley, J. R., 1960, Liver Cirrhosis as a Means to Estimate the Prevalence of Alcoholism, Publication No. 8, Finnish Foundation for Alcohol Studies, Helsinki.

Burch, G. E., and Giles, T. D., 1974, Alcoholic cardiomyopathy, in: *The Biology of Alcoholism*, Volume 3 (Kissin, B., and Begleiter, H., eds.), pp. 435–460, Plenum Press, New York.

Chomet, B., and Gach, B. M., 1967, Lobar pneumonia and alcoholism: An analysis of 37 cases, *Am. J. Med. Sci.* **253:**300.

Ciompi, L., and Medvecka, J., 1976, Étude comparative de la mortalité à long terme dans les maladies mentales, *Arch. Suisses Neurol. Neurochir. Psychiatrie* **118:**111.

Dahlberg, G., 1952, Dodlighet bland Alkoholister, *Svenska Lakartidningen* **49:**112.

Dahlgren, K. C., 1951, On death rates and causes of death in alcohol addicts, *Acta. Psychiatr. Neurol. Scand.* **26:**297.

Dahlgren, L., and Myhred, M., 1977, Alcoholic females II. Causes of death with reference to sex difference, *Acta Psychiatr. Scand.* **56:**81.

Davies, K. M., 1965, The influence of alcohol on mortality, *Proc. Home Office Life Underwriters Assoc.* **46:**159.

de Lint, J., and Levinson, T., 1975, Mortality among patients treated for alcoholism: A 5-year follow-up, *Can. Med. Assoc. J.* **113:**385.

de Lint, J., and Schmidt, W., 1968, The distribution of alcohol consumption in Ontario, *Q. J. Stud. Alc.* **29:**968.

Doll, R., and Hill, A. B., 1964, Mortality in relation to smoking: Ten years' observation of British doctors, *Br. Med. J.* **1:**1399, 1460.

Dorpat, T. L., and Ripley, H. S., 1960, A study of suicide in the Seattle area, *Comp. Psychiatry* **1**:349.

Dreher, K. F., and Fraser, J. G., 1967, Smoking habits of alcoholic outpatients. I, *Int. J. Addict.* **2**:259.

Dreher, K. F., and Fraser, J. G., 1968, Smoking habits of alcoholic outpatients. II, *Int. J. Addict.* **3**:65.

Dussault, P., Burford, R., and Chappel, C., 1972, *Studies of Alcoholic Beverages: Physiological Effects*, Brewers Association of Canada, Bio-research Laboratory, Montreal.

Dyer, A. R., Stamler, J., Paul, O., Berkson, D. M., Lepper, M. H., McKean, H., Shekelle, R. B., Lindberg, H. A., and Garside, D., 1977, Alcohol consumption, cardiovascular risk factors and mortality in two Chicago epidemiological studies, *Circulation* **56**:1067.

Dyer, A. R., Stamler, J., Oglesby, P., Berkson, D. M., Shekelle, R. B., Lepper, M. H., McKean, H., Lindberg, H. A., Garside, D., and Tokich, T., 1981, Alcohol, cardiovascular risk factors and mortality: The Chicago experience, *Circulation* **64**(Suppl. III):20.

Eckhardt, M. J., Harford, T. C., Kaelber, C. T., Parker, E. S., Rosenthal, L. S., Ryback, R. S., Salmoiraghi, G. C., Vanderveen, E., and Warren, K. R., 1981, Health hazards associated with alcohol consumption, *JAMA* **246**:648.

Emirgil, C., and Sobol, B. J., 1977, Pulmonary function in former alcoholics, *Chest* **72**:45.

Emirgil, C., Sobol, B. J., Heymann, B., and Shibutani, K., 1974, Pulmonary function in alcoholics, *Am. J. Med.* **57**:69.

Engeset, A., Lygren, T., and Idsoe, R., 1963, The incidence of peptic ulcer among alcohol abusers and non-abusers, *Q. J. Stud. Alcohol* **24**:622.

Fallon, H. J., and Lesesne, H. R., 1978, Medical complications of excessive drinking, in: *Drinking—Alcohol in American Society* (J. A. Ewing and B. A. Rouse, eds.), pp. 63–69, Nelson-Hall, Chicago.

Feingold, A. O., 1976, Association of tuberculosis with alcoholism, *South. Med. J.* **69**:1336.

Gabriel, E., 1935, Über die Todesursachen bei Alkoholikern, *Zentralbl. Ges. Neurol. Psychiatr.* **153**:385.

Giffen, P. J., Oki, G., Lambert, S., Timothy, E., and Maxwell, J., 1971, The Chronic Drunkenness Offender: Ages and Causes of Death of the Chronic Drunkenness Offender Population, A.R.F. Substudy No. 418, Addiction Research Foundation, Toronto.

Gillis, L. S., 1969, The mortality rate and cause of death of treated alcoholics, *S. Afr. Med. J.* **43**:230.

Goldberg, L., 1955, Drunken drivers in Sweden, in: *Proceedings of the 2nd International Conference on Alcohol and Road Traffic*, pp. 112–127, Garden City Press, Toronto.

Gorwitz, K., Bahn, A. K., Klee, G., and Solomon, M. S., 1966, Release and return rates for patients in state mental hospitals of Maryland, *Public Health Rep.* **81**:1095.

Hudson, P., 1978, The medical examiner looks at drinking, in: *Drinking—Alcohol in American Society* (J. A. Ewing and B. A. Rouse, eds.), pp. 71–92, Nelson-Hall, Chicago.

Ignev, I., 1979, Reasons for suicides and attempted suicides among chronic alcoholics, in: *Proceedings of the 10th International Congress for Suicide Prevention and Crisis Intervention*, p. 224, International Association of Suicide Prevention, Ottawa.

Ipsen, J., 1950, An epidemic of infectious hepatitis, predominantly of adults, and highly fatal for elderly women, *Am. J. Hyg.* **51**:255.

Israel, Y., Kalant, H., Orrego, H., Khanna, J. M., Videla, L., and Phillips, J. M., 1975, Experimental alcohol-induced hepatic necroses suppression by propylthiouracil, *Proc. Natl. Acad. Sci. USA* **72**:1137.

Jolliffe, N., and Jellinek, E. M., 1941, Vitamin deficiencies and liver cirrhosis in alcoholism. VII, Cirrhosis of the liver, *Q. J. Stud. Alcohol* **2**:544.

Kalant, H., LeBlanc, A. E., Wilson, A., and Homatidis, S., 1975, Sensorimotor and physiological effects of various alcoholic beverages, *Can. Med. Assoc. J.* **112**:953.

Keller, A. Z., and Terris, M., 1965, The association of alcohol and tobacco with cancer of the mouth and pharynx, *Am. J. Public Health* **55**:1578.

Kendall, R. E., and Staton, M. C., 1966, The fate of untreated alcoholics, *Q. J. Stud. Alcohol* **27**:30.

Kissin, B., and Begleiter, H. (eds.), 1974, *The Biology of Alcoholism, Clinical Pathology*, Plenum Press, New York.

Kissin, B., and Kaley, M. M., 1974, Alcohol and cancer, in: *The Biology of Alcoholism*, Volume 3 (B. Kissin and H. Begleiter, eds.), pp. 481–511, Plenum Press, New York.

Klatsky, A. L., Friedman, G. D., and Siegelaub, A. B., 1979, Alcohol use, myocardial infarction, sudden cardiac death, and hypertension, *Alcohol. Clin. Exp. Res.* **3**(1):33.

Knott, D. H., and Beard, J. D., 1972, Changes in cardiovascular activity as a function of alcohol intake, in: *The Biology of Alcoholism*, Volume 2 (B. Kissin and H. Begleiter, eds.), pp. 345–365, Plenum Press, New York.

Krasner, N., Davis, M., Portman, B., and Williams, R., 1977, Changing pattern of alcoholic liver disease in Great Britain: Relation of sex and signs of autoimmunity, *Br. Med. J.* **1**:1497.

Lasserre, O., Flamant, R., Lellouch, J., and Schwartz, D., 1966, Alcool et cancer: Etude de pathologie géographique portant sur les Départements Français, *Bull. Inst. Nat. Santé* **22**:53.

Lebowitz, M. D., 1981, Respiratory symptoms and disease related to alcohol consumption, *Am. Rev. Respir. Dis.* **123**:16.

Ledermann, S., 1956, *Alcool, Alcoolisme, Alcoolisation: Données Scientifiques de Caractère Physiologique Economique et Social*, Institut National d'Etudes Démographiques. Travaux et Documents, Cahier No. 29. Presses Universitaires de France, Paris.

Ledermann, S., 1964, *Alcool, Alcoolisme, Alcoolisation: Mortalité, Morbidité, Accidents du Travail*, Institut National d'Etudes Démographiques, Travaux et Documents, Cahier No. 41. Presses Universitaires de France, Paris.

Lee, F. I., 1966, Cirrhosis and hepatoma in alcoholics, *Gut* **7**:77.

Leevy, C. M., Gellene, R., and Ning, M., 1964, Primary liver cancer in cirrhosis of the alcoholic, *Ann. N.Y. Acad. Sci. USA* **114**:1026.

Lelbach, W. K., 1974, Organic pathology related to volume and pattern of alcohol use, in: *Research Advances in Alcohol and Drug Problems*, Volume 1 (R. J. Gibbins, Y. Israel, H. Kalant, R. E. Popham, W. Schmidt, and R. G. Smart, eds.), pp. 93–198, John Wiley and Sons, New York.

Lelbach, W. K., 1976, Epidemiology of alcoholic liver disease. *Prog. Liver Dis.* **5**:494.

Lindelius, R., and Salum, I., 1972, Mortality, in: Delirium tremens and certain other acute sequels of alcohol abuse. *Acta Psychiatr. Scand.* (Suppl. 235):86.

Lindelius, R., Salum, I., and Agren, G., 1974, Mortality among male and female alcoholic patients treated in a psychiatric unit, *Acta Psychiatr. Scand.* **50**:612.

Lindenbaum, J., 1974, Hematologic effects of alcohol, in: *The Biology of Alcoholism*, Volume 3 (B. Kissin, and H. Begleiter, eds.), pp. 461–480, Plenum Press, New York.

Lorber, S. H., Dinoso, V. P., and Chey, W. Y., 1974, Diseases of the gastrointestinal tract, in: *The Biology of Alcoholism*, Volume 3 (B. Kissin and H. Begleiter, eds.), pp. 339–357, Plenum Press, New York.

Lundquist, G. A. R., 1970, Alcohol Dependence and Depressive States, Paper presented at the 16th International Institute on the Prevention and Treatment of Alcoholism, Lausanne.

Lyons, H. A., and Saltzman, A., 1974, Diseases of the respiratory tract in alcoholics, in: *The Biology of Alcoholism*, Volume 3 (B. Kissin and H. Begleiter, eds.), pp. 403–434, Plenum Press, New York.

Mäkelä, K., 1971, Concentration of alcohol consumption, *Scand. Stud. Criminol.* **3**:77.

McDermott, P. H., Delaney, R. L., Egan, J. D., and Sullivan, J. F., 1966, Myocardosis and cardiac failure in men, *JAMA* **198**:253.

McLean, N. J., and Campbell, I. M., 1979, The drinking driver: A personality profile, in: *Proceedings of the 7th International Conference on Alcohol, Drugs, and Traffic Safety* (I. R. Johnson, ed.), Australian Government Publishing Service. Canberra.

McMichael, A. J., 1979, Laryngeal cancer and alcohol consumption in Australia, *Med. J. Aust.* **1**:131.

Menge, W., 1950, Mortality experience among cases involving alcohol habits, *Proc. Home Office Life Underwriters Assoc.* **31**:70.

Newman, A. F., 1965, Alcoholism in Frontenac County: A Survey of the Characteristics of an Alcoholic Population in its Native Habitat, Ph.D. Thesis, Queen's University, Kingston, Ontario, Canada.

Nicholls, P., Edwards, G., and Kyle, E., 1974, Alcoholics admitted to four hospitals in England. II. General and cause-specific mortality, *Q. J. Stud. Alcohol* **35**:841.

Olin, J. S., 1966, "Skid row" syndrome: A medical profile of the chronic drunkenness offender, *Can. Med. Assoc. J.* **95**:205.

Parker, R. G. F., 1957, The incidence of primary hepatic carcinoma in cirrhosis, *Proc. R. Soc. Med.* **50**:145.

Pearl, R., 1923, Alcohol and mortality, in: *Action of Alcohol on Man* (E. H. Sterling, ed.), pp. 213–286, Longmans, London.

Pearl, R., 1926, *Alcohol and Longevity*, Alfred A. Knopf, New York.

Pell, S., and D'Alonzo, A., 1973, A five-year mortality study of alcoholics, *J. Occup. Med.* **15**:120.

Péquignot, G., 1974, Les problèmes nutritionnels de la société industrielle, *Vie Mèd. Canad. France* **3**:216.

Péquignot, G., Chabert, C., Eydoux, H., and Courcoul, M. A., 1974, Augmentation de risque de cirrhose en fonction de la ration d'alcool, *Rev. Alcool.* **20**:191.

Popham, R. E., 1956, The Jellinek alcoholism estimation formula and its application to Canadian data, *Q. J. Stud. Alcohol* **17**:559.

Popham, R. E., 1970, Indirect methods of alcoholism prevalence estimation: A critical evaluation, in: *Alcohol and Alcoholism* (R. E. Popham, ed.), pp. 294–306, University of Toronto Press, Toronto.

Popham, R. E., and Schmidt, W., 1978, The biomedical definition of safe alcohol consumption: A crucial issue for the researcher and the drinker, *Br. J. Addict.* **73**: 233.

Popham, R. E., Schmidt, W., and de Lint, J., 1976, The effects of legal restraint on drinking, in: *The Biology of Alcoholism*, Volume 4 (B. Kissin and H. Begleiter, eds.), pp. 579–625, Plenum Press, New York.

Purtilo, D. T., and Gottlieb, L. S., 1973, Cirrhosis and hepatoma occurring at Boston City Hospital 1917–1968, *Cancer* **32**:485.

Rankin, J. G., Hale, G. S., Wilkinson, P., O'Day, D. M., Santamaria, J. N., and Babarzy, G., 1969, Relationship between smoking and pulmonary disease in alcoholism, *Med. J. Aust.* **1**:730.

Riddick, L., and Luke, J. L., 1978, Alcohol associated deaths in the District of Columbia—a postmortem study, *J. Forensic Sci.* **23**:493.

Robinette, C. D., Hrubec, Z., and Fraumeni, J. F., 1979, Chronic alcoholism and subsequent mortality in World War II veterans, *Am. J. Epidemiol.* **109**:687.

Robins, E., Murphy, G. E., Wilkinson, R. J., Gassner, S., and Kayes, J., 1959, Some clinical considerations in the prevention of suicide based on a study of 134 successful suicides, *Am. J. Public Health* **49**:888.

Room, R., and Day, N., 1974, Alcohol and mortality, in: *Second Special Report to the U.S. Congress: Alcohol and Health* (M. Keller, ed.), pp. 104–123, U.S. Government Printing Office, Washington, D.C.

Rothman, K. J., 1978, The effect of alcohol consumption on risk of cancer of the head and neck, *Laryngoscope* **88**(1)(Suppl. 8):51.

Rothman, K. J., 1980, The proportion of cancer attributable to alcohol consumption, *Prev. Med.* **9**:174.

Rothman, K., and Keller, A., 1972, The effect of joint exposure to alcohol and tobacco on risk of cancer of the mouth and pharynx, *J. Chron. Dis.* **25:**711.

Rubin, E., and Lieber, C. S., 1974, Fatty liver, alcohol hepatitis and cirrhosis produced by alcohol in primates, *N. Engl. J. Med.* **290:**128.

Rushing, W. A., 1969, Suicide and interaction of alcoholism (liver cirrhosis) with the social situation, *Q. J. Stud. Alcohol* **30:**93.

Schmidt, W., 1973, Analysis of alcohol consumption data: The use of consumption data for research purposes, in: *Proceedings of the Epidemiology of Drug Dependence: Report on a Conference,* pp. 57–66, World Health Organization Regional Office for Europe, Copenhagen.

Schmidt, W., 1974, Unpublished report, Addiction Research Foundation, Toronto.

Schmidt, W., 1975, Agreement, disagreement in experimental, clinical and epidemiological evidence on the etiology of alcoholic liver cirrhosis: A comment, in: *Alcoholic Liver Pathology* (J. M. Khanna, Y. Israel, and H. Kalant, eds.), pp. 19–30, Addiction Research Foundation, Toronto.

Schmidt, W., 1977, Cirrhosis and alcohol consumption: An epidemiological perspective, in: *Alcoholism: New Knowledge and New Responses* (G. Edwards and M. Grant, eds.), Croom Helm Ltd. Publishers, London.

Schmidt, W., and de Lint, J., 1970, Estimating the prevalence of alcoholism from alcohol consumption and mortality data, *Q. J. Stud. Alcohol* **31:**957.

Schmidt, W., and de Lint, J., 1972, Causes of death of alcoholics, *Q. J. Stud. Alcohol* **33:**171.

Schmidt, W., and de Lint, J., 1973, The mortality of alcoholic people, *Alcohol Health Res. World* **Experimental Issue:**16.

Schmidt, W., and Kornaczewski, A., 1975, The effect of lowering the legal drinking age in Ontario on alcohol-related motor vehicle accidents, in: *Proceedings of the 6th International Conference on Alcohol, Drugs, and Traffic Safety* (S. Israelstam, and S. Lambert, eds.), pp. 763–770, Addiction Research Foundation, Toronto.

Schmidt, W., and Popham, R. E., 1980, Sex differences in mortality: A comparison of male and female alcoholics, in: *Research Advances in Alcohol and Drug Problems,* Volume 5 (O. J. Kalant, ed.), pp. 365–384, Plenum Press, New York.

Schmidt, W., and Popham, R. E., 1981, The role of drinking and smoking in mortality from cancer and other causes in male alcoholics, *Cancer* **47:**1031.

Schmidt, W., and Smart, R. G., 1963, Drinking-driving mortality and morbidity statistics, in: *Alcohol and Traffic Safety* (B. H. Fox and J. H. Fox, eds.), pp. 27–43, Public Health Service Publication No. 1043, U.S. Government Printing Office, Washington, D.C.

Schmidt, W., Smart, R. G., and Popham, R. E., 1962, The role of alcoholism in motor vehicle accidents, *Traffic Safety Res. Rev.* **6:**21.

Schoenberg, B. S., Bailar, J. C., and Fraumeni, J. F., 1971, Certain mortality patterns of osophageal cancer in the United States, 1930–67, *J. Nat. Cancer Inst.* **46:**63.

Schwartz, D., Lasserre, O., Flamant, R., and Lellouch, J., 1966, Alcool et cancer: Etude de pathologie géographique portant sur 19 pays, *Eur. J. Cancer* **2:**367.

Seixas, F. A., Williams, K., and Eggleston, S. (eds.), 1975, Medical consequences of alcoholism, *Ann. N.Y. Acad. Sci.* **252.**

Selzer, M. L., 1975, Alcoholics and social drinkers: Characteristics and differentiation, in: *Proceedings of the 6th International Conference on Alcohol, Drugs, and Traffic Safety* (S. Israelstam and S. Lambert, eds.), pp. 13–20, Addiction Research Foundation, Toronto.

Skog, O-J., 1971, *The Distribution of Alcohol in a Population,* National Institute for Alcohol Research, Oslo.

Skog, O-J., 1980, Liver cirrhosis epidemiology: Some methodological problems, *Br. J. Addict.* **75:**227.

Sulkunen, P., 1973, *On International Alcohol Statistics, Reports from the Social Research Institute of Alcohol Studies,* No. 72, Finnish Foundation for Alcohol Studies, Helsinki.

Sundby, P., 1967, *Alcoholism and Mortality,* Publication No. 6, National Institute for Alcohol Research, Universitetsforlaget, Oslo.

Takala, M., Pihkanen, T. A., and Markkanen, T., 1957, The Effects of Distilled and Brewed Beverages, Publication No. 4, Finnish Foundation for Alcohol Studies, Helsinki.

Thaler, H., 1977, Alcohol consumption and diseases of the liver, *Nutr. Metab.* **21:**186.

Thorarinsson, A. A., 1979, Mortality among men alcoholics in Iceland 1951–74, *J. Stud. Alcohol* **40:**704.

Tibblin, G. T., Wilhelmsen, L., and Werko, L., 1975, Risk factors for myocardial infarction and death due to ischemic heart disease and other causes, *Am. J. Cardiol.* **35:**514.

Tuyns, A. J., 1970, Cancer of the oesophagus: Further evidence of the relation to drinking habits in France, *Int. J. Cancer* **5:**152.

Tuyns, A. J., 1979, Epidemiology of alcohol and cancer, *Cancer Res.* **39:**2840.

Tuyns, A. J., and Audigier, J. C., 1976, Double wave cohort increase for oesophageal and laryngeal cancer in France in relation to reduced alcohol consumption during the second world war, *Digestion* **14:**197.

Tuyns, A. J., Jensen, O. M., and Péquignot, G., 1977, Le cancer de l'oesophage en Ille-et-Vilaine en fonction des niveaux de consommation d'alcool et de tabac: Des risques qui se multiplient, *Bull. Cancer* **64:**45.

Tuyns, A. J., Péquignot, G., and Jensen, O. M., 1978, Nutrition, alcohol and oesophageal cancer, *Bull. Cancer* **65:**59.

Tuyns, A. J., Péquignot, G., and Abbatucci, J. S., 1979, Oesophageal cancer and alcohol consumption: Importance of type of beverage, *Int. J. Cancer* **23:**443.

Wallgren, H., Kosunen, A., and Nikander, S., 1970, *Alkoholhaittojen Kytkeytyminen Juoman Laatuun*, Alkon Kesjuslaboratorio, Seloste 8065, Helsinki.

Weir, J. M., and Dunn, J. E. Jr., 1970, Smoking and mortality: A prospective study, *Cancer (Brussels)* **25:**105.

Whitlock, F. A., 1974, Liver cirrhosis, alcoholism and alcohol consumption, *Q. J. Stud. Alcohol* **35:**586.

Williams, P., Robinson, D., and Bailey, A., 1979, High-density lipoprotein and coronary risk factors in normal men, *Lancet* **1:**72.

Wilkinson, P., Santamaria, J. N., and Rankin, J. G., 1969, Epidemiology of alcoholic cirrhosis, *Australas. Ann. Med.* **18:**222.

Wynder, E. L., 1978, The epidemiology of cancers of the upper alimentary and upper respiratory tracts, *Laryngoscope* **88:**50.

Wynder, E. L., and Mabuchi, K., 1972, Etiological and preventive aspects of human cancer, *Prev. Med.* **1:**300.

Wynder, E. L., Bross, I. J., and Feldman, R. M., 1957, A study of etiological factors in cancer of the mouth, *Cancer (N.Y.)* **10:**1300.

6

Assessing Alcohol Use by Patients in Treatment

HARVEY A. SKINNER

1. INTRODUCTION

The monitoring of alcohol use by patients undergoing treatment is vital to ensure good clinical care, and to provide data on the effectiveness of therapy. However, many treatment programs do not have systematic procedures for the regular assessment of a patient's alcohol consumption during treatment and follow-up. Part of the difficulty is due to skepticism about the validity of alcoholics' self-reports, as well as resource constraints that limit the use of objective measures of alcohol use, such as tests of liver functioning. Although the notion that alcoholics deny the extent of their drinking is widely held, recent studies have found that various populations of alcohol abusers may provide fairly accurate information about their drinking and alcohol-related problems. An important instance where self-reports tend to underestimate drinking is when the patient has a positive blood alcohol level while undergoing the assessment.

    This chapter provides a critical evaluation of different methods for monitoring alcohol use. First, a conceptual framework is introduced that differentiates among alcohol consumption levels, symptoms of alcohol dependence, and disabilities related to drinking. This framework is useful for understanding *what* variables should be monitored with a given patient. The next three sections concern *how* one should monitor alcohol use with patients in treatment. A distinction is drawn among (1) retrospective methods, such as self-reported consumption for a given follow-up period; (2) prospective methods, such as daily self-monitoring of alcohol use; and (3) objective indicators, such as verification

HARVEY A. SKINNER ● Addiction Research Foundation, Toronto, Ontario, Canada.

by collateral sources, sequential blood tests, and use of official records. Finally, a practical strategy is described for monitoring alcohol use that integrates the relative merits of the three approaches.

This chapter focuses on research with clinical populations published up to the end of 1982. A comprehensive review of the validity of self-reports that includes general population surveys and special groups has been published recently by Midanik (1982a).

Measurement Accuracy

The goal of assessment is to obtain "valid" measurement of a patient's alcohol use and related problems. There are two principal sources of invalidity: measurement error and systematic biases. Both constrain or set limits upon the confidence one may have when interpreting a measure of alcohol use.

According to classic reliability theory (Wiggins, 1973), the observed scores of a measure (X) consist of a reliable or true score component (R) plus a random error component (E), that is, $X = R + E$. The degree of random error sets an upper bound on the reliability of a measure. Reliability may be defined as the consistency of an individual's reporting of drinking behavior both within a single assessment (internal consistency) as well as between two assessment occasions (test–retest). Thus, reliability addresses only the reproducibility or stability of indices related to alcohol consumption. Numerous factors influence the extent of measurement error (Cronbach, 1970), such as the patient's general ability to comprehend instructions, health status, and fluctuations in memory or concentration. These factors are especially germane to monitoring patients with alcohol-related problems, given the deleterious effects of prolonged alcohol intake upon memory (Rankin, 1975). The patient's status with respect to treatment can also influence the extent of measurement error. For instance, conditions such as an alcohol-withdrawal syndrome can produce relatively temporary alterations in functioning (e.g., heightened anxiety level, poor concentration). Various studies have demonstrated recoverability of cognitive, sensory, and motor disabilities following the cessation of drinking in chronic alcoholics (Bean and Karasievich, 1975; Lishman, 1981; Wilkinson and Carlen, 1981). As a patient progresses in treatment, the potential for obtaining more reliable information increases.

In addition to measurement error, a second major issue is systematic biases that can obscure the "correct" interpretation of a measure. For example, because of defensiveness sometimes observed among alcoholics, a patient may deny or minimize problem areas related to drinking. Since this bias is consistent it will contribute to the reliability of a measure, but one may draw an incorrect interpretation of the scores. For instance, an alcoholic may *consistently* deny drinking when in fact heavy alcohol consumption had occurred. Skinner (1981) reviews other sources of systematic biases in the assessment of alcohol problems. The best way to establish confidence in the assessment of alcohol use is to measure it by alternative modes (Campbell and Fiske, 1959; Sobell and Sobell, 1980). To the extent that different measurement methods converge, the validity of assessment is empirically established.

2. DEFINING ALCOHOL ABUSE

A persistent problem in alcoholism treatment and research has been the lack of consensus on concepts. In the present context, one is interested in how much a particular patient is drinking, as well as whether the patient is experiencing problems that are either a direct or indirect consequence of drinking. A three-dimensional framework is presented in Table 1 that is based upon recent classifications by the World Health Organization (Edwards et al., 1977) and

Table 1. Three-Dimensional Framework for Monitoring
Alcohol Use

Dimension	Status
I. Drinking history	
Frequency of drinking	0 = abstinence
Quantity consumed	1 = light drinking
Drinking style (e.g., binge)	2 = moderate drinking
Time of day	3 = heavy drinking
Drinking situation	
Antecedents to drinking	
II. Dependence syndrome	
Impaired control over drinking	0 = no symptoms
Increased tolerance	1 = mild symptoms
Withdrawal symptoms	2 = moderate symptoms
Compulsive drinking style	3 = severe symptoms
Drinking to relieve withdrawal symptoms	
III. Disabilities related to drinking	
Biomedical	0 = no problems
Traumatic injury	1 = mild problems
Liver disease	2 = moderate problems
Hypertension	3 = severe problems
Pancreatitis	
Gastrointestinal symptoms	
Psychosocial	
Absenteeism at work	
Poor work performance	
Family problems	
Anxiety/depression	
Intellectual impairment	
Legal problems	

American Psychiatric Association (1980). Although the three dimensions are empirically interrelated, they are conceptually distinct and may not be present simultaneously or to the same degree. Thus, all three dimensions should be assessed when monitoring a patient's progress in treatment and aftercare.

The first dimension concerns the patient's drinking behavior. Key variables include the frequency of drinking, quantity consumed per drinking day, variability in consumption (minimum and maximum amounts consumed), types of beverages, time of day when alcohol is consumed, and situations where alcohol is consumed. With these data, one could calculate various indices for a given time period, such as the number of abstinent days, light or moderate drinking days (less than 60 g absolute alcohol), and heavy drinking days (greater than 60 g absolute alcohol). Also, these data would allow one to document antecedent situations in which a patient tended to drink excessively, which could be important for determining treatment.

The second dimension refers to the alcohol dependence syndrome described by Edwards and Gross (1976), and further elaborated in the World Health Organization Task Force Report (Edwards et al., 1977). The cardinal symptom of this syndrome is an impaired control over alcohol intake. Other aspects include increased tolerance to alcohol, withdrawal symptoms following cessation of drinking, awareness of a compulsion to drink excessively, reinstatement of the syndrome after abstinence, and salience of drink-seeking behavior. The alcohol dependence syndrome is viewed as existing along a continuum of severity, and there is empirical evidence that supports quantitative variance among individuals with this syndrome (Skinner and Allen, 1982).

The third dimension concerns various biomedical and psychosocial disorders that stem either directly or indirectly from excessive drinking (Edwards et al., 1977). Estimates suggest that approximately 5-7% of men in North America may meet criteria for alcohol dependence, indicated by symptoms of alcohol withdrawal and impaired control over drinking (Polich, 1982). However, many individuals have "drinking problems" without symptoms of alcohol dependence. The prevalence of nondependent alcohol abusers has been estimated between 15 and 35% of men (Cahalan, 1970). Problem drinkers tend more to be young men; alcohol-related disabilities (e.g., traumatic injury) are often linked to acute episodes of intoxication.

The three-dimensional framework provides a basis for evaluating treatment outcome. For example, a patient's alcohol consumption (dimension I) may decrease substantially during treatment, but his or her level of depression (dimension III) may be exacerbated. Hart and Stueland (1979) found that different types of alcoholic patients achieved varying levels of success as determined by seven measures of rehabilitation (e.g., economic security, physiological problems, family relationships). However, patients failed to be differentiated on measures of drinking. This study is a good example of the need to monitor several criteria, since an emphasis on drinking status alone (dimension I) would have concealed differential rehabilitation in other areas (dimension III). The framework in Table I could serve as a diagnostic "shorthand" similar to the TNM classification of malignant tumors (Harmer, 1978) that is widely used in the staging of cancer. Recommendations for the operational definition of the dimensions are given by Skinner (1983).

3. RETROSPECTIVE METHODS

Retrospective methods involve the assessment of drinking behavior for a specified time period by the use of a structured interview or self-report questionnaire. These methods have advantages in that they are relatively inexpensive, yield immediate information on the patient's alcohol use history, allow the clinician to probe and gain more personal knowledge about a patient, and allow assessment of drinking behavior over a longer time period than is possible using objective methods (e.g., blood alcohol concentration). However, the accuracy of self-reported drinking histories has generated much controversy. Clearly, there are situations where one would expect an individual to have more motivation to distort his or her alcohol use in order to portray a positive level of functioning. Examples of these situations are a criminal justice system where abstinence is a condition of parole, and an employment context where the employee's work performance is being scrutinized due to "suspected" drinking. On the other hand, with patients in treatment a number of studies over the past few years have demonstrated that reliable and valid self-reports can be collected under a fairly broad set of conditions.

Reliability Studies of Clinical Populations

Reliability is the reproducibility or consistency of indices related to alcohol use. Although issues of reliability have attracted attention for some time (e.g., Bailey et al., 1966; Guze and Goodwin, 1972), methodologic problems have been noted with this research (Skinner and Sheu, 1982). In response to these criticisms, recent research has focused on shorter test–retest intervals and on more representative samples of treatment populations.

Two studies have focused on lifetime patterns of alcohol use. Rohan (1976) designed an interview format in which a patient's lifetime drinking history was analyzed in annual segments since the onset of regular drinking. The interview was conducted on two occasions (2 weeks apart) by different interviewers with two independent samples of 40 male alcoholics in a rehabilitation program. The lifetime duration of regular drinking was quite reliable with estimates above 0.97 for the two samples. Furthermore, the reliability of the lifetime total quantity consumed was highly consistent with reliability estimates above 0.90. Skinner and Sheu (1982) evaluated the reliability of a lifetime drinking history with patients initially assessed upon entry to treatment and then retested on an average of 4.8 months later. Aggregate indices achieved reasonably high reliability: 0.94 for lifetime duration of drinking, 0.80 for lifetime total volume consumed, and 0.68 for lifetime daily average. Recall accuracy was lower for the drinking phase that immediately preceded entry to treatment. Inconsistent reporting of alcohol consumption was correlated with a lower level of social stability and a greater degree of neuropsychological impairment.

Sobell and associates (1979b) conducted a detailed drinking history of male outpatients over the previous 12 months (test–retest interval of 6 weeks). Daily drinking disposition measures were recorded for pretreatment time periods covering: 30 days, 90 days, 180 days, and 360 days. Correlations ranged from 0.79 for the number of days abstinent 30 days prior to treatment to 0.98 for the number of days incarcerated in the 30-, 180-, and 360-day period prior to treatment. They concluded that outpatient male alcoholics' self-reports of daily drinking behavior during the 12 months prior to treatment were highly reliable. Armor and co-workers (1978) found that the internal consistency reliability for self-reported quantity of alcohol consumption ranged from 0.57 for data obtained from an alcoholic population to 0.28 for a general population. The internal consistency reliability of self-reported frequency of drinking ranged from 0.71 for the alcoholics to 0.78 for the general population. Thus, Armor et al. (1978) found that reliability of *frequency* of drinking was quite satisfactory. However, self-reports on the actual *quantity* of alcohol consumed tended to be less consistent.

Concerning drinking-related problems, Skinner and Sheu (1982) included the Michigan Alcoholism Screening Test (MAST) in their study. Both test–retest (0.84) and internal consistency reliability estimates (0.85 and 0.88) for the MAST total score were quite high. However, a detailed analysis revealed that patients were more consistent in reporting objective events (e.g., arrested for drunk driving), than they were in reporting their subjective recognition of alcohol misuse. Similarly, in a study of self-reports from 72 skid-row alcoholics, Annis (1979) found relatively high consistency on reinterview for demographic items, but less consistency was noted for items related to drinking patterns and social functioning. With respect to drinking variables, 85% agreement was observed for usual frequency of drinking, 54% agreement was evident for usual alcoholic beverage, and 56% agreement was observed for usual daily alcohol consumption. This finding corroborates the Armor et al. (1978) study in demonstrating that more reliable estimates are generally given for frequency of drinking than for the actual quantity consumed.

In measuring symptoms of alcohol dependence, Skinner and Allen (1982) found that a brief 29-item Alcohol Dependence Scale (ADS) exhibited substantial internal consistency reliability (0.92) using a large (n = 225) outpatient sample. An estimate of the test–retest reliability of the ADS was 0.92 with 76 inpatients who were retested after 1 week of therapy (Skinner and Horn, 1984). Using a different self-report questionnaire for assessing the degree of alcohol dependence, Stockwell and colleagues (1983) examined the test–retest reliability with 45 inpatients (2 week interval between testing). The reliability coefficient for the total score was 0.85. Various subscales of the dependence syndrome also achieved excellent reliability estimates: 0.81 physical withdrawal signs, 0.85 affective withdrawal, 0.77 withdrawal relief drinking, 0.79 alcohol consumption, and 0.76 reinstatement of withdrawal symptoms following abstinence.

In review, these studies indicate that reliable information can be collected by patient self-reports in clinical settings. When carefully designed interviewing techniques and questionnaires are used the reliability estimates are often substantial. With regard to dimension I (Table 1), reliable drinking histories have been obtained both for the recent past and for lifetime patterns. However, the frequency of drinking may be assessed more reliably than the actual quantity consumed. The degree of alcohol dependence (dimension II) has been measured by self-report questionnaires with excellent reliability. Finally, problems related to drinking (dimension III) have also been assessed with good reliability, although patients tend to give the most consistent responses when reporting discrete events (e.g., hospitalizations) rather than subjective symptoms.

Validity Studies

Validity is the accuracy or verisimilitude of the patient's reporting of drinking behavior. In order to establish the validity of self-reports, they must be compared with an independent and objective criterion representing the same behavior. External criteria have included laboratory tests (e.g., blood alcohol concentration), collateral reports, official records (e.g., hospitalizations), and behavioral observations.

Collateral Reports.   Information from collateral sources has been the most widely used method of verifying patients' self-reports. One of the first studies (Guze et al., 1963) compared information from 90 male criminals (39 had been previously diagnosed as alcoholics) with reports by their relatives, and found 74% agreement. Of particular interest was the finding that most of the disagreements (80%) involved a positive answer from the patient and a negative response from the relative, which suggested that patients were more likely than collaterals to report drinking problems. This trend has been supported by subsequent research (e.g., Maisto et al., 1982a).

More recent studies have found a trend towards increased accuracy of patients' self-reports as treatment progresses. McCrady et al. (1978) interviewed 44 alcoholics and their spouses on three occasions: during hospitalization, 6–8 weeks after discharge, and 5–7 months after discharge. These interviews concerned the patient's drinking behavior during the previous 30-day period. The agreement on alcohol use category (e.g., 0–1, 1–3 oz per day) was 56% for pretreatment drinking, 84% agreement at first follow-up, and 96% at second follow-up. With respect to serious consequences from drinking (tremors, black-outs, missed meals, morning drinking, drunkenness, or missing work), agreement on the presence or absence of at least three of these consequences was 67% at pretreatment, 84% at first follow-up and 87% at second follow-up. Although this study suggests a trend towards increasing accuracy of patient self-reports as treatment progresses, 19 patients dropped out of this study before the first follow-up. Thus, the increased accuracy during the two follow-ups may reflect the possibility that "less accurate" patients dropped out from treatment. Miller et al. (1979) compared self-reported alcohol consumption from patients in four treatment studies with collateral data. The correlation between self-report and collateral for the patient's highest alcohol consumption was 0.48 at intake to treatment, 0.66 at the termination of treatment, and 0.79 at 3-month follow-up. This finding was not influenced by patient dropouts since the sample size was consistent across the three occasions ($n = 127$, $n = 131$, $n = 128$). Moreover, this study found that approximately equal numbers of collaterals overestimated and underestimated the patients' self-reports.

Other studies have demonstrated that discrepancies between patient self-reports and collateral reports vary as a function of the drinking behavior measured. Maisto et al. (1979) collected 6-month posthospitalization data for 52 alcoholics and their collaterals. The correlations between patient and collateral reports ranged from 0.97 for days hospitalized, 0.82 for days drunk, 0.81 for days abstinent, 0.49 for days of limited drinking, to 0.46 for days jailed. A similar trend was found by Polich (1982) in a study of 128 alcoholics interviewed 4 years after treatment (Table 2). A very high degree of consensus (91%) was found with respect to whether or not the patient drank during the previous 6 months. In the few cases where disagreements occurred, the collaterals tended to understate the patient's drinking. Likewise, excellent agreement was evident for alcohol-related problems, such as being jailed or hospitalized due to drinking. However, much less consensus was evident for alcohol dependence symptoms (tremors or shakes, morning drinking), where the alcoholics were more likely than collateral to report symptoms of alcohol dependence.

Table 2. Agreement between Patient and Collateral Reports at 4-Year Follow-Up

Drinking behavior (past 6 months)	Agreement (percent)	Overreports[a] (percent)	Underreports (percent)
Any drinking	91	9	1
In jail, related to drinking	89	6	6
In hospital, related to drinking	88	8	4
Any tremors or "shakes"	55	29	15
Any morning drinking (on awakening)	48	45	8

[a] Overreports: patient's response was affirmative but collateral's was negative. Underreports: patient's response was negative but collateral's was affirmative.
From Polich (1982).

Official Reports.   Another approach to validating patient self-reports is to compare them with official records. Most comparisons show good agreement and when disagreements occur they are usually due to overreporting by the patient (Sobell and Sobell, 1975, 1978; Cooper et al., 1980, 1981). For instance, Cooper et al. (1981) examined the validity of patients' self-reports against official records for the number of days (duration) associated with each jail, hospital, or residential treatment episode over different pretreatment time intervals. In general, fairly good agreement was found between self-reports and the official record data. The correlation coefficients with 24 outpatients for number of days in jail ranged from 0.61 to 0.89 for different pretreatment intervals. With 24 inpatients, the correlations for residential and hospital days ranged from 0.73 to 0.94, and these validity coefficients were higher and more consistent across pretreatment intervals than were the correlations for days in jail. Discrepancies most frequently resulted from patients' overreporting days incarcerated. With the inpatient sample, the self-reported number of days incarcerated differed from official records on 28% of reported occasions. However, in the majority of these discrepancies (74%), the patients actually reported more days incarcerated than were reflected on the official records. Similarly, the outpatient self-reports differed from official records on 39% of reported occasions, and in 85% of the discrepancies the patients had overreported days incarcerated. Thus, patients tended to depict themselves more negatively than the official record data.

Laboratory Tests.   A third way of validating patients' self-reports is to compare them with results of appropriate laboratory tests. Two studies have examined alcoholic patients attending a liver clinic. Orrego et al. (1979) questioned 37 patients over a 6-month period regarding their drinking during the previous week. Patients were told that their daily urine was being monitored for compliance in taking medication, but the patients were not aware that alcohol content in their urine was also being assessed. For patients with alcohol in their urine, drinking was denied 52% of the time they were interviewed. However, patients who always admitted to drinking had a significantly greater average urine alcohol value than those who denied drinking every time. Furthermore, intermittent drinkers had significantly greater urine alcohol values when admitting drinking than when they denied drinking. Deniers appeared to consume less alcohol than those who admitted drinking. The Orrego et al. (1979) study is notable in that urinary alcohol was measured *daily* over a 6-month follow-up period. Iber and Miller (1982), on the other hand, compared self-reported drinking with urine screening for alcohol only at the time of outpatient visits, and found that very few patients (7%) had been drinking. However, this finding may reflect a higher abstinence rate among cirrhotic patients in anticipation of their attendance at the liver clinic.

Another approach to validating self-reports is through a measure of blood alcohol concentration (BAC). A general tendency toward underreporting by patients has been found in several settings (Armor et al., 1978; Jalazo et al., 1978; Polich, 1982; Sobell et al., 1979). For instance, Armor et al. (1978)

compared self-reported alcohol consumption with blood alcohol concentrations from alcoholics tested at treatment admission. The BAC computed from self-reports was about 25% lower than the actual BAC. Polich (1982) conducted further research with this clinical population at the 4-year follow-up. Among those patients reporting a drink in the past 24 hr ($n = 220$), 59% had a positive BAC. In contrast, of 412 patients who reported not drinking in the past day, only 16 patients (4%) had a positive BAC. These findings indicate that the patient self-reports of abstinence were highly valid. Among the patients who reported any drinking in the previous 24 hr, the self-reported amount consumed was compared with the BAC levels. The majority (58%) of self-reports were consistent with the BAC, although underreporting was revealed in 35% of cases whereas overreporting was found in only 7%. However, one recent study found that a substantial number of alcoholics on admission to treatment overreported their consumption levels compared with the BAC (Midanik, 1982b).

Since results of both BAC analysis and urinary screening are useful only for verifying alcohol consumption in the recent past (e.g., 24 hr), considerable research has been directed towards the establishment of biomedical indices that can measure total alcohol consumption over longer time intervals (Holt et al., 1981). Two of the more promising laboratory tests are gamma-glutamyl transpeptidase (GGT) and mean corpuscular volume (MCV). However, abnormal results on these tests are not specific to alcohol use, and may reflect various non-alcohol-related conditions. This research is examined in Section 5.

Behavioral Observations.   A fourth approach for validating patient self-reports is to compare them with behavioral observations. This can take place either in controlled settings where the purpose of the experiment may be disguised, or in natural settings where drinking behavior is observed in everyday conditions, such as a bar (Bridell and Nathan, 1975). Harford et al. (1976) interviewed patrons in three bars about their recent drinking behavior. The self-report data were compared with direct observations of their alcohol consumption made 3–7 days previously. Of 85 individuals who agreed to participate in the study, 54% remembered the exact number and type of drinks, 21% underreported, and 9% overreported by only one drink. Although overall frequency and quantity of drinking throughout the week before the interview did not affect recall, the number of drinks during the interview date did affect accuracy.

In review, patients tend to provide fairly valid self-reports of their alcohol use (Table 1, dimension I) if they are free from alcohol at the time of assessment. When patients have been drinking, there is a general tendency toward underreporting, although the extent of underreporting may not seriously affect outcome status (Polich, 1981). Little research has been conducted to date on the validity of patient self-reports regarding symptoms of alcohol dependence (dimension II). Polich (1982) found that collaterals were often unsure about whether or not the patient had experienced such symptoms. However, the accuracy of the collateral information must be questioned since the collaterals may not have observed enough of the patient's behavior to make a valid report. Stockwell et al. (1983) reported significant correlations between self-reported symptoms of alcohol dependence and physician's assessment of withdrawal severity in a detoxication unit. With respect to alcohol-related problems (dimension III), patient self-reports are often in exact agreement with official records such as driving infractions, hospitalizations, and employment records. When discrepancies occur, it is generally found that the patient has reported an incident that has not been recorded in the official records.

4. SELF-MONITORING

Self-monitoring is a behavioral assessment technique that involves having an individual record the occurrences of a target behavior (e.g., alcohol consumption), as well as the conditions that precede and follow the behavior (Nelson, 1977; Kazdin, 1974). Self-monitoring offers considerable potential for increasing the accuracy of assessment. However, the very act of self-monitoring can alter the frequency of the target behavior. This reactivity often produces behavioral changes that are in a therapeutic direction, such as a reduction in smoking or alcohol use. Thus, self-monitoring has both assessment and therapeutic implications. Despite the growing popularity of self-monitoring techniques in behavior therapy and research, there is a curious lack of research on these methods in the alcoholism treatment literature. One explanation is the fact that most alcoholism treatment programs emphasize total abstinence for patients with serious drinking problems, in which case there would be no drinking behavior to record if the patient complies with the abstinence goal. However, self-monitoring techniques would appear to be particularly relevant for therapies with early-stage problem drinkers who are attempting to moderate or control their alcohol use.

Self-monitoring techniques or "daily diaries" have been used in several research studies of drinking practices in nonalcoholic populations (Uchalik, 1979; Fuller et al., 1972). One of the first clinical applications of self-monitoring described a simple self-feedback technique to monitor drinking behavior in an outpatient treatment program (Sobell and Sobell, 1973). Patients were instructed to record their daily alcohol consumption using a specially prepared Alcohol Intake Sheet, which included the date, specific type of drink, percentage alcohol content, time of day, number of sips per drink, amount of drinks consumed, and situation where the drinking occurred. Daily self-monitoring of alcohol intake was also emphasized in two self-help programs for problem drinkers (Miller and Munoz, 1976; Amit et al., 1977). In an evaluation of behavioral and traditional treatment for problem drinkers, Pomerleau and Adkins (1980) used daily records of drinking with patients in the behavioral treatment condition.

In one of the few studies that have evaluated self-monitoring techniques and alcohol use, Kennedy and Gilbert (1978) tested the hypothesis that self-monitoring would increase control over drinking and perceived control over internal/external pressures. Twenty patients in an alcohol treatment program were randomly assigned to either the self-monitoring ($n = 10$) or alcohol education ($n = 10$) control condition. No significant differences were found in postdischarge alcohol use. However, most of the patients were reported to be abstinent after discharge, which means that there was little variability in posttreatment drinking behavior. Patients with generalized expectations of internal control in the self-monitoring condition did show an increase in perceived control over external stresses. Kennedy and Gilbert (1978) concluded that internally oriented (locus of control) patients appear to benefit most from the use of self-monitoring.

More recently, Sanchez-Craig and Annis (1982) compared the validity of self-monitoring and retrospective measures of alcohol consumption with 40 male problem drinkers. During treatment patients were trained in daily self-monitoring of drinking behavior, and following discharge they were instructed to monitor their alcohol use and bring their records at the time of the follow-up assessment. Eighteen patients complied with the self-monitoring regime, whereas 22 patients did not. The total number of drinks consumed over the 3-week period prior to the day of follow-up assessment was obtained either by self-monitoring methods (18 patients who complied) or recall method (22 patients who failed to maintain daily records). For the recall patients, the drinking history was taken retrospectively using a standardized interview format. The accuracy of each method was corroborated by two biochemical indicators of drinking: gamma-glutamyl transpeptidase (GGT) and high-density lipoproteins (HDL). Considering 6- and 12-month follow-up data, the recall group scored significantly higher than the self-monitoring group on both biochemical measures. Both GGT and HDL failed to correlate significantly with self-reported alcohol consumption in the recall group. However, with the self-monitoring group, GGT correlated 0.54 with alcohol consumption and HDL correlated -0.55 with alcohol consumption.

These findings suggest that self-monitoring yielded a more accurate picture of alcohol consumption than the retrospective drinking history. However, Sanchez-Craig and Annis (1982) cautioned that the difference could be due to motivational variables rather than the assessment method, since patients who complied with self-monitoring may have been more motivated to observe their drinking and give more accurate reports.

Clearly, there is pressing need for further research on the issues of self-monitoring of alcohol use with patients in treatment. Self-monitoring would appear to be especially important for interventions aimed at the early-stage problem drinker, where moderation in drinking is a realistic goal. In her comprehensive review of self-monitoring research, Nelson (1977) identified a number of variables that can affect the accuracy of self-monitoring. For instance, self-monitoring tends to be more accurate when individuals are aware that accuracy is being monitored, and when an individual is reinforced for accurate self-records. With respect to reactivity, the effect of self-monitoring appears to attenuate with time. In the treatment of early-stage problem drinkers, one might investigate the therapeutic value of self-monitoring for shaping positive as opposed to negative behaviors. For instance, rather than having a patient record only the number of instances of "heavy" drinking days (negative results), one might emphasize the self-monitoring of positive events such as successfully controlling an urge to drink, or the maintenance of a prescribed abstinence schedule.

5. OBJECTIVE METHODS

Because of concerns over the validity of self-reported drinking and alcohol-related problems, considerable interest has been generated in the use of objective methods, including official records of alcohol-related arrests, measures of blood alcohol concentration, and reports by a collateral observer. Although these methods have the potential of providing more valid information than patient self-reports, objective data are generally most costly to obtain and may pertain only to a fairly limited time frame, such as alcohol use in the past 24 hr. Thus, within any treatment setting constraints of both time and resources may limit the use of objective techniques. This section provides a synopsis of the advantages and limitations of these methods.

Laboratory Tests

Various laboratory tests and biochemical markers have been proposed for the assessment and diagnosis of alcohol abuse (Holt et al., 1981). The only true indicator of alcohol consumption is the detection of alcohol or one of its metabolites in the patient's body fluids. However, these compounds have a relatively short half-life which limits their application to drinking in the previous 24 hr. Also, the detection of alcohol use (dimension I) does not necessarily imply that the patient is experiencing symptoms of dependence (dimension II) or alcohol-related disabilities (dimension III). On the other hand, a positive blood alcohol level or results of a urine screening test may be quite helpful when monitoring a patient's compliance with a treatment goal of total abstinence.

Detecting Alcohol Use in the Previous 24 Hr.   Kalant (1971) provides an extensive review of studies which have shown that the urine alcohol curve parallels the blood alcohol curve when serial collections of urine are obtained. Thus, the amount of alcohol in urine is a good indicator of alcohol concentration in plasma water at the time when urine is formed. Factors that determine the blood alcohol concentration (BAC) will also influence urine alcohol concentrations (UAC). According to Sellers and Kalant (1976), a 70-Kg person of average build metabolizes 7–10 g of absolute alcohol/hr. Table 3 provides a guide to the number of hours required for the metabolism of alcohol for a given quantity of alcohol consumed. For example, an intake of one standard drink may take 2 hr

Table 3. Metabolism of Alcohol in a 70-kg Person

Alcohol consumed in 1 hr		Approximate blood alcohol concentration[b] (mg/100 ml)	Approximate time required for complete metabolism (hr)
Standard drinks[a]	Absolute alcohol (g)		
1	13.6	26	1.4–2
2	27.2	52	2.7–3.9
4	54.4	104	5.4–7.8
6	81.6	155	8.2–11.7
8	108.8	207	10.8–15.5
10	136.0	259	13.6–19.4
12	163.2	311	16.3–23.3

[a] One standard drink is approximately 12 oz Canadian beer, $1^1/_2$ oz liquor, 5 oz wine, or 3 oz fortified wine (e.g., sherry).
[b] BAC (mg/100 ml) = $(80 \times g)/(f \times w)$, where 80 is used because blood absorbs about 80% of alcohol in body tissue, g = grams of absolute alcohol consumed, f = subject's body water expressed as a decimal fraction of body weight/type (lean/muscular = 0.7, average = 0.6, fat = 0.5), and w = subject's weight in kg.

to be completely metabolized; 12 drinks will require up to 24 hr for metabolism. Any urine formed during these periods will contain a detectable concentration of alcohol.

Blood alcohol concentrations are reduced by approximately 20 mg/100 ml per hour due to the metabolism of alcohol in the nonalcoholic person (Sellers and Kalant, 1976). However, a number of factors can influence alcohol absorption and disposition in body fluids as well as its subsequent metabolism, including: amount and timing of alcohol consumed, body type, gastric emptying rate, use of other drugs, individual differences in ethanol metabolism by the liver, type and amount of food consumed, and presence of certain disease states, especially liver disease (Sellers and Kalant, 1976; Holt, 1981). A consideration of these factors is important when interpreting a BAC or UAC level for a given patient.

A urine specimen could be requested routinely as part of the patient's attendance at a therapy session or medical clinic. The urine could be analyzed not only for alcohol concentration, but also for the presence of other drugs or for monitoring the patient's compliance with medication (Orrego et al., 1979). Depending upon the treatment setting, the patient may or may not be told that the urine specimen is being analyzed for alcohol content. If the patient is aware that his or her urine is being so scrutinized, then one would expect the patient to provide more valid self-reports of alcohol consumption in the immediate 24 hr. Also, the determination of whether the patient is alcohol-free at the time of assessment is important for corroborating the validity of self-reports (Sobel et al., 1979). Urine or breath samples may be used to differentiate patients according to those for whom self-reported drinking behavior may be expected to be valid (negative UAC or BAC) or invalid (positive UAC or BAC).

An especially promising development involves the use of a dipstick for rapid determination of ethanol in body fluids (Kapur and Israel, 1983). The methodology is based on a competitive enzyme inhibitor of alcohol dehydrogenase. Different concentrations of ethanol produce quantifiable differences in color intensity of the dipstick. This procedure could be used in a variety of settings to provide a convenient objective indicator of alcohol consumption.

Blood alcohol concentrations assessed by a portable breath analyzer have been used for the infield testing of patients. For example, Miller (1975) and Miller et al. (1974) randomly administered breath alcohol analyses to chronic drunkenness offenders as part of a behavioral contingency treatment. Patients were contacted on a random basis and instructed to visit one of the agencies within an hour for an alcohol test. If the patient was unable to attend, a research assistant was sent to administer the blood alcohol test in the natural environment. This approach would appear to be particularly effective for ensuring patient compliance with a drinking goal of abstinence.

Inaccuracies in the use of breath alcohol analyzers have prompted studies of saliva (McColl et al., 1979) and sweat (Phillips and McAloon, 1980). The sweat patch test monitors the amount of alcohol consumed over a number of days. A method for rapid assay of ethanol in the sweat patch using a handheld electrochemical detector has been developed (Phillips, 1982). Thus, it may be possible to obtain an objective estimate of the amount of alcohol consumed over a longer period of time than the previous 24 hr by analyzing sweat. However, this instrument has yet to be evaluated outside of Phillips' developmental work. For instance, research is needed on how the sweat patch test may be influenced by an intermittent drinking style and variability in the amount of alcohol consumed per drinking session.

Indicators of Alcohol Use Over Longer Periods.   Ideally, a biochemical marker should be sensitive to quantitative changes in drinking over a 2- or 3-week period, and be highly specific to alcohol use. To date, the many laboratory tests evaluated for this purpose have tended to be relatively nonspecific to alcohol use (Holt et al., 1981), and they have demonstrated only moderate sensitivity in differentiating between alcoholic and nonalcoholic patients (Bernadt et al., 1982). However, two emerging trends deserve note. Several studies have shown that diagnostic accuracy can be increased by the use of laboratory tests in combination. Evidence also suggests the need to establish a within-subject baseline on results of a given laboratory test so that subsequent deviations from this baseline more closely reflect changes in the individual's alcohol consumption.

Three of the more promising biochemical markers include measurement of the liver enzyme gamma-glutamyl transpeptidase (GGT), mean corpuscular volume (MCV), and high-density lipoprotein (HDL). Studies have found a graded response between these biochemical indices and recent alcohol use in populations of heavy drinkers (e.g., Sanchez-Craig and Annis, 1982), psychiatric patients (Bernadt et al., 1982), and healthy subjects (e.g., Chick et al., 1981; Papoz et al., 1981). Although the reported correlations are statistically significant, they generally are of low magnitude. This calls into question the usefulness of a particular biochemical test for indicating alcohol consumption at the individual level. Recent studies have shown that the predictive accuracy can be increased by the *combined* use of laboratory tests (Chalmers et al., 1981; Eckardt et al., 1981; Ryback et al., 1982a,b). For instance, Chick et al. (1981) found that an abnormal GGT or MCV has 62% sensitivity for detecting individuals who consumed more than 60 g ethanol per day (more than four standard drinks), and this combination was greater than the sensitivity of either test considered singly. Similarly, Sanchez-Craig and Annis (1981) found a composite index of GGT and HDL to be superior to either test alone in differentiating between abstinent/light, moderate, and heavy drinkers. Thus, the use of composite indices for predicting an individual's level of alcohol consumption is a promising area of research.

At the same time, more needs to be learned about the various factors that can influence a given biochemical test. Devenyi et al. (1981) found that HDL can be a good indicator of alcohol use only in the absence of liver disease. With respect to drinking style, Skinner (1982) found in an outpatient sample with alcohol problems that GGT, MCV, and HDL showed a linear increase with the *frequency* of drinking (days per month) but not with the actual *quantity* consumed. MCV tended to become abnormally elevated around a threshold of 60 g ethanol/day (four drinks). At higher quantities the percentage abnormal MCV reached an asymptote, and even declined somewhat at very high consumption levels (more than 160 g absolute alcohol). On the other hand, GGT showed a marked increase in percent abnormal above the threshold of 160 g ethanol/day.

Morgan et al. (1981) have reported one of the few studies on the use of biochemical indices for monitoring alcohol consumption with patients in treatment. Aspartate transaminase (AST), GGT, and MCV were assessed biweekly for 3 months in 20 alcoholics who had precirrhotic, alcohol-related liver disease. These patients were chronic alcoholics who had abused alcohol on average for 20 years and had been consuming more than 20 standard drinks per day (260 g ethanol). After 3 months, the biochemical markers did not differentiate patients who abstained, decreased, or continued alcohol abuse. However, at the individual patient level Morgan et al. (1981) found that one or more biochemical markers tended to mirror changes in alcohol intake. The authors concluded that: (1) certain markers reflect alcohol intake more accurately than others in a particular patient; (2) individual differences are present in the threshold of alcohol intake below which laboratory measures remain normal; and (3) the biochemical indices tend to exhibit different dose–response curves. This study underscores the need to establish a baseline level of laboratory indices for each individual. Subsequent measurements relative to this baseline will then tend to mirror patterns of alcohol consumption. This within-subject relationship may be masked or washed in analyses that only consider group data. Further support for this relationship has been found by research currently in progress at the Clinical Institute (H. Orrego and Y. Israel, personal communication).

Other Techniques

Aside from laboratory tests, collateral reports and official records have been the most widely used sources of objective data on alcohol use and alcohol-related problems. This research, described in Section 3 above, suggests the following conclusions.

Collateral Reports.   As external criteria of alcohol use, these have tended to show little net bias in self-reported behavior, with disagreements indicating both overreporting and underreporting by patients. Indeed, some studies found that patients themselves are more likely to report drinking problems than the collateral observer and that collateral sources may yield less reliable data. With patients who lack social supports, difficulties often arise in finding a suitable collateral individual who can consistently observe the patient's behavior. Collaterals may provide fairly accurate data with respect to a patient's drinking frequency, but have only a gross estimate of the actual quantity consumed or of the presence of alcohol dependence symptoms. Thus, there are inherent limitations in the potential accuracy of collateral sources. Moreover, the issue of possible reactivity of collateral observers (especially, if a family member) on the patient's alcohol behavior has not been systematically explored. The use of collateral individuals may have more to offer in enhancing the patient's motivation for treatment than in providing detailed information about the patient's alcohol use.

Official Records.   These generally show at least two-thirds agreement with patient self-reports. When discrepancies occur, they are usually because the patient has admitted to an event that has not been recorded in the official record (e.g., arrest, hospitalization). This finding reflects inaccuracies or omissions in official recordkeeping as well as the possibility of the patient having changed names. Given the expense in finding official records, one must question the value in using this external criterion except in well-funded research studies (e.g., O'Farrell and Connors, 1982).

Physical Examination.   Clinical signs and symptoms of alcohol abuse are often nonspecific and are evident mainly in later stages of alcoholism, such as in patients with alcoholic liver disease (Holt et al., 1981). Thus, results of physical examinations are likely to be of limited value in providing objective evidence of continued alcohol use. On the other hand, heavy drinking has been associated with neuropsychological impairment among alcoholics and recent evidence suggests deficits in cognitive functioning even with nonproblematic social drinkers (Wilkinson and Carlen, 1981). Moreover, there is evidence of reversibility in cerebral dysfunction of alcoholics after a period of abstinence. Thus, the sequential application of brief neuropsychological tests, such as Digit Symbols (Wechsler, 1955) and the Trail-Making Test (Halstead, 1947), could provide corroborative evidence of abstinence or continued drinking.

For example, Wilkinson and Sanchez-Craig (1981) investigated changes in the results of the Digit Symbols Test over the first 3 weeks of treatment in 62 socially stable problem drinkers. They found that patients who abstained or consumed on average less than one drink per day showed a significant improvement of cognitive ability after 3 weeks. In contrast, patients who consumed four or more drinks per day showed no improvement in cognitive functioning. These findings suggest that results of neuropsychological tests may be sensitive indicators of patients' alcohol use, at least during the initial stages of treatment.

6.   RECOMMENDED STRATEGY

The ideal strategy for monitoring alcohol use with patients in treatment would contain the following components. First, breath or urine screening would be used at each clinic attendance to ensure that the patient is alcohol- (and drug-) free. Also, the patient should be relatively stable and not experiencing major symptoms of alcohol withdrawal or physical/mental distress. Good rapport should be established with the patient, and he or she should be assured of the confidentiality of any information provided. The best way to assess alcohol use would be by self-monitoring techniques. The patient would be requested to keep a daily log of the number and type of drinks consumed, as well as the particular situation and time when drinking occurred. Self-monitoring techniques are valuable not only for increasing assessment accuracy but also for their potential therapeutic effect in reducing the problem behavior. The patient should be informed that his or her self-reports will be checked against other sources, such as collateral individuals, official records, and results of laboratory tests. It is likely that this "awareness" factor in itself is important for enhancing the validity of self-reports. Finally, an initial baseline should be established for each patient on results of key biochemical tests (e.g., GGT and MCV) as well as brief neuropsychological tests (e.g., Digit Symbols, Trail-Making Test). Then, one may compare subsequent measurements to this within-subject baseline as an indicator of changes in the patient's alcohol use.

At the other extreme, the best way to ensure "invalid" self-reports is to use unstructured, general, or vague questions regarding the patient's alcohol use, provide minimal supportive counseling or safeguards about the confidentiality of information provided, assess patients when they have been drinking, are suspected of having a positive blood alcohol concentration, or when they are

Table 4. Factors Influencing the Validity of Self-Reports

Invalid	$\xleftarrow{\text{self-reports of}}$ alcohol use $\xrightarrow{}$	Valid
Patient has positive blood alcohol concentration at time of assessment		Patient is alcohol- (or drug-) free at time of assessment
Patient is experiencing withdrawal symptoms or acute physical/mental distress		Patient is stable, no major symptoms
Use of unstructured, general, or vague items in taking the drinking history		Use of structured, carefully developed drinking items (especially self-monitoring methods)
Patient not aware that self-reports will be checked against objective criteria		Patient aware that self-reports will be checked with other sources (laboratory tests, collaterals, records)
Minimal contact or supportive counseling with the patient		Good rapport has been established with the patient
Patient shows poor compliance with the treatment regime		Patient complies with other aspects of treatment
Patient has clear motives to dissimulate (e.g., abstinence is a condition of parole or continued employment)		Patient has no obvious reasons for distorting reports of alcohol use
Patient doubts the confidentiality of information provided		Patient is assured of confidentiality of information

experiencing major withdrawal symptoms, and not to inform the patient that his or her self-reports will be checked against objective criteria.

Table 4 summarizes factors that can influence the validity of assessments of alcohol use. This table may be used as a checklist for determining whether self-reported information in a specific program may be expected to be particularly valid or invalid. Many of the elements in Table 4 are of a policy or procedural nature that could be readily implemented in most treatment settings. Although self-monitoring is preferred over retrospective self-reports, patient compliance with keeping a daily drinking log may present problems, especially with chronic alcoholics who lack social supports. Special incentives or rewards could be given contingent on completion of the log as one mechanism for increasing compliance.

Of the many factors that can affect the quality of self-report data, the following issues deserve special note.

Blood Alcohol Concentration (BAC) at Assessment

Alcoholics who have been drinking (positive BAC) tend to underreport their recent alcohol consumption (Armor et al., 1978; Jalazo et al., 1978; Orrego et al., 1979; Polich, 1982; Sobell et al., 1979). For example, Sobell et al. (1979) found that when patients had not been drinking, their self-reports were valid. However, when patients had been drinking (positive BAC), they typically underreported their alcohol consumption for the previous 48 hr. Moreover, trained observers could identify only about two-thirds of the patients who were intoxicated but gave invalid self-reports. They concluded that breath analysis for BAC should be a routine component of assessment programs for monitoring patients' alcohol use during treatment and follow-up.

Confidentiality of Data and Awareness That Self-Reports Will Be Verified

Patients provide more accurate self-reports when they are assured of who will or will not have access to this information, and when they are informed that their self-reports will be checked against other sources, such as collateral reports or results of laboratory tests. Patients who are unaware that their self-reports are being corroborated by objective sources may give less accurate information (e.g., Orrego et al., 1979).

Assessment Context

Patients who are actively involved in treatment tend to provide fairly valid self-reports about their alcohol use and drinking problems. However, heavy drinkers who are not in treatment may significantly underreport their alcohol use (Popham and Schmidt, 1981; Pernanen, 1974). Thus, self-reports of alcohol use are more problematic when screening for heavy drinkers in a population than when patients are actively involved in treatment and follow-up.

Assessment Method

Several studies have compared different techniques for obtaining self-report data on alcohol use. Sobell and Sobell (1981) found little overall difference in the validity of patient self-reports under three conditions: (1) interview setting (group versus individual), (2) method of interview administration (self versus other), and (3) question type (alcohol versus nonalcohol versus demographic). Similarly, Skinner and Allen (1983) found no important differences in the reliability, level of drug abuse problems, or consumption patterns reported by patients who had been randomly assigned to either face-to-face interview, self-report questionnaire or computer-assisted interview. With carefully developed techniques, few differences have been observed in the method of assessing drinking behavior. However, quantity–frequency methods that derive an estimate of typical consumption levels may mask certain types of atypical drinking, such as infrequent binges (Sobell et al., 1982; Maisto et al., 1982b).

Drinking Interval

Alcoholics tend to provide less reliable and accurate information when assessing alcohol use during the period *before* entry to treatment. More valid information on self-reports is generally obtained while patients are in treatment itself or during subsequent follow-up (Cooper et al., 1980; McCrady et al., 1978; Miller et al., 1979).

7. CONCLUSION

Self-reports are the most widely used method for assessing alcohol use and alcohol-related problems. This approach is less costly than other techniques, has been shown to be generally valid with varying populations of alcohol abusers, often provides less "noisy" information than many objective measures, and is the only way to assess certain variables such as drinking or alcohol problems that occurred in the distant past. In particular, self-monitoring of alcohol use offers the greatest potential for obtaining accurate data. This technique may also have therapeutic value in reducing the patient's alcohol consumption. However, compliance with self-monitoring may prove problematic for some patients. Objective methods such as collateral reports and official records can provide important data for corroborating self-reported information, but when discrepancies occur it is often found that the patient has provided the more accurate information. Breath and urine screenings for alcohol are excellent for confirming self-reported abstinence; however, these laboratory tests are relevant only to drinking behavior in the past 24 hr. The search for other biochemical indices that would detect alcohol use over longer periods (e.g., 2 or 3 weeks) has yet to provide a procedure that possesses both high sensitivity and specificity. Thus, the establishment of a "gold standard" objective measure of alcohol use remains elusive.

Confidence in the assessment of alcohol use is established by convergence among alternative measurement methods. This is a basic mechanism of science. Although practical constraints often limit the extent of multimodal assessment possible in a given setting, a *minimum* standard should be the use of a carefully developed self-report technique (preferably self-monitoring), administered under conditions that enhance its validity (Table 4), corroborated by breath or urine screening for recent alcohol use, and interpreted by the treatment staff with a healthy level of caution.

ACKNOWLEDGMENT

The author thanks Ms. Sheila Henderson for her help in the preparation of this chapter.

REFERENCES

American Psychiatric Association, 1980, *Diagnostic and Statistical Manual of Mental Disorders*, Third Edition, Washington, D.C.

Amit, Z., Sutherland, E. A., and Weiner, A., 1977, *Guide to Intelligent Drinking*, Fitzhenry and Whiteside, Don Mills, Ontario.

Annis, H. M., 1979, Self-report reliability of skid-row alcoholics, *Br. J. Psychiatry* 134:459–465.

Armor, D. J., Polich, J. M., and Stambul, H. B., 1978, *Alcoholism and Treatment*, Wiley, New York.

Bailey, M. B., Haberman, P. W., and Scheinberg, J., 1966, Identifying alcoholics in population surveys: A report on reliability. *Q. J. Stud. Alcohol* 27:300–315.

Bean, K. L., and Karasievich, G. O., 1975, Psychological test results at three stages of inpatient alcoholism treatment, *J. Stud. Alcohol* 36:838–852.

Bernadt, M. W., Munford, J., Taylor, C., Smith, B., and Murray, R. M., 1982, Comparison of questionnaire and laboratory tests in the detection of excessive drinking and alcoholism, *Lancet* 1:325–328.

Bridell, D. W., and Nathan, P. E., 1975, Behavior assessment and modification with alcoholics: current status and future trends, in: *Progress in Behavior Modification*, vol. 2 (M. Hersen, R. M. Eisler, and P. M. Miller, eds.), pp. 1–51, Academic Press, New York.

Cahalan, D., 1970, *Problem Drinkers*, Jossey-Bass, San Francisco.

Campbell, D. T., and Fiske, D. W., 1959, Convergent and discriminant validation by the multitrait-multimethod matrix, *Psychol. Bull.* 56:81–105.

Chalmers, D. M., Rinsler, M. C., MacDermott, S., Spicer, C. C., and Levi, A. J., 1981, Biochemical and haematological indicators of excessive alcohol consumption, *Gut* 22:992–996.

Chick, J., Krietman, N., and Plant, M., 1981, Mean cell volume and gamma-glutamyl transpeptidase as markers of drinking in working men, *Lancet* 1:1249–1251.

Cooper, A. M., Sobell, M. B., Maisto, S. A., and Sobell, L. C., 1980, Criterion intervals for pretreatment drinking measures in treatment evaluation, *J. Stud. Alcohol* 41:1186–1195.

Cooper, A. M., Sobell, M. B., Sobell, L. C., and Maisto, S. A., 1981, Validity of alcoholics' self-reports: Duration data, *Int. J. Addict.* 16:401–406.

Cronbach, L. J., 1970, *Essentials of Psychological Testing*, Third edition, Harper and Row, New York.

Devenyi, M. D., Robinson, G. M., Kapur, B. M., and Roncari, D. A. K., 1981, High density lipoprotein cholesterol in male alcoholics with and without severe liver disease, *Am. J. Med.* 71:589–594.

Echardt, M. J., Ryback, R. S., Rawlings, R. R., and Graubard, B. I., 1981, Biochemical diagnosis of alcoholism, *JAMA* 246:2707–2710.

Edwards, G., and Gross, M. M., 1976, Alcohol dependence: Provisional description of clinical syndrome, *Br. Med. J.* 1:1058–1061.

Edwards, G., Gross, M. M., Keller, J., Moser, J., and Room, R., 1977, *Alcohol-Related Disabilities*, World Health Organization Offset Publication Number 32, Geneva, Switzerland.

Fuller, R. K., Beebe, H. T., Littell, A. S., Houser, H. B., and Witschi, J. C., 1972, Drinking practices recorded by a diary method, *Q. J. Stud. Alcohol* 33:1106–1121.

Guze, S. B., and Goodwin, D. W., 1972, Consistency of drinking history and diagnosis of alcoholism, *Q. J. Stud. Alcohol* 33:111–116.

Guze, S. B., Tuason, V. B., Stewart, M. A., and Picken, B., 1963, The drinking history: A comparison of reports by subjects and their relatives, *Q. J. Stud. Alcohol* 24:249–260.

Halstead, W., 1947, *Brain and Intelligence*, University of Chicago Press, Chicago.

Harford, T. C., Dorman, N., and Feinhandler, S. J., 1976, Alcohol consumption in bars: Validation of self-reports against observed behavior, *Drink. Drug Pract. Surv.* 11:13–15.

Harmer, M. H., 1978, *TNM Classification of Malignant Tumors*, International Union against Cancer, Geneva, Switzerland.

Hart, L. S., and Stueland, D., 1979, An application of the multidimensional model of alcoholism to program effectiveness, *J. Stud. Alcohol* **49**:645–655.

Holt, S., 1981, Observations on the relation between alcohol absorption and the rate of gastric emptying, *Can. Med. Assoc. J.* **124**:267–278.

Holt, S., Skinner, H. A., and Israel, Y., 1981, Early identification of alcohol abuse: Clinical and laboratory indicators, *Can. Med. Assoc. J.* **124**:1279–1295.

Iber, F. L., and Miller, P. A., 1981, Alcohol among stable cirrhotic patients in a liver clinic, *Hepatology* **2**:692.

Jalazo, J., Steer, R. A., and Fine, E. W., 1978, Use of breathanalyzer scores in the evaluation of persons arrested for driving while intoxicated, *J. Stud. Alcohol* **39**:1304–1307.

Kalant, H., 1971, Absorption, diffusion, distribution, and elimination of ethanol: Effects on biological membranes, in: *The Biology of Alcoholism*, (B. Kissin and H. Begleiter, eds.), pp. 1–63, Plenum Press, New York.

Kapur, B. M., and Israel, Y., 1983, A dipstick methodology for rapid determination of alcohol in body fluids (abst.), *Clin. Chem.* **29**:1178.

Kazdin, A. E., 1974, Self-monitoring and behavior change, in: *Self Control: Power to the Person* (N. J. Mahoney and C. E. Thorsen, eds.), pp. 218–246, Brooks-Cole, Monterey, California.

Kennedy, R. W., and Gilbert, G. S., 1978, A self-control program for drinking antecedents: The role of self-monitoring and control orientation, *J. Clin. Psychol.* **34**:238–243.

Lishman, W. A., 1981, Cerebral disorder in alcoholism: Syndromes of impairment, *Brain* **104**:1–20.

Maisto, S. A., Sobell, L. C., and Sobell, M. B., 1979a, Comparison of alcoholics' self reports of drinking behavior with reports of collateral informants, *J. Consult. Clin. Psychol.* **47**:106–112.

Maisto, S. A., Sobell, M. B., Cooper, A. M., and Sobell, L. C., 1979b, Test-retest reliability of retrospective self-reports in three populations of alcohol abusers, *J. Behav. Assess.* **1**:315–326.

Maisto, S. A., Sobell, M. B., and Sobell, L. C., 1982a, Reliability of self-reports of low ethanol consumption by problem drinkers over 18 months of follow-up, *Drug Alcohol Depend.* **9**:273–278.

Maisto, S. A., Sobell, M. B., Cooper, A. M., and Sobell, L. C., 1982b, Comparison of two drinking techniques to obtain retrospective reports of drinking behavior from alcohol abusers, *Addict. Behav.* **7**:33–38.

McColl, K. E. L., Whiting, B., More, M. W., and Goldberg, A., 1979, Correlation of ethanol concentrations in blood and saliva, *Clin. Sci.* **56**:283–286.

McCrady, B. S., Paolino, T. J., and Longabaugh, R., 1978, Correspondence between reports of problem drinkers and spouses on drinking behavior and impairment, *J. Stud. Alcohol* **39**:1252–1257.

Midanik, L., 1982a, The validity of self-reported alcohol consumption and alcohol problems: A literature review, *Br. J. Addict.* **77**:357–382.

Midanik, L., 1982b, Overreports of recent alcohol consumption in a clinical population: A validity study, *Drug Alcohol Depend.* **9**:101–110.

Miller, P. M., 1975, A behavioral intervention program for chronic public drunkenness offenders, *Arch. Gen. Psychiatry* **32**:593–596.

Miller, P. M., Hershen, M., Eisler, R. M., and Watts, J. G., 1974, Contingent reinforcement of lowered blood alcohol levels in an outpatient chronic alcoholic, *Behav. Res. Ther.* **12**:261–263.

Miller, W. R., and Munoz, R. F., 1976, *How to Control your Drinking*, Prentice-Hall, Englewood Cliffs, New Jersey.

Miller, W. R., Crawford, V. L., and Taylor, C. A., 1979, Significant others as corroborative sources for problem drinkers, *Addict. Behav.* **4**:67–70.

Morgan, M. Y., Colman, J. C., and Sherlock, S., 1981, The use of a combination of peripheral markers for diagnosing alcoholism and monitoring for continued abuse, *Br. J. Alcohol Alcoholism* **16**:167–177.

Nelson, R. O., 1977, Assessment and therapeutic functions of self-monitoring, in: *Progress in Behavior Modification*, (M. Hersen, R. M. Eisler, and P. M. Miller, eds.), pp. 264–308, Academic Press, New York.

O'Farrell, T. J., and Connors, G. J., 1982, Obtaining driver's license records for use in evaluating alcoholism treatment, *J. Stud. Alcohol* **43**:1046–1052.

Orrego, H., Blendis, L. M., Blake, J. E., Kapur, B. M., and Israel, Y., 1979, Reliability of assessment of alcohol intake based on personal interviews in a liver clinic, *Lancet,* **2**:1354–1356.

Papoz, L., Warnet, J. M., Pequignot, G., Eschwege, E., Claude, J. I., and Schwartz, D., 1981, Alcohol consumption in a healthy population: Relationship to gamma-glutamyl transferase activity and mean corpuscular volume, *JAMA* **245**:1748–1751.

Pernanen, K., 1974, Validity of survey data on alcohol use, in: *Research Advances in Alcohol and Drug Problems,* vol. I (R. J. Gibbons, Y. Israel, H. Kalant, R. E. Popham, W. Schmidt, R. G. Smart, eds.), pp. 355–374, Wiley, New York.

Phillips, M., 1982, Sweat-patch test for alcohol consumption: Rapid assay with an electrochemical detector, *Alcohol. Clin. Exp. Res.* **6**:532–534.

Phillips, M., and McAloon, M. H., 1980, A sweat-patch test for alcohol consumption: Evaluation in continuous and episodic drinkers, *Alcohol. Clin. Exp. Res.* **4**:391–395.

Polich, M. J., 1981, Epidemiology of alcohol abuse in military and civilian populations, *Am. J. Public Health* **71**:1125–1132.

Polich, M. J., 1982, The validity of self-reports in alcoholism research, *Addict. Behav.* **7**:123–132.

Pomerleau, O., and Adkins, D., 1980, Evaluating behavioral and traditional treatment for problem drinkers, in: Evaluating Alcohol and Drug Abuse Treatment Effectiveness (L. C. Sobell, M. B. Sobell, and E. Ward, eds.), pp. 93–128, Pergamon Press, New York.

Popham, R. E., and Schmidt, W., 1981, Words and deeds: The validity of the self-report data on alcohol consumption, *J. Stud. Alcohol* **42**:355–368.

Rankin, J. G., 1975, *Alcohol, Drugs and Brain Damage,* Addiction Research Foundation, Toronto.

Rohan, W. P., 1976, Quantitative dimensions of alcohol use for hospitalized problem drinkers, *Dis. Nerv. Syst.* **37**:154–159.

Ryback, R. S., Eckardt, M. J.; Felsher, B., and Rawlings, R. R., 1982*a*, Biochemical and hematologic correlates of alcoholism and liver disease, *JAMA* **248**:2261–2265.

Ryback, R. S., Eckardt, M. J., Rawlings, R. R., and Rosenthal, L. S., 1982*b*, Quadratic discriminant analysis as an aid to interpretive reporting of clinical laboratory tests, *JAMA* **248**:2342–2345.

Sanchez-Craig, M., and Annis, H. M., 1981, Gamma-glutamyl transpeptidase and high density lipoproteins cholesterol in male problem drinkers: Advantages of a composite index for predicting alcohol consumption, *Alcohol. Clin. Exp. Res.* **5**:540–544.

Sanchez-Craig, M., and Annis, H. M., 1982, Self-monitoring and recall measures of alcohol consumption: Convergent validity with biochemical indices of liver function, *Br. J. Alcohol Alcoholism* **17**:117–121.

Sellers, E. M., and Kalant, H., 1976, Alcohol intoxication and withdrawal, *N. Engl. J. Med.* **294**:757–762.

Skinner, H. A., 1981, Assessment of alcohol problems: basic principles, critical issues and future trends, in: *Research Advances in Alcohol and Drug Problems,* Vol. 6, (Y. Israel, F. B. Glaser, H. Kalant, R. E. Popham, W. Schmidt, and R. G. Smart, eds.), pp. 319–369, Plenum Press, New York.

Skinner, H. A., 1982, Alcohol abuse: Strategy for early detection and brief intervention, workshop sponsored by the Association for Medical Education and Research in Substance Abuse and by the World Health Organization, November 15–16, 1982, Berkeley, California.

Skinner, H. A., and Allen, B. A., 1982, Alcohol dependence syndrome: measurement and validation, *J. Abnorm. Psychol.* **91**:199–209.

Skinner, H. A., and Allen, B. A., 1983, Does the computer make a difference? Computerized versus face-to-face versus self-report assessment of alcohol, drug and tobacco use, *J. Consult, Clin. Psychol.* **51**:267–275.

Skinner, H. A., 1983, Screening for alcohol dependence with drinking-driver offenders, unpublished manuscript.

Skinner, H. A., and Horn, J. L., 1984, Guidelines for Using the Alcohol Dependence Scale (ADS), Addiction Research Foundation, Toronto.

Skinner, H. A., and Sheu, W. J., 1982, Reliability of alcohol use indices: Lifetime drinking history and MAST, *J. Stud. Alcohol* **42:**1157–1170.

Sobell, L. C., and Sobell, M. B., 1973, A self-feedback technique to monitoring drinking behavior in alcoholics, *Behav. Res. Ther.* **11:**237–238.

Sobell, L. C., and Sobell, M. B., 1975, Outpatient alcoholics give valid self-reports, *J. Nerv. Ment. Dis.* **161:**32–42.

Sobell, L. C., and Sobell, M. B., 1978, Validity of self-reports in three populations of alcoholics, *J. Consult. Clin. Psychol.* **46:**901–907.

Sobell, L. C., and Sobell, M. B., 1980, Convergent validity: An approach to increasing confidence in treatment outcome conclusions with alcohol and drug abusers, in *Evaluating Alcohol and Drug Abuse Treatment Effectiveness*, (L. C. Sobell, M. B. Sobell, and E. Ward, eds.), pp. 177–183, Pergamon Press, New York.

Sobell, L. C., and Sobell, M. B., 1981, Effects of three interview factors on the validity of alcohol abusers' self-reports, *Am. J. Drug Alcohol Abuse* **8:**225–237.

Sobell, M. B., Sobell, L. C., and Vanderspek, R., 1979a, Relationships among clinical judgment, self-report, and breath analysis measures of intoxication in alcoholics, *J. Consult. Clin. Psychol.* **47:**204–206.

Sobell, L. C., Maisto, S. A., Sobell, M. B., and Cooper, A. M., 1979b, Reliability of alcohol abuser's self-reports of drinking behavior, *Behav. Res. Therapy* **17:**157–160.

Sobell, L. C., Cellucci, T., Nirenberg, T. D., and Sobell, M. B., 1982, Do quantity–frequency data underestimate drinking-related health risks? *Am. J. Public Health* **72:**823–828.

Stockwell, T., Murphy, D., and Hodgson, R., 1983, The severity of alcohol dependence questionnaire: Its use, reliability and validity, *Br. J. Addict.* **78:**145–155.

Uchalik, D. C., 1979, A comparison of questionnaire and self-monitored reports of alcohol intake in a nonalcoholic population, *Addict. Behav.* **4:**409–413.

Wechsler, D., 1955, *Manual for the Wechsler Adult Intelligence Scale*, Psychological Corporation, New York.

Wiggins, J. S., 1973, "Personality and Prediction: Principles of Personality Assessment," Addison-Wesley, Reading, Massachusetts.

Wilkinson, D. A., and Carlen, P. L., 1981, Chronic organic brain syndromes associated with alcoholism, in: *Research Advances in Alcohol and Drug Problems*, Vol. 6 (Y. Israel, F. B. Glaser, H. Kalant, R. E. Popham, W. Schmidt, and R. G. Smart, eds.), pp. 107–145, Plenum Press, New York.

Wilkinson, D. A., and Sanchez-Craig, M., 1981, Relevance of cognitive dysfunction to treatment objectives: Should alcohol-related cognitive deficits influence the way we think about treatment, *Addict. Behav.* **6:**253–260.

7

An Overview of Techniques and Problems in the Measurement of Alcohol Consumption

TIMO ALANKO

1. INTRODUCTION

This chapter presents an overview of the measurement of alcohol consumption in general population surveys. Throughout the literature on alcohol, various aspects of the subject have been treated, most often in connection with reports on empirical survey studies. Specific studies include those of Mäkelä (1971), Pernanen (1974), and Duffy (1982) among others. In recent years discussion has concentrated on the validity of survey results, especially on the coverage problem as discussed below.

Our approach is somewhat different from earlier attempts. We aim at a critical evaluation of the state of the art by emphasizing that assumptions concerning the actual drinking process of particular individuals should be made as explicit as possible and that various existing measurement techniques should be judged against this background. To do this, we start by reviewing what is known (or more-or-less guessed) about the drinking processes of individuals and how the processes can be described statistically. This review should also reveal the extremely complicated structure of the concept of alcohol consumption. We point out several factors that may distort the results obtained from sample surveys. We review various measurement techniques applied in consumption studies, devoting special attention to a technique that appears to have fewer shortcomings than others: the mapping out of actual drinking occasions within a fixed survey period. Finally, problems and questions requiring further research are discussed.

TIMO ALANKO ● Finnish Foundation for Alcohol Studies, Helsinki, Finland.

The scope of this chapter is restricted to quantitative alcohol consumption expressed in units of absolute alcohol; questions related to types of beverages, determination of their alcohol content, problems with official consumption statistics, etc., are not treated. The examples are based on primary data from Finnish consumption studies. For ease of exposition, formal statistical derivations are omitted. For a reader interested in technical details, these are available from the author on request.

2. THE DRINKING PROCESS

Even when merely quantitative aspects are of concern, it is evident that drinking is a highly complex and varying phenomenon, essentially because it involves the behavior of an individual in time. Furthermore, although there are numerous cross-sectional empirical studies on drinking, the lack of longitudinal studies is notorious. In fact, to our knowledge, the only available longitudinal study in which drinking by individuals over time is covered in detail is the drinking rhythm study conducted at the Finnish Foundation for Alcohol Studies in 1963 by Anders Ekholm. In the discussion that follows we rely heavily on this study despite some of its shortcomings. In the study, the drinking occasions, amounts consumed, and various other aspects were recorded by repeated interviews for a (nonrepresentative) group ($N = 94$) of Finnish men over a 1-year survey period (Ekholm, 1968; Pöysä and Niiranen, 1966). The two basic aspects of drinking that concern us here are the occurrence of drinking in time, the drinking frequency, and the amounts of alcohol consumed on the drinking occasions. The interplay between these elements can be illustrated by what is called a rhythmogram of drinking, an example of which is presented in Figure 1. As can be seen from the rhythmogram of this particular respondent, both the amounts

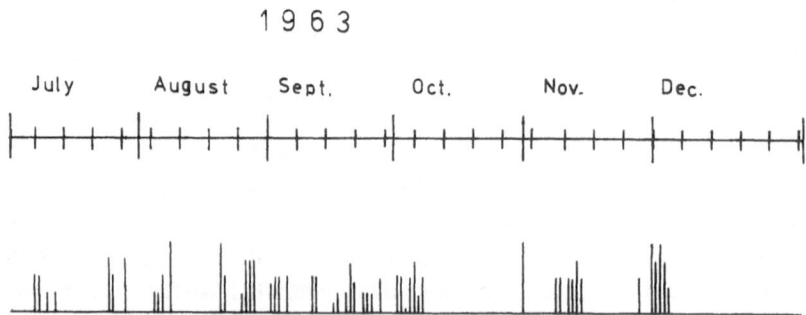

Figure 1. Drinking rhythmogram of a respondent in the 1963 Finnish drinking rhythm study. (From Pöysä and Niiranen, 1966). On the lower axis of the figure each drinking occasion is denoted by a vertical line, the length of which expresses the amount (in absolute alcohol) the respondent consumed. The upper axis indicates the months and weeks in the latter half of year 1963.

consumed and the occurrence of drinking show considerable irregularity. The drinking occasions follow each other neither with regular intervals nor with any determinate pattern, and the same applies to amounts consumed. However, drinking appears to be concentrated on weekends although, for this individual, periods of abstinence break even this pattern. There are certainly individuals whose drinking behavior is more pronouncedly regular than in our example and, in general, the regularity of drinking varies greatly in almost any population. To proceed, we next try to describe different drinking styles with respect to their regularity using a loose typology. The purpose of the typology is not to classify the *respondents* to a given category but to give a description of the different idealized types of *drinking behavior* expected to be found among a sample of respondents.

A Typology of Drinking Behavior

The following categories of drinking rhythm will be used in this chapter:

1. Abstaining

2a. Regular drinking with constant intervals between occasions

2b. Regular drinking with a periodic pattern governing the distances between occasions (for example, drinking each Friday and Saturday)

3a. Random drinking (in a statistical sense, specified below)

3b. Random drinking with a periodic pattern (see below)

4. Binge or spree drinking: periods of abstinence alternate (randomly or according to a pattern) with any of the types above

This crude typology is meant to be comprehensive enough for evaluation of various measurement techniques to be discussed later. For a given individual, any of the categories can be thought of as only an idealization of his or her true drinking pattern. Also, several of the categories may be present simultaneously. Similarly, in a given population, all of the types of drinking can be expected to be present but in differing proportions, reflecting the drinking culture in an interesting manner.

For the *amounts* consumed, a slightly less complicated characterization is necessary. In abstaining the amounts are of course zero. There may be drinking styles in which consumption is in constant quantities, but as a general rule the amounts consumed on different occasions may be expected to show considerable variation. A crude description of this variation is given below.

Statistical Description of the Drinking Process

As the drinking behavior of individuals is evidently characterized by many random features, an adequate description of drinking behavior requires probabilistic or statistical formulations. These will be discussed first for drinking frequency.

Statistical Formulations for Drinking Frequency.   The random occurrence of particular drinking occasions for an individual in time can be described as a stochastic point process, each occasion taken as a point on the time axis. Analyzing the data of the 1963 drinking rhythm study, Ekholm (1968) showed that for a majority of respondents the drinking process could be adequately described by a fairly simple mathematical model, the Poisson process. In particular he states that: "The structure of the drinking day process of c.50% of our interviewees is thus adequately characterized as a non-homogeneous or a simple Poisson process. Of the rest, the majority do not deviate seriously and we would classify only about 15% of the series studies as definite deviations from the Poisson hypothesis." Typical deviations from the Poisson hypothesis were drinking in sprees corresponding to our cateogry 4, long-term trends (see below) and hangover effects which, according to Ekholm "were usually not very pronounced, though significant." Within the Poisson hypothesis, the nonhomogeneity, which means that probability to drink varies according to some rule, was due to a periodicity factor: the concentration of drinking occasions to weekends. Thus we can assume for the moment that our category 3a can be described by a homogeneous, and our category 3b by a nonhomogeneous, Poisson process. The delightful thing about this result is that the Poisson process is characterized by a single parameter: the rate of the process. For the nonhomogeneous case, the rate is dependent on time, which complicates the matter. However, since we can assume that nonhomogeneity is in most cases due to a periodic factor, the weekly drinking pattern, it is possible even in this case to develop suitable estimation procedures for the drinking frequency.

A property of the Poisson process that is useful in the measurement of drinking frequency is that the number of drinking occasions falling in a fixed time period is Poisson-distributed. The mean of this distribution is, in our application, just the mean drinking frequency of the individual measured in time units equal to the fixed time period. In the nonhomogeneous case it can be shown that the same property holds on the condition that the time period is fixed to be equal to the periodic pattern of the drinking rate; in our case typically to 1 week.

Statistical Properties of the Amounts Consumed.   The theory of the amounts consumed on particular occasions is less well formulated than that of the drinking rhythm. On the basis of recent work by the author, statistical description of the variability of the amounts in the drinking rhythm study can be resumed as follows. For all the respondents, the empirical distribution of the amounts consumed on different occasions appears to arise from a positively skewed distribution reminiscent of the well-known lognormal distribution and consequently can satisfactorily be described by two parameters. At present, no definite theoretical model for this variability is offered although Alanko (1982) suggested that the inverse Gaussian distribution (see Folks and Chhikara, 1978) would be a convenient tool for analyzing the amounts. No analysis of how the amounts consumed might be dependent on the drinking frequency is at present available.

Measurement as an Inference Problem

The usual goal of measurement in drinking surveys is to determine the drinking frequency, the amounts consumed per occasion and/or composite measure, the total alcohol consumption of an individual per time unit. As we have seen, usually none of these objects of measurement are fixed but show considerable random variation in time and thus the inference concerns unknown parameters governing the drinking process. In typical surveys the data available for inference are based either on the personal judgment of the respondent or on recording the actual drinking of the respondent during a short period of time (typically 1 week). Given the random nature of drinking, the inference problem concerns estimation of the parameters governing the drinking process on the basis of a very limited number of observations. In practical terms, the parameters to be estimated can be taken to describe drinking only in a limited section of a person's life history: a typical inference problem is to estimate the annual consumption of the respondent. As the measurement of alcohol consumption is essentially a statistical inference problem, the measurement techniques developed in the alcohol literature can and should be evaluated using statistical criteria such as unbiasedness and efficiency (see below). Before that, we must turn to another important factor that causes problems in the measurement of alcohol consumption, namely the respondent.

Response Problems

The drinking process fluctuates randomly in time and the very nature of drinking causes complications in the definition of the object of measurement. This analysis was based on the drinking rhythm study data in which response error, according to Pöysä and Niiranen (1966), was very insignificant. In large surveys one must also question the ability and will of the respondents to provide truthful and reliable answers to the questions put to them in a questionnaire or interview. Basically, two sources of misinformation are suspected: that the respondents forget a part of the drinking occasions and/or the quantities consumed on recorded occasions, and that they conceal a portion of their drinking. Evidence about response error is mainly indirect but very convincing (see below). Evidence for forgetting as a function of time has been directly analyzed by Pernanen (1974) and Mäkelä (1971), for example. What makes concealing and forgetting particularly troublesome is that there is no feasible way of obtaining independent information about true drinking behavior so as to locate the respondents with response errors, or to model more generally the forgetting/concealment mechanism. A proposal to use modern "breathalyzer" techniques (Wechsler, 1981) in surveys is also surrounded by enormous and costly practical difficulties. It therefore seems survey research has to live with response error for at least some time to come.

3. EVALUATION OF EXISTING MEASUREMENT TECHNIQUES

In this section, we present an overview of the main methods of measurement that have been used in drinking surveys, and discuss their statistical and other properties against the background developed in previous sections. The first main variant of alcohol consumption measures includes questions mapping out customary drinking habits of the respondents, of the type "How often do you usually . . ." or "How many drinks do you usually . . ." The other main variant is based on the actual recent consumption history of the respondent. We will first discuss the customary drinking habit variant.

Measurement of Customary Drinking Habits

Measurement of customary drinking habits has the general feature that the respondent is required to evaluate his or her own drinking pattern. The major use of this approach has been in connection with indexes of drinking patterns, typical in Anglo-Saxon survey tradition, and it is convenient to review these indexes first.

We will describe the classic quantity–frequency index (Q–F) and its later refinements, and then discuss elements of these indexes relevant to the measurement problem.

Quantity–Frequency Index.   The Q–F index was introduced by Straus and Bacon (1953) and is an attempt to provide a unidimensional classification of the existing drinking patterns of the population. Respondents are typically asked about their customary drinking frequency of three types of beverages—beer, wine, and spirits—and about the usual quantity of beverage consumed on a drinking occasion. The classifications employed for the frequency of drinking and the usual quantities consumed are fairly coarse. The Q–F index is formed by combining the essentially two-dimensional quantity–frequency classes into a unidimensional scale ranging roughly from heavy drinkers to abstainers.

As the aim of this index is to provide a taxonomy of existing drinking patterns, it is unjustified to evaluate it merely as a measurement instrument for quantitative alcohol consumption. However, because many studies whose aim is not merely taxonomic also inquire about customary drinking habits, it is worthwhile to study the general approach also from the measurement point of view which is done below.

Quantity–Frequency–Variability Index.   In 1969, Cahalan and co-workers used a modification of the Q–F index, the quantity–frequency–variability index, to provide also for variability of drinking habits in the classification. For each beverage type the proportion of occasions was determined when five or more, three to four, or one to two units of alcohol (defined suitably for different beverages) were consumed. The proportions were defined on a four-point scale from "nearly every time" to "never." Further questions concerned the frequency of drinking of any alcoholic beverage (lacking in Q–F). On the basis of the answers to these questions a complicated classification of individuals was constructed.

Volume–Variability Index.   A measure of the total consumption of alcohol per time unit can be attempted on the basis of customary drinking habits indexes. Such a measure is included in the volume–variability index by Cahalan and Cisin (1968). In the index, one of the classification criteria is the average daily volume, estimated by multiplying the customary frequency of consumption (evaluated by the respondents as number of drinking occasions for each beverage per thirty days) by the customary amount consumed for each beverage. Thus, on the basis of essentially Q–F questions (although not on the basis of Q–F categories), it is possible to construct an essentially quantitative measure of total consumption. All the critical remarks noted above apply to this composite measure and, furthermore, the possible biases noted above have multiplicative effects in the total consumption measure.

Evaluation of the Customary Drinking Habits Approach.   The ability of individuals to evaluate their consumption, expressed in fairly abstract terms such as frequency, can be questioned on psychological grounds. This point has recently been strongly emphasized by Gregson and Stacey (1982). On the other hand, as respondents are able to evaluate a longer period of time than possible in surveys, they should also be able to eliminate the effects of lapses of memory and temporary fluctuations in their drinking rhythm from the estimates, at least in principle.

A more technical question has recently been pointed out by Duffy (1982). The problem is with the word "usual" or "customary" employed in Q–F. Duffy suggests that most respondents tend to interpret "usual" as the modal value whereas in drinking surveys the mean or average values of quantities and frequencies are really of interest. As we have pointed out above, it can be assumed that both the distribution of quantities consumed on single separate occasions and the distribution of frequencies (expressed as drinking occasions per time unit) over time (over separate time periods of equal length) are positively skewed for the majority of consumers. Even if the respondents were aware of the problem, the algebraic task of calculating the average value of a skew distribution in their minds would be difficult. Thus we strongly agree with Duffy. The conclusion is that, because the mode of a positively skewed distribution is generally lower than the mean, an underestimation of both the average drinking frequency and the average amount consumed can be expected when questions related to customary drinking habits are used. The importance of this bias depends naturally on the particular drinking pattern of the individual and is difficult to separate from other sources of error, notably deliberate or unconscious concealment of consumption.

The Q–F categories, formed by combining frequency and quantity classes into a single category, do not provide measures of total consumption and attempts to discuss, for example, distributions of consumption by means of the Q–F index are surrounded by difficulties.

All the critical remarks noted above apply also to the measure of total consumption included in v–v and, furthermore, the possible biases in frequency and amounts estimates have multiplicative effects for this measure.

Measurement Based on Actual Consumption

In contrast to questions in which respondents are required to calculate the parameters of customary drinking habits themselves, we are in a better position to evaluate measures based on recent consumption history. Whereas with the customary drinking habits approach the calculations of the estimates are left to the shadows of respondents' minds, estimation is done in the actual consumption approach by the researcher on the basis of information concerning a number of fairly recent drinking occasions. The psychological advantages of this approach are evident and a more detailed picture about sociologically interesting aspects of drinking occasions and drinking contexts becomes available (Bruun, 1972). Furthermore, the actual consumption approach can be legitimately judged as a measure of quantitative alcohol consumption.

There are two basic variants of this approach. Each subject may be asked about a specified number of drinking occasions (preceding the interview or questionnaire): the "when last" approach. Alternatively, an effort may be made to obtain information about all the drinking occasions within a specified time period defined as the survey period.

The "When Last" Approach.   The first of these approaches was used in a number of Finnish studies in the 1950s and 1960s (e.g., Kuusi, 1957; Sariola, 1954; Bruun and Hauge, 1963) and has recently been revived in a New Zealand study (Gregson and Stacey, 1980).

The usual procedure has been to obtain data about one or two most recent drinking occasions, particularly about the quantities of alcohol consumed and the occurrence time of these occasions. Then an estimate of drinking frequency has been based on the interval between the two occasions or on the interval between the interview and the second occasion (or the first if only one occasion has been recorded). The quantities consumed on the occasions have been taken as estimates of intake per occasion. An estimate of total consumption per time unit is formed by multiplying the two estimates. In the example of the New Zealand study, t_1 can be taken to denote the time of the last, and t_2 the time of the next-to-last drinking occasion. The amount consumed on the last occasion is denoted by q_1. Gregson and Stacey report $1/(t_1 - t_2)$ as the estimate of drinking frequency and q_1 as the estimate of intake per occasion. The total consumption estimate is thus $q_1/(t_1 - t_2)$. For alternative but theoretically similar formulae see Mäkelä (1971).

Evaluation of the "When Last" Approach.   It is possible to evaluate this measurement technique by referring to our categories of regularity of drinking rhythm. For the formula $1/(t_1 - t_2)$ it is possible to show that only for our category 2a, perfectly regular drinking, is the estimate unbiased. Any periodicity, even if of determinate type 2b, causes an obvious bias. Skog (1981) showed that the estimate is biased for our category 3a by using a discrete time version of the Poisson process as an example, and by giving numerical examples showing how the size of the bias varies with the true drinking frequency. The situation is aggravated even more for our category 3b with a periodic pattern in the nonhomogeneous stochastic process, not to mention spree drinking of category 4. For all these categories it is easy to prove the bias formally by fairly elementary mathematical methods but the proofs are omitted here.

The amount consumed on a single occasion provides, theoretically, an unbiased estimate of the mean amount, but an estimate based on a single or at best two occasions tends to have very high sample variance. Naturally the sampling variance of the estimate depends directly on the distribution of the amounts of the respondent: in the extreme case where the respondent always consumes the same amount the sampling variance reduces to zero. As indicated by the drinking rhythm study, this is a rare case.

The estimate of total consumption per time unit formed by multiplication suffers naturally from the bias in the estimate of drinking frequency. This measure may furthermore suffer from an association between the drinking frequency and the amounts consumed. For example, some respondents may consume larger amounts when their drinking frequency temporarily increases. At present, practically no empirical evidence is available for assessing such associations, with the exception described later. To make a bad case worse, all sorts of response errors, from lapses of memory concerning the time of drinking occasions to conscious concealing, may have considerable additional effects on the estimates described in this section. Although some of these effects could be expected to counteract the biases in the drinking frequency estimates, this would be a very small consolation.

In conclusion, this approach to measurement of alcohol consumption is extremely unsatisfactory on theoretical grounds and is expected systematically to give too high estimates of drinking frequencies and total consumption.

The Survey Period Approach.   Several studies base the measurement of consumption on the drinking occasions falling within a period of specified length, typically 1 week. This approach has several advantages to the "when last" approach, especially for the measurement of drinking frequency. In the survey period approach the number of drinking occasions occurring in the period is taken as the estimate of drinking frequency. An estimate of the amount consumed per occasion is obtained using the average amount over the recorded occasions. Finally, the measure of total consumption per time unit (the length of the survey period) is simply the sum of the amounts consumed on separate occasions.

Evaluation of the Survey Period Approach. Consider first the estimate of drinking frequency. For our category 2a, regular drinking, it can be shown that if the survey period is chosen randomly, the number of occasions falling in the period is an unbiased estimate of the drinking frequency. For this category, the number of occasions estimate is only unbiased in the statistical sense but has sample variation related to sampling the survey period from the time axis. We saw that with the "when last" approach with two recorded occasions, the frequency could be determined exactly from the interval between occasions and, consequently, the survey period approach is inferior for the rare category 2a. For category 2b with a deterministic periodic pattern of drinking occasions, the number of recorded occasions provides an unbiased estimate. On the condition that the length of the survey period is the same or a multiple of the periodicity in the drinking rhythm, the frequency is determined exactly both in categories 2a and 2b. Because for most respondents the basic length of periodicity is 1 week, the number of occasions estimated with a 1-week survey period works very well, at least when compared with the "when last" approach. For categories 3a and 3b, it is possible to show the unbiased character of the number of occasions estimate using the assumptions about the drinking process described previously. (The Poisson assumption for the number of occasions holds true in category 3b only on the condition used above for category 2b.) Assumptions needed to make the result applicable also for category 4, spree drinking, are too restrictive to be of any practical value. Special techniques should be employed to measure drinking frequency in this drinking category which is often thought to be typical of alcoholics (see Ahlström-Laakso, 1975). In general surveys it is, of course, difficult to identify this drinking style, and this may be thought to be one of the reasons for the coverage problem discussed below.

The variances of the number of occasions estimators, expressing the magnitude of random measurement error due to time-variation of drinking, and reflecting the capability to predict the true drinking frequency of an individual, can be formally derived using a number of assumptions but are omitted here. In general these variances are relatively high: for example, in categories 3a and 3b, one could describe the situation as an attempt to estimate the mean of the number of occasions distribution (here assumed to be Poisson) on the basis of a *single* observation; the number of occasions in a single period.

The problem is accentuated for infrequent drinkers who very often have no occasions in the survey period and thus rate zero as the estimate of frequency.

To overcome this difficulty, a modification to the survey period approach was introduced in the 1968/1969 Finnish survey, described by Mäkelä (1971). The modification was that the length of the survey period was determined on the basis of self-reported drinking frequency, measured by a customary drinking frequency question. Thus the survey period varied from 1 week for those who reported drinking every day to 1 year for the most infrequent drinkers. This technique ensured that on the average at least four drinking occasions could be covered.

The unbiased result rests also on the assumption that drinking remains stable in the long term: in practice, often the annual drinking is estimated by 1 week's drinking so that the stability requirement is extended only to 1 year. This means, for example, that the seasonal variation of drinking should remain in the limits allowed by the random models of categories 3a and 3b. In survey design, this is often done by avoiding untypical seasons such as summer holidays, when the survey is made. However, the results of Ekholm (1968) showed that for some respondents the stability requirement may not hold.

The mean amount consumed per occasion is estimated by the average of the amounts consumed on separate survey period occasions. The variance of this unbiased estimate depends naturally on the number of occasions available and on the variation in the amounts drunk by the respondent. The modification described above is helpful also in this case.

An unbiased estimate of total consumption per time unit is obtained by summing the amounts consumed on the survey period occasions. It is theoretically possible that an association between realized drinking frequency (the particular number of drinking occasions in the survey period) and the amounts consumed on the drinking period occasions may cause additional variance in the estimate. The author has recently reanalyzed the total consumption measures based on 1 week's drinking from the 1963 drinking rhythm study by considering the distribution of this measure over the 52 weeks of the year for each respondent. The general conclusion is that this longitudinal distribution of the total consumption estimate has the same characteristic form and positive skewness as many distributions familiar from the cross-sectional level. This means that in general the variance of the total consumption estimate is fairly high and increases with consumption level.

The general concern for response error is as serious for the survey period approach as for every other measurement approach. Especially with long survey periods, slips of memory and concealment in recalling drinking occasions, amounts consumed, etc., are very likely to occur but are extremely difficult to detect or analyze.

Measures other than those related to average consumption are easily formed in the actual consumption approach: these may include the proportion of drinking occasions leading to intoxication or maximum intake on an occasion within a specified period (Mäkelä, 1971; Kreitman and Duffy, 1982).

General Conclusions

Two basic types of measurement have been treated: the techniques based on mapping out customary drinking habits and those based on recording actual recent consumption. The indexes formed by using customary drinking approaches do not directly aim to provide measures of quantitative alcohol consumption of individuals. Attempts to form such measures on the basis of questions about customary amounts and frequencies have serious pitfalls. The "when last" type of approach about actual consumption was seen to lead to seriously biased drinking frequencies and, consequently, also to biased measures of total consumption for almost all drinker groups. The survey period approach gives unbiased estimates for most existing drinking styles and is thus the one to be preferred in actual survey work.

All the existing techniques suffer from response error which, as discussed in the next section, has serious effects on the reliability of survey results.

In many studies the customary drinking habits approach and the actual drinking approach have been used side by side. An example of this was the 1968/1969 Finnish survey. An interesting theoretical possibility would be to use both types of measures simultaneously to estimate the reliability of the measurement variables, much as large test batteries are used in psychometric studies or multiple indicators in sociology. It seems, however, that the assumption of independent measurements cannot be met in alcohol consumption studies because of concealment. These considerations lead us to consider the measurements obtained as sample or population level variables.

4.   ALCOHOL CONSUMPTION MEASUREMENTS AS VARIABLES

A common procedure in survey variables is to treat the measures obtained from individual respondents as observations of a sociologic variable. These variables often have interest in themselves. For instance, in the discussion about distribution of alcohol consumption in a population, their parameters, such as the mean and measures of dispersion, can be used to describe and compare different populations or subgroups within a population. Often, the variables are used as explanatory variables for consequences of drinking, for example. Furthermore, the average of the total consumption variable provides the only widely used validity check for a survey as this statistic can be compared with official sales records or other corresponding sources.

Formally, three types of variables are obtained using the measurement techniques in Section 3. The Q–F and related indexes are categorical variables and are usually treated accordingly with the methods appropriate for frequency data. A special case is the dichotomous variable abstainer/consumer. The drinking frequency variable can also be a categorical variable if only a crude classification is used. Depending on the measurement technique, it can be a discrete or a continuous one. In the actual consumption approach, the variables related to amounts consumed or total consumption are usually treated as continuous ones. These distinctions are of interest when the measurements are related to other variables, for instance to social consequences of drinking or when drinking in substrata of a population is compared. We will not digress into the general methodology of drinking variables in empirical research, but will give a brief account of the properties of the frequency and total consumption variables obtained by using the survey period approach.

The variability in an alcohol consumption variable is generally taken to express differences between drinking habits of different respondents. We have, however, seen that each observation already involves variability caused by time variation of the drinking process and probable response errors, and thus all variability cannot be taken to express individual differences. At the moment, it is not known how to evaluate and separate the contributions of these factors; response errors are particularly problematic. Using distributional statistical models, the author has recently tried to analyze the effects of time-variation on the distribution of drinking frequency and total consumption variables under the assumption of no response error in a survey period approach to measurement. Already this part of the problem involves complicated statistical modelling (Alanko, 1982). We will review some of the findings having relevance for the use of these variables and the measurement problem.

The Drinking Frequency Variable

Let us assume that the number of occasions per week is Poisson-distributed over the weeks for each respondent, but that parameters (expressing mean long-term frequency) vary from respondent to respondent as indicated by the drinking rhythm study. How, then, is the across-the-population frequency variable distributed? In addition to individual variance expressed by the Poisson assumption, one must allow for interindividual differences. Typically, it is assumed that the distribution of individual mean frequencies can be expressed by a theoretical distribution such as the lognormal, gamma, or (as in the recent work by the author) inverse Gaussian distribution. Then the distribution of drinking frequency in a population can be expressed by a theoretical distribution of mixed Poisson type. The best known of these distributions is the negative binomial distribution, often employed in, for example, purchase behavior studies. (For some empirical examples from alcohol studies, see Alanko, 1981, 1982.) What is the practical use of these theoretical developments? First, inference concerning the parameters (e.g., mean drinking frequency of a population) of the distribution can be improved by using properties of the theoretical distribution, for instance, in comparison between subgroups of population. Secondly, a theoretical basis is given to estimate the reliability of the frequency variable: the concept of reliability of a measurement frequently used in psychometric studies is defined as the square of the correlation between the "true" frequency and the measured frequency. In practice, the reliability of the frequency measure obtainable from the theoretical model overestimates the true reliability in a survey context because no allowance for response errors is made. Thirdly, the theoretical models can be used to "improve" individual frequency estimates by considering also the obtained population mean frequency (for technical details see Morrison and Schmittlein, 1981).

The Total Consumption Variable

The distribution and various other properties of this variable have been at the center of a long and persistent debate in alcohol literature. As indicated above, the total consumption measure is a complicated composite measure involving an estimate of both drinking frequency and the amounts consumed. A general finding is that the distribution of the total consumption variable in almost any population is positively skewed and tends to have a very long tail. Furthermore, a theoretical model, the lognormal distribution, has been found to give a fairly good fit to many sets of empirical data. Many pages in the alcohol literature have been devoted to the question of whether or not alcohol consumption truly is lognormally distributed and what conclusions this finding implies for alcohol policy. The debate about lognormality is not particularly relevant here because the measures of total consumption have been taken as given in the discussion and no attention has been paid to the various random and other components involved in the composition of this measure. Alanko (1982) presents an attempt to analyze the structure and distribution of this variable, and the attempt shows clearly how complicated the stochastic structure of the variable is. Despite the difficulties, the three advantages of the theoretical model for drinking frequencies noted above can be extended theoretically also to the total consumption variable, but at present only on the assumption of no response errors.

Dependency between Drinking Frequency and the Amounts Consumed

Alanko (1982) reported an association between drinking frequency and the average amount per occasion. Briefly stated, the average intake per occasion increases linearly with the number of drinking occasions per week on the basis of data from a 1976 Finnish survey study (reported in Simpura, 1981). The finding was formally ascertained by using a generalized least-squares model accounting for the heteroscedasticity in the calculation of the average amounts. The finding has two possible explanations. One is that in the individual level drinking process, a temporary increase of drinking frequency brings about also larger intakes per occasion or vice versa. The other explanation is that frequent drinkers generally consume larger amounts per occasion. The first possibility is more serious from the measurement point of view because this sort of association decreases the reliability of the total consumption measure. It would be hasty to generalize on the basis of a single survey; the association may be present in only this study or may reflect only the peculiarities of Finnish drinking habits. A more detailed account will be available in a forthcoming paper by the author.

Mean of the Total Consumption Variable and the Coverage Problem

A rather more concrete problem of the total consumption variable is related to the single parameter on which independent information is available: the mean consumption. It is well-known that the average (per capita) consumption obtained from sample surveys consistently deviates downwards from the per capita consumption obtained from other sources, such as sales records or other aggregate statistics. The ratio between the survey estimate of mean per capita consumption and the known per capita consumption is often called the coverage of the survey and tends to vary between 40 and 60% if the survey period approach is used. Better coverages have been reported in several studies but often either biased measurement techniques have been used (Gregson and Stacey, 1981) and/or populations with exceptionally regular drinking habits have been studied (Ledermann's early studies in rural France). Generally, the low coverages have been taken as serious indicators of the lack of validity of consumption surveys. Basically, two lines of explanation for the coverages have been proposed. Among others, Wilson (1981) and Mulford and Fitzgerald (1981) have suggested that low coverage reflects sample frame defects; in particular, selective nonresponse of alcoholics. On the other hand, Mäkelä (1971) showed that in a study with a very high response rate, nonresponse could maximally explain 6.2% of the discrepancy and concluded that deliberate underreporting or concealment of consumption was the main reason for the low coverage. The same conclusion was reached by Popham and Schmidt (1981a). The arguments by Wilson and by Mulford and Fitzgerald have been questioned by Duffy (1982) and Popham and Schmidt (1981b). Our conclusion is that the evidence of low coverage points strongly to the presence of individual level response errors.

The low coverage percentages indicate only the magnitude of the problem but tell nothing about how response errors are distributed among respondents, whether and how they are associated with consumption levels, drinking patterns, particular drinking occasions, heavy drinking (cf. Popham and Schmidt, 1981). A tacit assumption in many studies seems to be that response errors affect all respondents in the same way, typically by reducing the total consumption estimate by the same proportion. The assumption is certainly convenient because it preserves rank-order validity, but not particularly plausible although almost no empirical evidence exists. The fair degree of stability in the coverage percentages in different studies is a consolation for this assumption. One could replace this assumption by the weaker assumption that response errors depend on a large number of psychosocial characteristics of respondents which are not strongly associated with alcohol consumption over the sample and are in practice random. This would save, at least in theory, the usefulness of sample surveys as regards sample level characteristics, for example, in monitoring changes in drinking habits by repeated surveys or in comparisons between populations and subgroups within a population. As neither of these assumptions is particularly plausible, serious doubt always remains when the individual level measurements are used directly in analyses or when inference is based on subgroups with a limited number of respondents.

The conclusion is that the coverage problem calls for further research. To gain even approximate knowledge about the mechanisms behind response error, a large number of costly and difficult empirical studies are needed, covering such aspects as forgetting, concealment, interview techniques, etc. At present, even preliminary suggestions concerning the design of such studies seem overwhelmingly difficult to give.

The magnitude of the coverage discrepancy points strongly to the conclusion that response errors are the primary source of error in the measurement of alcohol consumption. That in turn means that unless response error problems can be managed to some extent, the other problems treated in this paper are of secondary practical interest. This is, of course, no excuse for neglecting the known sources of bias in measurement techniques nor should it discourage future research that is not directly related to response error.

5. CONCLUSION

So many defects, sources of bias, and unreliability are seen to surround measurement of alcohol consumption in surveys that it is easy to ask as Duffy (1982) does: "Why are surveys performed at all?" One answer is that, despite all the shortcómings on the level of individual measurements, no better way of monitoring drinking habits of a population is at present available. Awareness of the problems should stimulate further research leading to improved methodology; many research areas have already been pointed out. These include research concerning not only the coverage problem and response error, but also the drinking process of individuals in time, both in relatively short periods (from some months up to a year), and also on the lifetime level. Thus, a multitude of research approaches and efforts is needed to make consumption surveys withstand critical examination. A small consolation is that many of the problems of alcohol consumption surveys are shared by consumption surveys in general, ranging from marketing research on toothpaste consumption to the use of public libraries.

REFERENCES

Ahlström-Laakso, S., 1975, *Drinking Habits among Alcoholics*, Finnish Foundation for Alcohol Studies, Forssa.

Alanko, T., 1981, A Statistical Framework for Describing the Measurement of Alcohol Consumption from Survey Data. 27th International Institute on the Prevention and Treatment of Alcoholism, Vienna, Austria.

Alanko, T., 1982, A Statistical Model for Describing Drinking Behaviour and Distribution of Alcohol Consumption in Sample Surveys. Unpublished licentiate thesis, Department of Statistics, University of Helsinki.

Bruun, K., 1972, Surveys of drinking and abstaining: Urban, suburban and national studies—introduction, *Q. J. Stud. Alcohol* (Suppl. 6).

Bruun, K., and Hauge, R., 1963, *Drinking Habits among Northern Youth*, Finnish Foundation for Alcohol Studies, Helsinki.

Cahalan, D., and Cisin, I. H., 1968, American drinking practices: Summary of findings from a national probability sample. I. Extent of drinking by population subgroups, *Q. J. Stud. Alcohol* **29**:130–151.

Cahalan, D., Cisin, I. H., and Crossley, H., 1969, *American Drinking Practices*, Rutgers Center of Alcohol Studies, New Brunswick, N.J.

De Lint, J., Hyman, M. M., Mulford, H. A., and Fitzgerald, J. L., Wechsler, H., 1981, "Words and deeds;" responses to Popham and Schmidt, *J. Stud. Alcohol* **42**(3):359–368.

Duffy, J. C., 1982, The Measurement of Alcohol Consumption in Sample Surveys. Alcohol Epidemiology Section of the International Council of Alcohol and Addictions, Helsinki, July 1982.

Ekholm, A., 1968, A Study of the Drinking Rhythm of Finnish Males. 28th International Congress on Alcohol and Alcoholism, Washington, D.C.

Folks, J. L., and Chhikara, R. S., 1978, The inverse gaussian distribution and its statistical application, *J. R. Stat. Soc.* [B] **40**:263–289.

Gregson, R. A., and Stacey, B. G., 1980, Distribution of self-reported alcohol consumption in New Zealand, 1978–1979, *Psychol. Rep.* **47**:159–170.

Gregson, R. A., and Stacey, B. G., 1982, Self-reported alcohol consumption: A real psychophysical problem, *Psychol. Rep.* **50**:1027–1033.

Kreitman, N., and Duffy, J., 1982, Beyond Consumption: The Effect of Drinking Patterns on the Consequences of Drinking. Alcohol Epidemiology Section of the International Council of Alcohol and Addictions, Helsinki.

Kuusi, P., 1957, *Alcohol Sales Experiment in Rural Finland*, Finnish Foundation for Alcohol Studies, Helsinki.

Mäkelä, K., 1971, *Measuring the Consumption of Alcohol in the 1968–1969 Consumption Study*, Oy Alko Ab, Helsinki.

Morrison, D. G., and Schmittlein, D. C., 1981, Predicting future events based on past performance, *Management Sci.* **27**(9):1006–1023.

Mulford, H. A., and Fitzgerald, J. L., 1981, cited in De Lint et al.

Pernanen, K., 1974, Validity of survey data on alcohol use, in: *Research Advances in Alcohol and Drug Problems*, Vol. 1, (R. J. Gibbins, Y. Israel, H. Kalant, R. E. Popham, W. Schmidt, and R. G. Smart, eds.), pp. 355–374, John Wiley, New York.

Popham, R. E., and Schmidt, W., 1981*a*, Words and deeds: The validity of self-reported data on alcohol consumption, *J. Stud. Alcohol* **42**(3):355–358.

Popham, R. E., and Schmidt, W., 1981*b*, "Words and deeds": A rejoinder, *J. Stud. Alcohol* **42**(5):533–535.

Pöysä, T., and Niiranen, S., 1966, Kokemuksia samojen henkilöiden toistuvista haastatteluista (Experiences from repeated interviews with the same persons), *Alkoholipolitiikka* **2**:73–80.

Sariola, S., 1956, *Drinking Patterns in Finnish Lapland,* Finnish Foundation for Alcohol Studies, Helsinki.

Simpura, J., 1981, Drinking habits in Finland and Scotland: A comparison of survey results, *Int. J. Addict.* **16:**1129–1141.

Skog, O-J., 1981, Distribution of self-reported alcohol consumption: Comments on Gregson and Stacey, *Psychol. Rep.* **49:**771–777.

Straus, R., and Bacon, S. D., 1953, *Drinking in College,* Yale University Press, New Haven.

Wechsler, H., 1981, cited in De Lint et al.

Wilson, P., 1981, Improving the methodology of drinking surveys, *Statistician* **30:**159–167.

8

Studies of Driver Impairment and Alcohol-Related Collisions

A Methodologic Analysis

ROBERTA G. FERRENCE and PAUL C. WHITEHEAD

1. INTRODUCTION

The study of driving impairment and alcohol-related collisions is complicated by the fact that relevant data are derived from a variety of sources and are based on different groups of drivers. In order to discover changes in the magnitude of driving impairment and collisions and to assess the effects of countermeasures of various types, it is essential to understand the nature of these differences and to establish the validity and reliability of the findings. The purpose of this chapter is to assess systematically the different sources of information that are available on this topic, indicate the advantages and disadvantages of each, specify the uses to which they can best be put, and identify the major gaps in our knowledge.

In this chapter, the term "driving impairment" refers to the impairment by alcohol of certain functions involved in operating a vehicle. Theoretically, impairment can only be measured for each individual separately, comparing performance with and without the prior ingestion of alcohol. In reality, impairment of an individual is often inferred on the basis of comparison with other individuals who are sober.

The very serious problems of validity and reliability encountered when trying to measure impairment among several individuals have led to the use of the concentration of alcohol in the blood (BAC) as an operational measure of

ROBERTA G. FERRENCE ● Addiction Research Foundation, Toronto, Ontario, Canada, and Queen's University, Kingston, Ontario, Canada.   PAUL C. WHITEHEAD ● University of Western Ontario, Addiction Research Foundation, London, Ontario, Canada.

behavioral impairment. Although BAC is a reliable measure of the presence of alcohol in the body, it is not a valid measure of behavioral impairment for all individuals, particularly at low levels. Nevertheless, the legal limits of BAC, at least in North America, are associated with at least some degree of impairment among most of the driving population. The term "legal impairment" will refer to the occurrence of BACs that exceed the legal limit. Legal, in this context, means "in terms of the law" rather than "authorized by the law" (Webster's, 1962).

Data on driving impairment and collisions can be applied to two areas of interest. The first is a description of the prevalence of driving after drinking and alcohol-related collisions for different social and demographic categories. The second is the etiologic significance of the blood alcohol level of the driver in the occurrence of impairment and collisions. These general areas of consideration can be expressed more precisely. Who drinks and drives and under what conditions? Of this group, who is apprehended? Who is charged and convicted? Do those impaired drivers who become involved in collisions differ from those who do not? Is driver impairment more likely to be reported for some categories of drivers than for others? What are the risks of collisions of varying severity among different sectors of the population with different levels of impairment? How are high levels of blood alcohol related to the occurrence of collisions?

These questions are investigated by examining a range of epidemiologic data on impaired drivers and drivers involved in alcohol-related collisions. Our focus is on methodologic considerations that may affect the validity and reliability of findings from entirely different types of investigations and from studies that are similar in design (cf. Edelman and Walker, 1977). We will try to account for discrepancies between findings in terms of methodologic factors such as techniques of reporting, sampling, data collection, and measurement of blood alcohol concentrations.

2. METHODOLOGY OF STUDIES ON DRIVER IMPAIRMENT AND COLLISIONS

Research Strategies

Populations.   Studies of the impairment of drivers and alcohol-related collisions can be categorized in four ways, depending on the population that is examined. The first type focuses on the general driving population. Some of these studies use breath-testing equipment to measure the blood alcohol level of drivers who are selected randomly at the roadside. Others rely on self-reports by drivers, either at roadside or in other locations such as their homes, about previous incidents of combining drinking and driving. Such studies usually involve interviews and report demographic characteristics of the drivers and some information about the situations in which the events occur.

A second type of study examines persons convicted of driving while impaired (DWIs). Sometimes the entire population of offenders is included, while in other cases a subgroup, such as recidivists or those referred for treatment, is isolated for study. Comparisons may be made between offenders and their neighbors, recidivists and first offenders, or between offenders who are considered to be alcoholics and those who are "social drinkers." The initial sample may be drawn from police records or clinical populations. These studies often report extensive psychological and personal history data in addition to more basic information.

The third type of study investigates the characteristics and circumstances of collisions in which the driver may or may not be deemed to be impaired. Data are collected from police records or medical examiners' reports. Drivers at zero or very low BACs may serve as a control group, or drivers at varying BACs may be compared.

The process of investigation by legal agents differs for nonfatal collisions that involve personal injury or property damage and those in which the driver dies. Alcohol involvement in nonfatal collisions can be measured by breath tests, but it is frequently based on the observations of the attending police officer. In fatal collisions, blood tests are usually performed when feasible, and the investigation of the collision may be much more intensive.

To establish the level of risk associated with the consumption of alcohol, comparisons can be made between the different types of alcohol-involved collisions or between collisions and noncollision controls matched on the basis of the time and place of the crash. Samples of different types can also be compared, for example, drivers selected at random in roadside surveys with drivers involved in fatal collisions.

The types of studies discussed to this point are all field studies in the sense that they focus on behavior that occurs in the real world and over which the investigator has little or no control. The fourth type of study is conducted within a laboratory setting that is controlled by the investigator. Such studies usually rely on volunteer subjects, often students, who ingest specified amounts of alcohol and are then tested to determine their responses to situations considered representative of driving in real life or of certain components of driving.

Scientific Design.   If we also wish to assess the etiologic role of alcohol and determine what gaps there are in our knowledge of this subject, some attention must be paid to the scientific design of these investigations. With the exception of most laboratory studies of the effects of alcohol on specific behavioral responses and some closed-circuit driving studies, the bulk of the studies on driving impairment and collisions cannot be termed true experiments. Many, however, incorporate controls to the extent that they can be classified as preexperimental or quasiexperimental studies (cf. Campbell and Stanley, 1963). Case–control studies, for example, have been carried out in which previous exposure to an agent among cases characterized by a particular condition is compared with that of a similar group without the condition. In these studies, drivers involved in collisions are usually matched with drivers passing the same point where the accident occurred at the same time of day.

Results of most roadside surveys are presented as descriptive accounts of the characteristics and BACs of the drivers sampled. Yet many are conducted for the purpose of gathering baseline data with which to evaluate the efficacy of specific countermeasures or more general prevention programs or to conduct legal impact studies (cf. Lemert, 1966). In this context, they constitute the first observation in a pretest/posttest design or a more sophisticated time series design, which incorporates a series of periodic observations. In some cases, the strength of these designs has been greatly enhanced by the addition of a nonequivalent control group, such as persons from a nonexposed age group or geographic area.

A fundamental difficulty with research on drinking and driving is that true field experiments are difficult to carry out and few have been attempted, whereas laboratory studies may lack validity because they cannot duplicate real-life situations on the road (Chapanis, 1967; Hurst, 1974; Browning and Wilde, 1975; O.E.C.D., 1978). Factors that may be associated with impairment and collisions, such as the social and demographic characteristics of the driver, the driver's motivation to avoid detection, the context of driving, and the condition of the vehicle, are usually controlled in laboratory studies, so they cannot be examined as independent variables. Furthermore, the choice of dependent variables may not adequately represent actual behavior on the road. Even field experiments require the imposition of certain controls, and these may result in problems similar to those experienced in laboratory settings. Browning and Wilde (1975) contend that by studying real populations of drinking drivers who are involved in collisions, new knowledge on this subject can best be gained. They conclude that some precision must be sacrificed to achieve validity. Collisions are rare events, however, and may involve a number of factors unrelated to impairment, so that the most fruitful approach is likely to be one that combines the results of both laboratory and field studies.

Useful information can be obtained without benefit of true experimental designs. For example, much epidemiologic knowledge is acquired using case–control studies, which can provide information that is sufficiently detailed to support strongly, though not necessarily confirm, a hypothesized causal relationship. Laboratory studies may provide information about the mechanism or mediating processes involved, but this knowledge is not usually considered essential for the institution of preventive measures.

Impairment of drivers and collisions are comparable to many diseases and other phenomena that can profitably be studied using the methods of epidemiology. They are widely but unevenly distributed in the population; a large number of factors contribute to their occurrence and distribution; and they are potentially amenable to preventive measures. Within this context, the best approaches to the study of impaired driving and collisions would involve studies that preserve the validity of the variables under study while attempting to avoid confounding factors as much as possible.

Assessing Research Strategies

Sources of data on impairment of drivers and collisions can be evaluated according to the manner in which populations are selected for study and the advantages and disadvantages of the methods used to collect data (Table 1). As a general rule, the most deviant levels of behavior are more likely to be included in records of social control and service agencies and they are least likely to be reported in surveys of individuals (Warren, 1980). Thus, police, medical examiners', and hospital records will underrepresent drivers apprehended or involved in collisions at low BACs, whereas household and other types of surveys that rely on self-reports and permit refusals will underrepresent those who drive while impaired at the highest BACs and with the greatest frequency.

Roadside Surveys.   Roadside surveys are unique in that drivers are selected at random, or at least systematically (for example, every tenth driver), at sites that are selected randomly from feasible survey locations. To the extent that randomization is successful, the study sample will be as representative as possible of the population at risk during the times that surveys are conducted, generally late at night and on weekends. The large samples of passing drivers that are usually obtained in such studies (some as high as 10,000) are often sufficient to carry out detailed examinations of several variables simultaneously. Large samples are necessary, however, if drivers at higher BACs are to be examined separately. In many cases, all drivers whose BACs are over the legal limit must be grouped together and separate analyses for those at very high BACs cannot be done.

The concentration of alcohol in the blood can be measured directly with blood tests or indirectly using breath tests. The use of either type of testing results in much greater precision than human estimates or self-reports of previous consumption (Zusman and Huber, 1979). BACs are usually expressed as the weight of alcohol per unit volume of blood, for example, 0.08 g/100 ml or 0.08%. The convenience of using breath tests for live subjects has led to the development of a number of devices based on various principles that have become increasingly reliable with improvements in technology (Dubowski, 1975). Screening devices, such as the Alcohol Level Evaluation Roadside Tester (A.L.E.R.T.), that provide a quantitative measure of blood alcohol concentration, are easily employed at roadside.

In general, both types of devices achieve a level of accuracy that is acceptable for screening or charging impaired drivers (Perrine, 1971; Lovell, 1972;

Table 1. Assessment of Data Sources on Driver Impairment and Alcohol-Related Collisions

Population	Source of data	Selection	Ascertainment	Advantages	Disadvantages
Drivers	Roadside surveys	Random	Breath tests	Large samples Random selection Controls for exposure Interval data on population at risk Reasonable cost Breath tests unbiased	Nonresponse may bias Breath tests may not correspond to actual impairment Not generalizable to all times and places
Drivers	Household and other surveys	Random	Self report	Can measure period prevalence Purpose may be hidden In-depth interview possible	May exclude high-risk individuals Problems of recall, underreporting Costly Samples may not be large enough for analysis
DWIs	Police records	Observed infractions Erratic driving Collisions Routine checks	Breath tests (Observed infractions; erratic driving if below legal limit)	Data available High-risk population Breath tests unbiased	Small proportion caught Police judgment basis of stopping, charging Patrols not randomly deployed

DWIs (clinical)	Originally police records; Clinical admissions	Referral to treatment facility	As above	Captive population; Can conduct intensive interviewing	As above; Not representative of DWIs
Nonfatal collisions	Police records	Reportable collisions; Suspicion of officer	Police observations (May use breath tests)	Data available; Most collisions included; Large numbers; Can use surrogate measure of alcohol involvement	Police judgment used; Dichotomous data on impairment; Exclude no or low property damage
Fatal collisions	Coroners' reports	Crashes involving deaths within 6 hr where testing feasible	Blood tests	Data available; Most fatal collisions included; Most cases tested	Not representative of all collisions; Small numbers; Those not tested may differ
Volunteers	Laboratory experiments	Usually volunteer students	Blood/breath tests	Experimental design; Can control for many variables	Questionable validity; May not apply to real-life traffic situations
Volunteers or driving public	Field experiments	Usually volunteers	Blood/breath tests	Experimental design; Can control for some variables; May approximate real-life traffic situations	Questionable validity; Limited number of variables can be examined; No risk involved

Levett and Karras, 1975). There is some variation between instruments, but the correlation coefficients between breath tests and blood tests and between repeated breath tests generally exceed 0.95 (Noordzij, 1975; Bonnichsen and Goldberg, 1981). Errors are often in the direction of underestimating blood alcohol concentration (Harger, 1975), and failure to obtain a maximum breath sample can reduce the reading further (Levett and Karras, 1975). Some variability is also associated with the phase of alcohol metabolism, with BACs slightly overestimated during the absorption phase and underestimated during the elimination phase (Jones, 1976). As a matter of practice, it appears that police officers rarely charge a driver whose test result is less than 25% above the legal limit.

For epidemiologic purposes, the devices appear to provide satisfactory but conservative results. Most studies report grouped data so that minor variations are less likely to make a difference. Obviously, evaluation studies should employ the same devices and procedures in repeated surveys to permit inferences to be drawn on the basis of small changes in rates of impairment. Surveys that use different devices and procedures are not strictly comparable, but most of the surveys conducted prior to the acceptance of a standardized methodology developed by the Organization for Economic Cooperation and Development (Carr et al., 1974) vary in so many other ways that meaningful comparisons are difficult or impossible. Minor variations in BAC readings are mainly a legal issue at this point and have been challenged in the courts where the issue is a false-positive reading. False-negative results, however, are more likely to occur. Significant proportions of false-positives have been reported, but these are largely due to the time lapse between the administration of breath and blood tests and to the deduction made at prosecution to account for possible analytical errors (Goldberg, 1981). Incorrect administration of tests or the presence of other substances such as cigarette smoke can also produce false-positive readings.

Although breath tests are fairly accurate measures of blood alcohol content, the degree of impairment at any particular BAC can vary greatly among individuals. Heavy drinkers, for example, may develop considerable tolerance to alcohol so that they exhibit much less impairment than those who usually drink less (Smart and Schmidt, 1969; O.E.C.D., 1978). These considerations are most relevant at fairly low BACs that are less than the legal limit (0.05% in the Netherlands; 0.08% in Canada; 0.10% in most of the United States). Even when BACs are at zero, the driver may still experience aftereffects of drinking that impair performance (Wilde, 1974). Alcoholics are also overrepresented in accidents when sober, which may be due to chronic effects of heavy drinking (Smart, 1969) or to personality factors (Zelhart, 1972; Selzer and Vinokur, 1974). Most of those actually charged with driving over the legal limit display rather high BACs at which most or all drivers exhibit at least some signs of impairment.

Disadvantages associated with particular sources of data derive to some extent from the purposes to which the data are applied and often involve tradeoffs in which certain desirable features are foregone to allow for others that are considered more important. Drivers tested in roadside surveys are not representative of the total driving population or of drivers on the road at all times of the day. They are not even necessarily representative of drivers on the road at the time of the survey because drivers who are on the road for longer periods of time are more likely to be sampled. According to Wolfe (1975), this is not a serious problem, however, because length of trip is only weakly related to impairment. Generalizability to all times of day and often to other geographic areas is sacrificed in order to reduce cost and effort by sampling at times when the proportion of drivers on the road who have elevated BACs is likely to be highest, that is, late at night, on weekends. These data are frequently collected in order to evaluate interventions designed to reduce the proportion of drivers with BACs above the legal limits so that, in these cases, nighttime roadside surveys are clearly preferable.

If one is more interested in describing the prevalence of legal impairment among different categories of the driving population, however, such samples are seriously biased and inadequate. The rate of driving with a BAC that exceeds the legal limit is comparatively low among drivers during the day and early evening, but the numbers involved are substantial because there are many more drivers on the road. For example, in a roadside survey carried out in Canberra, Australia, between 1000 and 0200 hours (Duncan, 1979), late-night drivers who were legally impaired accounted for only 32% of all those who drove over the legal limit. The late-night drivers are clearly at higher risk than those surveyed at other times, but their social and demographic profile is not generalizable to the total group. The types of drivers and the context of driving, particularly with regard to origin and destination, differ considerably from those at other times: male drivers, and young men in particular, are greatly overrepresented. Thus, most of the existing data on legal impairment among drivers on the road are inadequate for establishing age- and sex-specific rates for the prevalence of driving over the legal limit. When survey data for a 24-hr period are not available, estimates of the true distribution of legal impairment can still be calculated using data from other sources, such as collisions or records of driving habits, to correct for variations in exposure among different categories of drivers.

A potentially serious problem that affects roadside survey data, regardless of application, is the rate of refusals or nonresponse among those selected for testing. Refusal rates often reach 6% or more, which may be as great as the proportion of legally impaired drivers detected. Attempts have been made to estimate legal impairment among the nonresponse group (Wolfe, 1975), but these probably underestimate the proportion that are legally impaired. For example, interviewers in a Canadian survey correctly identified 88% of those with BACs less than 0.02% but only 44% of those who were legally impaired (Smith et al., 1976). Hurst and Darwin (1977) reanalyzed United States roadside survey data and estimated that the rate of impairment (over 0.10%) would increase from 5.4 to 10.3% when the nonresponse group was included. Warren and Simpson

(1981) used the Hurst and Darwin estimates to produce weights for Canadian data. This analysis indicated that impairment (over 0.08%) would increase from 5.9 to 8.3% with the addition of nonrespondents.

If drivers with BACs above the legal limit are overrepresented in the non-response category, the estimates of legal impairment among drivers will be low. This is not usually a problem if the data are to be used for evaluation, as long as the rate of underrepresentation is stable over time. The composition of the refusal group, however, in terms of who has been drinking and who has not, may change as a result of a countermeasure. For purposes of measuring prevalence, any underrepresentation will result in underestimates of the magnitude of the problem and any variation in the extent to which different categories of the population (e.g., age, sex, and socioeconomic categories) are underrepresented will produce biased pictures of legal impairment.

Household and Other Surveys.   A few surveys have been carried out in which respondents complete mailed questionnaires or are interviewed at home, in person, or by telephone, regarding their drinking and driving behavior (e.g., Wolfe and Chapman, 1973). This type of study shares some of the advantages of roadside surveys, such as random sampling and large sample size. In addition, anonymity can be preserved in surveys conducted by mail or telephone, and the purpose of the survey may be camouflaged by combining questions about drinking and driving with questions unrelated to this topic.

The population sampled in household surveys differs in many ways from drivers on the road. Many household surveys use the household rather than the individual as the unit of sampling. Thus, households with children over the age of 16 might provide three or more drivers, whereas single person households would provide only one. If only one driver were selected from each household, older drivers and women would be overrepresented because they are more likely to be at home. Young men are most often missed in household surveys, and persons of any age with drinking problems are more likely to be omitted because they are more often institutionalized or have no fixed address.

Surveys that incorporate questions on drinking and driving are also carried out among special populations such as high school and college students (e.g., Ennis, 1978; Wechsler and McFadden, 1979). These surveys share most of the characteristics of household surveys with the exception of the sampling method employed.

The nature of the data collected in household and other surveys is different from those obtained in roadside surveys. The investigator must rely on retrospective reports of drinking and driving behavior and collisions. Problems of recall are a major source of inaccurate reporting. Information such as the number of miles driven annually and the number of years of driving experience should be obtained to control for variations in exposure, but this is not always done.

In addition to problems of recall, deliberate underreporting of deviant behavior is a substantial problem. Whereas underreporting of undetected drinking and driving may be due largely to difficulties of recall, it is likely that most underreporting of charges for impaired driving and collisions is intentional, given the fact that the event is relatively rare and potentially more stigmatizing than undetected drinking and driving.

Bradburn and Sudman (1979) used four different interview techniques to investigate response effects to a series of threatening questions that included having "been charged for driving under the influence of liquor during the last 12 months." The rate of completed interviews ($n = 249$) was highest for those carried out by telephone (78%) and lowest for a self-administered questionnaire left at the home (48%). About 80% of the noninterviews with impaired drivers were due to a failure to locate the respondent. Thus, the actual refusal rate ranged from 4 to 10% of the total sample.

The investigators were able to determine the extent of underreporting by using police files to validate responses. Almost half of the impaired drivers failed to report charges made against them in the past year. Variation in the amount of distortion for the different techniques ranged from 35% for a random response technique to 54% for the self-administered questionnaire. The value of these findings is that they provide correction factors that can be used to produce estimates of true rates of legally impaired driving. Unfortunately, the size of the sample is not large enough to establish distortion rates for different age and sex categories.

In summary, household and other surveys produce data on legal impairment that are probably less accurate and less representative than those obtained in roadside surveys. Moreover, BAC levels cannot be obtained. There are certain advantages to this method of collecting data, however, that would make it preferable in some instances. Since respondents are not in transit, more extensive information can be obtained. The relevant questions can be included in a more general survey at considerable savings. Finally, the household survey can be used to determine drinking and driving histories of individual drivers and to measure the prevalence of legally impaired driving in a community for a given period of time, such as 1 year. Some roadside surveys have collected such retrospective data in addition to BACs, but these data cannot be used to measure period prevalence.

DWIs. Drivers charged with driving while impaired (DWI) or driving under the influence of liquor (DUIL) are frequent subjects of investigations of driver impairment. Relevant data can usually be obtained from police records in which specific infractions, such as driving over the legal blood alcohol limit, driving while impaired by alcohol, or refusing to take a breath test, are recorded. The selection of drivers usually depends on the arresting officer's observation of driving infractions, erratic or unusual driving behavior, or the occurrence of a collision. Occasionally, impaired drivers are detected in routine checks for other purposes such as vehicle safety. Once the driver is suspected of being impaired, the officer may administer a simple breath test with a device used only for the purpose of screening and, if warranted, refer the driver for a further test on more accurate equipment that produces legally admissable results. The officer can exercise considerable discretion in deciding which drivers to stop, which to test, and which to charge.

Data on drivers charged with impairment cannot be considered representative of all those who drive while impaired (Whitehead, 1975). In the first place, they represent only a very small fraction of this population. In fact, the chances of being charged during a particular drinking and driving episode in Canada, for example, fall somewhere between those of encountering a snowstorm in June on the high side and winning the Olympic lottery on the low side (Bragg and Cousins, 1979). More precise estimates have been made ranging from one charge in 200 drinking-driving episodes to one in 2000, although the risk varies directly with BAC and is much greater at high levels of impairment. For example, a driver who makes one trip each week at a BAC ≥ 0.15 will be arrested once every 18 months on the average (Beitel et al., 1975). In Canada, one charge is laid per 26,000 km driven while the driver is legally impaired. Here, the probability of apprehension is much greater than in some European countries (Wilde, 1974).

Second, police manpower is deployed on the basis of the distribution of crime in general. Since crime tends to be concentrated in the core areas of cities where most drinking establishments are located, those who drink at these establishments are more likely to be detected. In most cases, drivers are stopped only when they do something to gain the police officer's attention, and this may be quite unrelated to impairment. Even among those drivers suspected of being impaired, only a small percentage are actually tested (Borkenstein, 1975). The relatively low priority given to the apprehension of impaired drivers and the reluctance of police officers to test drivers who appear to be impaired reflects their own view, and ultimately that of the larger society, that impaired driving is not a real crime (Willett, 1964; Boshier and Johnson, 1974).

Because of the severe limitations associated with these data, they cannot be used to measure the prevalence of legal impairment in a community. There are other situations, however, where data from this source are useful. In some cases, one is interested only in those drivers who have been formally charged with impairment. In other cases, the data on DWIs can usefully be combined with data from other sources, such as roadside surveys, or with data from well-chosen control groups in order to test specific research hypotheses.

Clinical DWIs.   Subsamples of clinical populations of persons who have been charged with impairment are frequently studied. These investigations are relatively easy to conduct because they involve captive populations of drivers referred for treatment of their alcoholism or admitted to educational programs in lieu of fines or incarceration. Subjects may be interviewed in depth about their history of drinking and driving and they frequently receive extensive psychological testing. These samples are often compared with samples of alcoholics or impaired drivers who are not referred for treatment. Data on deviant behavior that are collected in studies of clinical populations appear to be more accurate than those gathered among the general population (Sobell et al., 1974). They are probably even less representative of all impaired drivers than are total populations of impaired drivers who have been charged, however, because those referred for treatment are more likely to be alcoholics.

Nonfatal Collisions.   Drivers involved in collisions that are reported to the police may be observed by the officer for evidence of recent drinking. Most of these drivers are not given a breath test, but many jurisdictions require that the investigating police officer indicate on the collision report whether or not the driver appears to have been drinking or is impaired by alcohol (O.E.C.D., 1978).

Reports of drivers involved in collisions are fairly comprehensive, although cases in which little or no damage occurs are not likely to be included. Large numbers of collisions are available for study, and considerable information can be extracted from police records. These data can be used alone in legal impact studies or in combination with other sets of data. Were it not for some serious limitations, they would be adequate for examining the role of alcohol in the occurrence of collisions. These limitations primarily involve the way in which impairment is determined. Since the police officer may indicate only that the driver has been drinking, it is not clear what proportion of drivers are legally impaired or what range of BACs is involved. Furthermore, human judgment of blood alcohol levels is known to be unreliable, particularly at moderate levels of impairment when error can reach 50% (Wilde, 1974). In fact, one study has shown, albeit in a less threatening context, that drivers' reports of their own drinking are more accurate than estimates made by interviewers (Zusman and Huber, 1979). Of drivers with BACs $\geq 0.10\%$, interviewers estimated that only 45% had been drinking at all, whereas 68% of these drivers reported that they had been drinking. These problems become less critical if one is making a series of observations within a geographic area. Variability between geographic areas in the coding of impairment can considerably reduce the comparability of data from different areas (Douglass et al., 1974).

A different approach is to employ a surrogate measure of alcohol involvement, such as the proportion of late-night, single-vehicle collisions involving male drivers (cf. Douglass et al., 1974). This category is known to have a high proportion of legally impaired drivers, and their rates of collisions may be employed as an alternate measure of impairment or as an additional method of verifying results obtained using police observations.

The exclusion of nonreportable collisions may produce some bias, but the direction is not entirely clear. For reportable crashes, rates of alcohol involvement increase with increasing severity of the crash. Nevertheless, there may be a substantial proportion of very minor "fender benders" involving alcohol, in which the driver just leaves the scene. When more than one vehicle is involved, police attention may be more likely, even for cases where damage is minor.

The attribution of responsibility for a crash is often omitted from accident records, and even when included, it is generally based on the judgment of a police officer. These deficiencies are important when one is investigating the etiologic role of alcohol in crashes (O.E.C.D., 1978).

Fatal Collisions. Most fatally injured drivers are given blood tests to determine BAC when death occurs within 6 hr of the collision. Estimates of rates of legal impairment can be based either on the proportion tested or on the total category of fatalities. Some of the factors that affect the likelihood of a driver being tested, such as severe loss of blood, are fairly independent of other characteristics of the driver. Yet, it appears that those not tested tend to belong to those categories of drivers who are less likely to be legally impaired, for example, adolescents and older drivers, and those involved in multiple vehicle collisions (T.I.R.F., 1975). In addition, BACs are usually not measured until after death. Since this can be up to 6 hr after the collision, some of those who were legally impaired at the time of the collision will produce test results that are below the legal limit (T.I.R.F., 1975). Thus, rates that use the total number of fatalities as the population base are conservative, but probably more accurate than those based only on the proportion of those who are tested (Simpson et al., 1978).

There are several advantages in using populations of fatally injured drivers. Existing data based on medical examiners' reports can be obtained. Most fatal collisions in which the driver is killed are included. Most drivers are tested for blood alcohol concentration, and the assessment is based on results of blood tests, which are more reliable than human judgment.

The principal limitation is that fatally injured drivers are not representative of those involved in alcohol-related collisions generally. Some of the factors that increase the likelihood that the driver will die are unrelated or only weakly related to impairment, for example, the crash-worthiness of the vehicle and the proximity of emergency facilities. This unrepresentativeness is not a relevant issue when one is simply describing the prevalence of legal impairment among fatally injured drivers. It may become a problem when populations of these drivers are used to represent all drivers in alcohol-related collisions, as is often done in evaluation or legal impact studies.

Other factors that increase the risk of death following a collision are highly associated with the use of alcohol. For example, heavy drinkers who are physically deteriorated are less likely to survive serious collisions (Waller, 1979). The use of seatbelts is inversely related to BAC. The Ontario Roadside survey found that seatbelt use ranged from 36% of those who had BACs in excess of 0.15 to 62% of those with BACs of 0–0.02 (Interministerial Committee on Drinking–Driving, 1979). Furthermore, intoxicated persons who are involved in collisions receive less effective medical care than those who are sober (Waller, 1979). Such factors inflate the proportion of fatal crashes that are alcohol-related, if one considers only the role of alcohol in causing the collision.

An additional difficulty occurs when rates of legally impaired fatalities are calculated for specific age groups or geographic areas, because the numbers involved may be too small to produce rates that are stable over time. International comparisons pose problems as well because the definition of a fatality in different countries may range from those deaths that occur at the site of the crash to those occurring within 1 year as a result of injuries associated with the crash (Waller, 1975).

The biases associated with incomplete testing of fatally injured drivers and the absence of any tests for those who survive collisions in which passengers may have died are apparently minimal (T.I.R.F., 1975), especially when compared to the nonrepresentativeness of the entire category of fatalities. Furthermore, these biases could be reduced by using estimates based on attributes of cases in which testing was carried out (Kannemann and Warren, 1980).

Laboratory and Field Experiments.   Considerable information about the relationship between alcohol and behavioral responses has been generated by experimental studies carried out under controlled conditions in the laboratory and in the field. Some of the problems associated with this approach have been described above.

The issue of validity is raised throughout the experimental literature. For example, there appears to be no correlation between performance on a driving simulator and actual driving behavior (Edwards et al., 1969). Many studies of processes that are believed to be components of driving behavior provide no evidence that there is indeed a relationship (O.E.C.D., 1978). Subjects who are tested in laboratory experiments are often moderate drinkers who do not normally drink and drive and who may react quite differently from those who do. Furthermore, drivers on the road may be more likely to compensate for the effects of alcohol because they are more motivated to avoid detection (O.E.C.D., 1978).

Ideally, experimental studies should provide information about the effect of alcohol on specific behavioral functions so that the exact mechanism whereby impairment occurs can be understood. Under controlled conditions, it is found that many apparent contributors to driving impairment, such as response time, visual discrimination, and manual dexterity, are affected even at relatively low BACs. Complex tasks that require the simultaneous performance of several functions are affected to an even greater extent (Moskowitz, 1974). These results are sometimes viewed as positive indicators of the role of alcohol in causing impairment and collisions, but their most valid function is to suggest directions for field experiments in which conditions are made to approach those of actual driving experiences (cf. Wilde, 1974). Multivariate studies in real-life traffic situations are more difficult to carry out, but would provide a much clearer picture of the relative importance of psychological and social factors, as well as physiological factors, in the occurrence of collisions.

Closed-course studies of driver impairment provide a closer approximation to real-world driving situations and allow for a greater range of measurements.

For example, Attwood and co-workers (1980) used multivariate functions to differentiate between sober and intoxicated drivers in a closed-course experiment.

3. CURRENT KNOWLEDGE ABOUT DRIVING IMPAIRMENT AND ALCOHOL-RELATED COLLISIONS

We can describe what is currently known about driving impairment and alcohol-related collisions by answering the following three questions. Which drivers are at risk and under what conditions? Which drivers are apprehended? What is the role of alcohol in driving impairment and collisions?

Drivers at Risk

To determine the proportion and characteristics of drivers who drive while legally impaired, we will focus on data collected in roadside surveys. According to Stroh (1974), more than 20 roadside surveys were carried out between 1938 and 1974, but only the most recent used random selection of sites and drivers, and similar methodologies. With the exception of the well-known Grand Rapids study (Borkenstein et al., 1964), sample sizes for North American studies have, until recently, been too small to investigate characteristics of drivers differentially as a function of their blood alcohol levels.

Four major roadside surveys that meet the criteria of random selection, large sample size, and comparable methodology were carried out during the 1970s in Canada (Smith et al., 1976; Interministerial Committee, 1979), the United States (Wolfe, 1974), and the Netherlands (Institute for Road Safety Research, 1977). All of these were conducted during the late evening and early morning. The average consumption of alcoholic beverages in these countries is reasonably similar, which means that other factors being equal, for example, research design, drinking patterns, and so forth, we might expect similar rates of driving impairment. Some factors that could contribute to increased variability among countries are variations in drinking and driving patterns and in the legal BAC limit, changes over time, variations by time of year, size and characteristics of the nonresponse category, and the use of different testing instruments, in particular the A.L.E.R.T., which can vary by a BAC reading of 0.02%. Sampling error could be substantial in the U.S. study, which used a relatively small sample ($n = 3192$), particularly among subsets of impaired drivers. For example, only seven drivers aged 16–17 were found to have BACs of 0.10% or greater (Wolfe, 1975). The use of multistage sampling techniques in all four surveys also increased sampling error substantially.

Given these considerations, the proportions of drivers at various BACs and in different age and sex categories are remarkably similar (Table 2). (The BAC of 0.10% has been chosen to indicate impairment because it represents the lower level at which charges for impairment are generally made in North America.) One-fifth to one-quarter of all drivers sampled had been drinking; 4–5% were at or above 0.10%, and more than 1% were at or above 0.15%. The proportion of men who were legally impaired exceeds that of women by a factor of about 2 : 1. The rate of legal impairment among underage drinkers is about half that of those of legal drinking age. Rates generally increase with age, but decrease somewhat among older drivers.

These findings can be supplemented with retrospective reports of drinking and driving. Lehman et al. (1975) found that 53% of drivers reported previous driving after drinking when interviewed in roadside surveys in the United States. Smith et al. (1976) found that 72% of Canadian drivers had driven after drinking during the previous year and 29% had been "high." In a household survey carried out in Michigan in 1973, with a response rate of 77% (Wolfe and Chapman, 1973), 60% of respondents reported that they drove after drinking at least once in the past year, and 29% "drove while high." In a previously unreported mail survey of neighbors of convicted impaired drivers in London, Ontario, we found that 52% of respondents reported that they had driven when they believed their BACs were over 0.08% at least once in their lives. Seventy per cent of men aged 17–64 interviewed in a community survey in New South Wales reported that they frequently combine drinking and driving (Freedman et al., 1973). Despite the variation in sources and methods used, these data exhibit some consistency. Drinking and driving are combined at least occasionally by a majority of drivers, and driving while legally impaired is fairly common.

Extensive information on other characteristics of drivers, by BAC as well as by situational and environmental factors, was collected in two of the major North American surveys (Wolfe, 1974; Smith et al., 1976). Because of the small number of drivers with BACs at or above 0.15%, no distinction is made between individuals with different BACs within that category. By comparison, in Scandinavian countries, where the average consumption of alcohol and the legal blood alcohol limit are considerably lower ($\geqslant 0.05\%$), studies of legal impairment report rates of less than two per cent (Penttila et al., 1979, 1981; Goldberg, 1981).

When characteristics of driver are examined individually, certain categories are associated with elevated BACs. Drivers who are separated or divorced are more likely to exceed the legal limit than those who are married or single. Ratios for these categories are 1.8 : 1 and 2.3 : 1 (Wolfe, 1974; Smith et al., 1976). Those with a university education are less likely to exceed the legal limit than those with less education, but there is no clear inverse relationship across all levels of education. The impairment ratio for blue collar to white collar workers is 1.6 : 1 (Wolfe, 1974; Smith et al., 1976), and unemployed persons were found to be at slightly higher risk than employed persons: a ratio of 1.3 : 1 (Smith et al., 1976). On the basis of these findings, the sex of the driver is probably the best predictor of legal impairment, with age and marital status also important.

Table 2. Design and Results of Four Major Roadside Surveys Conducted between 1970 and 1979

	Netherlands	United States	Canada	Ontario, Canada
Source	Institute for Road Safety (SWOV), 1977	Wolfe, 1974	Smith et al., 1976; Schliewen, 1979	Interministerial Committee, 1979
Year	1970, 1971, 1973, 1975	1973	1974	1979
Season	Sept.–Nov.	Oct.–Dec.	April–June; Sept.–Oct. (2 provinces)	May–June
Days	Fri.–Sun.	Fri.–Sat.	Wed.–Sat.	Wed.–Sat.
Time	2200–0400	2200–0300	2200–2400; 0100–0300	2100–0300
Vehicles	Passenger cars	Passenger cars	Passenger cars, trucks, vans, motorcycles	Passenger cars, noncommercial light trucks
BAC test: Breath	Omicron Intoxilyzer and Field Crimper	Omicron Intoxilyzer and Field Crimper	Intoximeter Mark II	A.L.E.R.T. (digital readout)
Blood Sampling	Yes	No	No	No
Sites	Multistage random	Multistage random	Multistage random	Multistage random
Vehicles	Systematic random	Random	Random	Random
No. of sites	90	184	584	256
No. of respondents	2675/2967/2109/3544	3192	9029	10,000
Nonresponse rate (%)	14/13/18/11	12	6.9	6.1
BAC level	22 (mean)	22.6 (>0.02)	20.4	24.8
≥0.02				
≥0.05	13/17/15/NA	13.5[a]	11.2[a]	13.2 (>0.05)
≥0.10	5/8/5/3	5.0[a]	4.1[a]	4.7 (>0.10)
≥0.15	Not available	1.4[a]	1.3[a]	1.5 (>0.15)

Sex				
Percent males ≥0.05	13/18/16/11	14.5	12.1	14.8 (>0.05)
Percent females ≥0.05	4/6/6/2	8.4	5.9	6.8 (>0.05)
Percent males ≥0.10	5/8/5/4	5.4	4.4	5.5 (>0.10)
Percent females ≥0.10	0/2/3/1	2.6	2.3	1.6 (>0.10)
Sex ratio ≥0.10	3.7 : 1 (mean)	2.1 : 1	1.9 : 1	3.4 : 1
Age (percent ≥0.10)				
16–17	—	2.7	1.1	2.3 (>0.10; 16–18)
18–24	—	4.7	3.7	4.6 (>0.10; 19–24)
Total: 16–24	5 (<25; 1970–73)	4.4	3.4	4.1 (>0.10)
25–34	7 (1970–1973)	6.2	5.4	5.3
35–49	6 (1970–1973)	5.9 (35–44)	4.3	5.7
50+	4 (1970–1973)	3.7 (45+)	3.1	3.9
Legal BAC limit	0.05 after Nov. 1974	0.10%	0.08%	0.08%
Legal drinking age (yr)	16	18–21 (varies)	18	18/19 (increased in 1979)

[a] Weighted rate.

Several environmental and situational factors have also been examined. Legal impairment does not appear to be related to the time required for the trip, the distance travelled, or the number of passengers (Wolfe, 1974; Smith et al., 1976). Drivers with low annual mileage appear to be at higher risk (Wolfe, 1974). Not surprisingly, the origin and destination of the trip is highly related to the probability of legal impairment, with rates, among those leaving and even on their way to drinking establishments, above 8%. Even so, the range is somewhat less than one might expect, with the lowest rate at 2% for those on work-related trips (Wolfe, 1974; Smith et al., 1976).

It is important to differentiate between risk and magnitude of legal impairment because each concept has a different application. When one is most concerned with causation, risk is the appropriate measure to examine. For purposes of prevention, one is more interested in defining target groups so that magnitude may be of much greater interest, even for groups without elevated risk. Failure to differentiate between these two dimensions can result in misleading or at least unclear profiles. For example, the legally impaired driver is typically described as a separated or divorced man, working at a blue-collar job, who is returning from a drinking establishment. Yet, drivers who are separated or divorced account for only 9–14% of all legally impaired driving, so that their excessive level of risk is not that important numerically. By comparison, blue collar workers are at a somewhat lower level of excessive risk, but account for more than half of all legally impaired drivers. Only one-quarter of drivers with BACs above the legal limit have actually come from a bar, club, or tavern, whereas a higher proportion have come from homes of friends (Smith et al., 1976). Thus, although trips that originate in drinking establishments are more strongly associated with legal impairment than with other settings, they do not account for the greatest proportion.

Our description of individual factors associated with elevated risk of legal impairment has limited explanatory value because it ignores the effects of interaction between variables. Schliewen (1979) reanalyzed the Canadian national data (Smith et al., 1976) using a nonparametric segmentation technique in an attempt to identify the best predictors of variation in BAC level and to isolate subgroups based on a minimal number of variables. His analysis yielded six groups on the basis of three predictors each (Table 3). Interestingly, neither sex nor marital status is an important predictor when interaction effects are controlled. Time and origin of trip appear to be the best predictors of rates of impairment. Clearly, there is much scope for more sophisticated analyses of this type in clarifying the association between risk factors and driving impairment.

Some roadside surveys collect additional data on usual drinking patterns and previous drinking and driving incidents. In a series of roadside surveys in Vermont, Damkot et al. (1977) found that those who usually drink five or more beers at a sitting were three times as likely to have BACs ≥0.10% as those who usually drank one bottle (18.8% versus 6.1%). Daily drinkers were also many

Table 3. Rates of Legal Impairment and Relative Risk for Six Groups Identified by Best Predictors

Driver groups	Percent of all legally impaired	Percent of category ≥ 0.08 (N)	Percent of sample (N)	Relative risk of impairment[a]
A.M.: driver coming from bars, tavern, clubs	10.8	21.2	2.4	4.5
Completed education		(45)	(211)	
A.M.: driver coming from bars, taverns, clubs	5.8	10.8	2.5	2.3
Incomplete education		(24)	(220)	
A.M.: driver coming from other places	39.8	9.2	20.5	1.9
Aged 20 or more		(166)	(1805)	
P.M.: driver coming from other places	25.4	4.2	28.6	0.9
Blue-collar worker		(106)	(2519)	
P.M.: driver coming from other places	2.6	3.1	4.1	0.6
Aged 16–19		(11)	(361)	
P.M.: driver coming from other places	15.6	1.9	38.8	0.4
White-collar worker		(65)	(3417)	
Totals (percent)	100.0	4.9	96.9	1.0
(N)	(417)	(417)	(8533)	

Adapted from Schliewen, 1979.

[a] Relative to risk based on representation in sample.

times more likely to be legally impaired than those who drank rarely (19.7% versus 2.3%). Smith et al. (1976) found similar results in Canada where 8.7% of those who consumed at least 300 drinks per year were legally impaired compared to 1.0% of those who drank 1–23 drinks per year. Smith et al. (1976) also found that 11.8% of those who drove after drinking at least 209 times per year were legally impaired compared to 2.0% of those who did so 1–10 times per year. Similar proportions "drove while high" (11.2%, 41 or more times versus 4.4%, fewer than five times). Legal impairment is also related to previous alcohol-related convictions and collisions. Damkot et al. (1977) report that 5.1% of drivers with BACs ≥0.10% had two or more DWI convictions in the past 3 years compared to 2.4% of those who had lower BACs. Similarly, 6.5% of legally impaired drivers had two or more DWI collisions in the past 3 years compared to 2.2% of those who were not legally impaired.

Young male drivers aged 18–29 are overrepresented among men who have had one or more collisions of any type (Damkot et al., 1977). Ratios of the observed to expected rate of collisions are 1.4–1.0 for 18- to 24-year-olds and 1.2–1.0 for 25- to 29-year-olds. Previous DWI collisions are distributed somewhat differently by age, with 18- to 20-year-olds greatly underrepresented (0.3–1.0) and 30- to 39-year-olds overrepresented by 2.1–1.0. Drivers aged 21–29 are overrepresented to a lesser extent. The age distribution for those with one or more DWI convictions in the previous three years is similar, with 18- to 20-year-olds underrepresented (0.7–1.0) and 30- to 39-year-olds overrepresented (1.5–1.0). Young drivers in this sample drink liquor just as heavily as those aged 30–39 and they consume larger quantities of beer when they do drink, but the older drivers are twice as likely to drink these beverages very frequently. This may explain some of the difference between these two groups in rates of DWI collisions and convictions.

Only a small proportion of all drivers on the road at night and on weekends are at risk of being apprehended on the basis of their blood alcohol levels. This risk rises for certain groups of drivers according to their characteristics, their drinking patterns, and the nature of the trip. Nevertheless, meaningful profiles of legally impaired drivers must take into account the effect of interaction between these variables.

Drivers Who Are Apprehended

Studies of driver impairment and collisions customarily treat DWIs quite separately from drivers involved in alcohol-related collisions. While this seems reasonable in light of the differences in the characteristics of these populations and the context and outcome of their contact with agents of the law, there are good reasons for examining them together. The most important of these is that the circumstances that lead to arrest for DWI are frequently the same as those that result in collisions. Drivers are usually apprehended and charged with driving while impaired in one of three situations: following a collision, following an infraction of the laws governing motor vehicles, or while engaged in erratic driving behavior. Since none of these contexts is necessarily associated with alcohol, the attending police officer must decide whether or not a breath test is warranted. In at least one study of DWIs (Hickey et al., 1975), the mean BACs for the three different circumstances of arrest were essentially the same. The group of DWIs that is involved in collisions forms a part of the total group of drivers involved in collisions. DWIs have excessive rates of previous collisions (Perrine, 1970; Selzer and Vinokur, 1974; Maisto et al., 1979), and the occurrence of previous driving infractions and erratic driving is associated with an increased risk of collision.

Drivers charged with DWI are generally older, as a group, than drivers involved in collisions and those detected in roadside surveys. They usually include a higher proportion of men, divorced or separated persons, minority group members, and those from lower socioeconomic levels than either of the other two groups. Finally, they are heavier drinkers and have the highest BACs when apprehended of any group studied, averaging close to 0.20%, although

BACs for younger drivers tend to be lower (Hyman, 1968; Perrine, 1970; Chi et al., 1973; Yoder and Moore, 1973; Hickey et al., 1975; Foley et al., 1976; Shults and Layne, 1979).

With regard to BACs, Duncan and Vogel-Sprott (1978) report that impaired drivers referred for treatment did not drink much more on their day of arrest than they usually did, which suggests that the arrest, rather than the heavy drinking, was the unusual event. Furthermore, a comparison of first offenders with multiple offenders, using age as a covariate in covariance analyses, indicated that the mean dosages of alcohol taken prior to arrest did not differ significantly between the two groups. Although this study used a small sample that may not be typical of all impaired drivers ($n = 58$), it supports the notion that many of those charged consume large amounts of alcohol on a regular basis and drive while over the legal limit with some frequency.

The characteristics of DWIs outlined above are widely reported but have not been found in all studies. Van Ooijen (1979), for example, compared BACs and characteristics of three groups of DWIs in Rotterdam, The Netherlands: those involved in an accident, those detected because of their driving behavior, and those discovered in roadside surveys. The survey group had significantly lower BACs and were less likely to have used medicine or to be intoxicated by substances other than alcohol. There were no differences between the three groups with regard to age, marital status, social class, number of bars visited, criminal record, or refusal to take a breath test. These findings may not be generalizable to North America.

Since most studies of persons charged with or convicted of driving while impaired are carried out within relatively small jurisdictions, usually cities, it is not surprising that there is considerable variation in reported findings. Much of this is due to the prevailing level of enforcement. For example, the introduction of Alcohol Safety Action Programs in certain communities in the United States resulted in large increases in arrests for DWI in many cases (Oates, 1974). Zelhart and Schurr (1977) report that the proportion of DWIs detected as a result of an accident in Nashville, Tennessee, decreased from 50 to 15% between 1972 and 1976 when there was at least an eightfold increase in the frequency of citations, apparently due to the addition of mobile testing laboratories for measuring BACs in the field. Because of the small proportion of legally impaired drivers who are detected and charged, certain changes in enforcement procedures may produce a sizable change in both the rate of citations and the distribution of characteristics such as age, sex, and socioeconomic status among those cited.

In a study of DWIs in Kansas City, Missouri, Foley et al. (1976) found that driver characteristics were related to likelihood of arrest for DWI, disposition of the arrest, and referral after conviction. Nonwhites were less likely than whites to engage in plea-bargaining, probably because of their inability to hire a lawyer, and they were more likely to be placed in rehabilitation programs than to receive a fine. Compared to drivers judged responsible in fatal crashes, the DWIs were older, had lower incomes, and were more likely to be nonwhite. All of these findings are complicated by age differences among racial groups (most young persons arrested were white) and suggest that differences among age groups in both drinking patterns and driving patterns, together with the variation in BACs among drivers in crashes and DWIs, account for much of the variation in profiles of different groups of drivers.

Drivers involved in alcohol-related collisions are a minority of the total population of drivers who crash. Alcohol involvement varies directly with the severity of the collision, ranging from 5 to 10% of crashes involving only property damage to about half of serious injury and fatal crashes (O.E.C.D., 1978). Rates for nonfatal crashes involving personal injury range from 9 to 13% (Jones and Joscelyn, 1978). In a Canadian study of persons admitted to emergency rooms as a result of a motor vehicle accident, 26% of drivers tested (97% of the total) had BACs ≥ 0.08 (Rockerbie et al., 1981). Even higher rates of alcohol involvement are associated with both fatal and nonfatal late-night, single-vehicle crashes (Carlson, 1972; Rosenberg et al., 1974; Jones and Joscelyn, 1978). Alcohol-related fatal crashes constitute only about 1% of all alcohol-related crashes (BAC ≥ 0.10), whereas about one quarter involve personal injury and about three-quarters only property damage (Jones and Joscelyn, 1977).

Compared to matched controls or drivers in roadside surveys, young drivers, and to a lesser extent older drivers, are overrepresented in all types of crashes (Farris et al., 1976; Warren and Simpson, 1976), and young drivers involved in crashes have lower mean BACs than do older drivers (Ryan and Salter, 1979). Other high-risk groups include the following: those with previous convictions for impaired driving, speeding, and reckless driving; persons who are divorced or separated; and those with prison records (Lacey et al., 1979). Drivers with less education and unskilled occupations are also at higher risk (Farris et al., 1976). Jones and Joscelyn (1977) conclude, however, that marital status, education, occupation, and race are not important risk factors and that their relationship with the risk of crashes is frequently confounded by age. Women who have been drinking are overrepresented to a greater extent than men (Carlson, 1972; Farris et al., 1976), and their risk relative to men for a given BAC increases at higher BACs (Borkenstein et al., 1964). Since they are much less likely than men to drink and drive in the first place, however, their overall contribution to alcohol-related crashes is quite small.

The importance of age in explaining variability in rates of collisions is apparent when risk factors are calculated for individual age categories. In their study of drivers in personal injury crashes with matched controls, Farris et al. (1976) found that the risk ratio for young drivers under 25 years of age at BACs above 0.04% was 3.6 when compared to those at zero BAC. By comparison, drivers aged 25 and over had 2.2 times the risk of a crash at BACs over 0.08% compared to those at zero BAC.

When drivers involved in fatal crashes (T.I.R.F., 1975) are compared with drivers selected randomly at roadside (Smith et al., 1976), risk ratios within age categories at BACs ⩾0.08% are 4.2 for 16- and 17-year-olds and 2.1 for 18- and 19-year-olds. Ratios are about 1.0 or less for drivers aged 20–49 and rise to 1.4 for drivers aged 50 and over. The ratio for the total group is set at 1.0 and this is used as the basis for comparison.

Warren and Simpson (1976) attempted to account for some of the variability by age in rates of fatal crashes by calculating risk ratios for driving exposure and driving experience, both recent and cumulative, using a ratio of 1.0 for the total group as the basis for comparison. The risk factors for exposure and experience combined were 5.6 for adolescent drivers, 1.9 for drivers aged 20–24, and about 0.5 for drivers aged 25 and older. The relatively low level of exposure to night driving among drivers aged 50 and over would seem to compensate for their lack of recent driving experience. While exposure is clearly a risk factor for crashes, the role of driving experience needs further clarification, particularly since other factors such as the speed of the vehicle, the driver's motivation to drive safely, and the driver's tolerance for alcohol are also likely to be involved.

Data from case–control studies, in which BACs of drivers involved in collisions are compared with those of drivers selected randomly at the site of the crash or with those of drivers selected in roadside surveys, can be used to establish the relative hazard of various BACs. Farris et al. (1976) calculated "alcohol risk factors" for all drivers in crashes involving personal injury and separate factors for culpable drivers, that is, those judged responsible for the crash. For all drivers, risk is doubled at BACs of about 0.10% and is about four times as high at 0.15% when compared to drivers who have had little or no alcohol. Risk factors for culpable drivers are approximately double those for drivers in general, with about four times the risk at 0.10% and eight times at 0.15%. Lance and Romeder (1979) have calculated two measures of the risk associated with legal impairment (BAC ⩾ 0.08%) in fatal collisions occurring in Canada in 1974. They conservatively estimate an overall relative risk of 21 and an attributable risk of 37%. This means that impaired drivers are overrepresented in fatal crashes by a factor of 21 compared to nonimpaired drivers and that the exposure of the driver to amounts of alcohol that exceed the legal limit accounts for 37% of fatal crashes.

Hurst (1974) provides us with a thorough analysis of data from several case–control studies involving both fatal and nonfatal crashes and reports the findings listed below.

1. Total relative hazard increases monotonically with increasing BAC. The "dip" below the zero alcohol baseline found at BACs between 0.01% and 0.04% that was reported in the Grand Rapids study (Borkenstein et al., 1964) is an artifact of the data analysis accounted for by variations in relative hazard among subgroups distinguished by their drinking habits (Allsop, 1966).

2. Relative hazard is strongly related to BAC for drivers deemed to be at fault and appears to be unrelated to BAC for those not at fault. Pliner and Cappell (1977) found that student volunteers were somewhat more likely to attribute

responsibility for a crash to a driver who had been drinking excessive amounts of alcohol than to one who had consumed only coffee. The occurrence of bias in the attribution of responsibility for crashes should be investigated further among samples of law enforcement agents.

3. The driver's self-reported frequency of drinking is inversely related to the relative probability of a crash at any particular BAC (Fig. 1). The curve describing the relationship between relative hazard and BAC starts at a higher point and increases faster for drivers who drink less frequently. Despite their higher risk when drinking, these drivers are less likely than frequent drinkers to become involved in crashes because they drink less often and are less likely to combine drinking and driving. These variations in levels of risk can be attributed in part to acquired physical tolerance to alcohol and possibly to learned ability to compensate for impairment. There is probably a direct relationship between frequency of drinking and driving experience, which also contributes to this variation, because infrequent drinkers are at higher risk of crashes even when sober. Infrequent drinkers and abstainers may also differ in other ways that contribute to a higher risk of crashing: for example, they may have a higher incidence of health problems and may include former alcoholics.

The difference in degree of hazard between infrequent and daily drinkers is quite substantial. The infrequent drinker is at considerable hazard at 0.06%,

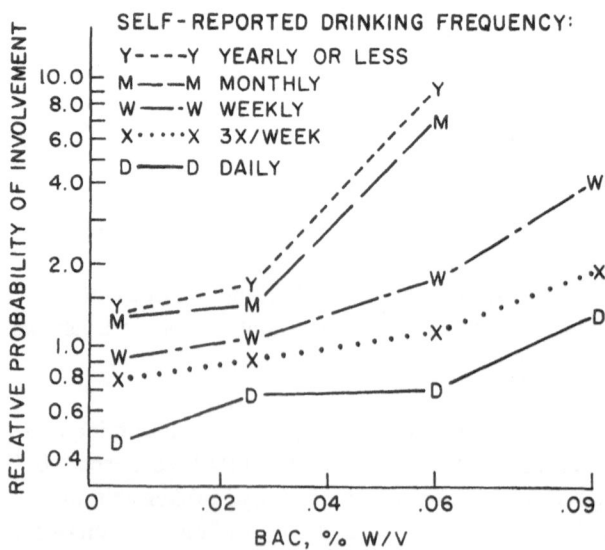

Figure 1. Relative probability of crash involvement (by drinking frequency subgroups) as a function of BAC, where 1.0 = relative probability of composite group at zero alcohol (Hurst, 1974).

whereas the daily drinker is almost as safe at 0.09% as a very infrequent drinker who has consumed no alcohol.

These findings provide one explanation for demographic differences between those receiving DWI citations and those involved in collisions (Hurst, 1974). Those charged with DWI are more likely to be frequent and heavy drinkers who would exhibit impairment, and therefore be noticed, only at high BACs. Less frequent drinkers, who might be just as impaired at lower BACs, are less likely to be charged, particularly if they are below the legal limit. Yet, this latter group, overrepresented with younger, less experienced drivers who drink less frequently, is at increased risk of becoming involved in collisions. The older drivers may manage to drive with sufficient caution that they avoid major collisions, but may still show sufficient erratic driving behavior to attract the attention of the police. This explanation derives support from Vingilis' (1979) study of suspected DWI cases. Women and young men were overrepresented among those suspected of driving while impaired, but a greater proportion were not charged because their BACs were below the legal limit.

Our review of the literature indicates that there are significant variations in the characteristics and BACs of drivers who are charged with impaired driving compared to those involved in collisions and in each group when compared to

Table 4. Characteristics of Drivers Involved in Crashes and Citations and Drivers with Clear Records[a]

	Crash		Citation		Clear record	
					Fatal–control ($N = 31$) percent	Hospital control ($N = 32$) percent
	Fatality ($N = 121$)[b] percent	Hospital ($N = 26$) percent	DWI ($N = 33$) percent	Other ($N = 30$) percent		
Male	86	77	97	97	77	69
Under 25 yr	42	73	27	73	26	22
Single	41	46	36	57	23	25
Wid.-Div.-Sep.	5	4	18	3	3	0
Socioeconomic level						
Upper	26	20	9	6	52	46
Middle	32	19	24	13	15	28
Lower	24	19	50	58	10	6
Not in labor force	18	42	16	22	22	19
Daily beer drinking	—	29	61	56	39	24
Heavy beer drinking	—	24	52	20	8	10
Heavy liquor drinking	—	6	28	16	0	4
1 + crash past 5 yr	—	43	52	77	0	0
1 + suspensions past 5 yr	24	8	58	36	6	9
1 + citations past 5 yr	27	35	78	90	0	0

[a] Adapted from Perrine, 1970.
[b] For some variables, N's are less than total given.

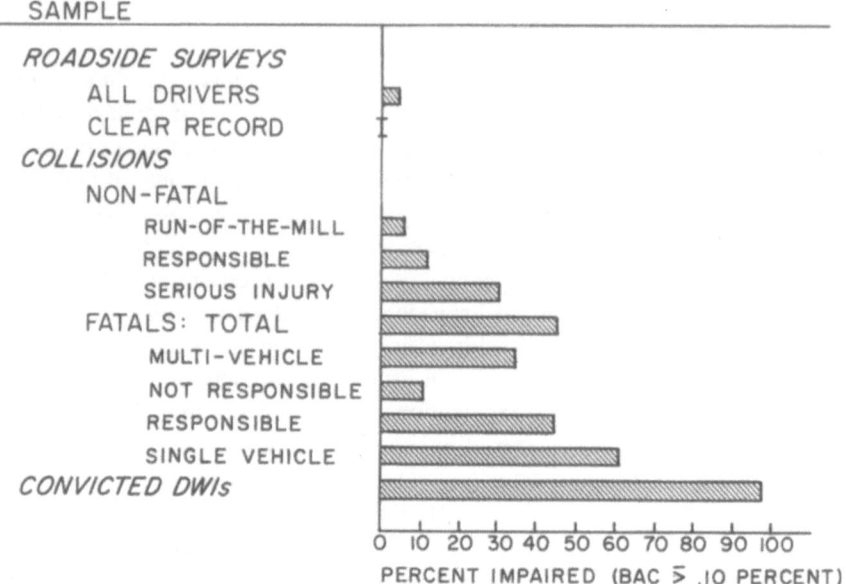

Figure 2. Percentage of impaired drivers in selected segments of the driving population. (Adapted from: Organisation for Economic Co-operation and Development, 1978. Based on average of relevant studies in Canada and in New York City; Grand Rapids, Michigan; Vermont; and other parts of the United States.)

controls or those selected at roadside. Perrine's (1970) comparison of drivers involved in crashes with those receiving citations for DWI or other reasons and those with clear records suggests the extent of the variation in driver characteristics and driving history (Table 4). The small sample sizes render some of the data suggestive rather than definitive, but a multiple discriminant analysis of twelve variables indicated significant differences between DWIs and clear-record drivers regarding convictions, occupational status, frequency of beer consumption, and quantity of liquor consumption (Perrine et al., 1971). The diversity of rates of impairment is illustrated by a comparison of selected segments of the driving population (Fig. 2). Here, the association of legal impairment with driving history, type of collision, severity of collision, and attribution of responsibility is very apparent.

One reason for these variations is the use of different definitions of impairment that result from different methods of selection. For example, DWIs are apprehended on the basis of highly visible signs of impairment that occur most often among drivers who are very intoxicated. When DWIs are compared with drivers selected at roadside who have very high BACs (≥0.15%), differences between the two groups, such as the sex and age distribution, are minimized (cf. Smith et al., 1976). Similarly, DWIs are most appropriately compared with drivers involved in collisions at very high BACs. The fact that young drivers have lower average BACs and are involved in collisions at lower BACs confounds many of the comparisons that are made between different populations of drivers, because age is associated with several variables, such as driving experience, tolerance to alcohol, and demographic characteristics. The restriction of most roadside surveys to high-risk times of the day and week when police are most likely to be patrolling the streets probably minimizes the differences between DWIs and impaired drivers detected at roadside. Because so few surveys have been carried out at other times of the day, we do not know whether impaired drivers in a comprehensive survey would look much different.

The issue of bias in the apprehension of impaired drivers is also complex. There is no evidence that police deliberately pick on drivers who are older, lower-class, or who belong to minority groups (Hyman et al., 1972; Zylman, 1972; Ward and Spronz, 1976). The fact that police patrols are concentrated in core areas of the city may put drivers with drinking patterns that increase their chances of being there at higher risk of apprehension (Whitehead, 1975). By comparison, young drivers, who are overinvolved in alcohol-related crashes, are more likely to be driving at excessive speeds, and this is easiest to achieve in semirural areas that are patrolled less intensively (Foley et al., 1976). A detailed analysis of age-specific data on drinking and driving patterns is required to provide definitive evidence for this type of selective bias.

To summarize, groups of drivers who are apprehended while impaired by alcohol differ in many ways that can be explained by the method of selection and the criteria for impairment. When these factors are controlled, the most important differentiating variables seem to relate to age, which is associated with variations in experience in driving, patterns of driving, psychological characteristics, risk of crashes unrelated to alcohol, patterns of alcohol consumption, and tolerance to alcohol.

The Role of Alcohol in Driving Impairment

The literature on driving impairment and alcohol-related collisions clearly indicates that there is an association between elevated BACs and increased risk of erratic driving and collisions. The extent to which alcohol plays a causal role in their occurrence is still undecided. The effect of personality factors, stress, and social role has been examined by a number of researchers (Pelz and Schuman, 1974; Selzer and Vinokur, 1974; Shaffer et al., 1974; Clark and Prolisko, 1979; Damkot, 1981), and there is evidence that drivers who are considered socially obstreperous or who have experienced unusual stresses in the recent past are at increased risk of becoming involved in collisions. Few studies are able to control for the use of other impairing drugs that might be associated with the use of alcohol (Waller, 1979). Other issues include the variability of degree of impairment for particular BACs, the role of driving experience, and the ability to compensate for the effects of alcohol.

It is clear that impairment by alcohol is neither a necessary nor sufficient cause of collisions (Warren and Simpson, 1978). Many fatal collisions and most nonfatal collisions are not associated with the use of alcohol by the driver. Nevertheless, there is stong evidence, both epidemiologic and experimental, that alcohol is a major risk factor for collisions.

Most laboratory and field studies show that some types of behavioral impairment occur even at low BACs (Moskowitz, 1974; Perrine, 1974; Laurell, 1979; Lovibond, 1979). In at least one field experiment, attempts were made to measure the impact of alcohol on driving responses in real life situations by using unobtrusive measures. Damkot et al. (1977) found that drivers with BACs between 0.08 and 0.15% stopped less smoothly than drivers at lower BACs. (Results for drivers at higher BACs were inconclusive because too few were sampled.)

Epidemiologic evidence that alcohol is an important risk factor for collisions comes from several sources. First, studies that examine culpability in multivehicle collisions report much higher blood alcohol levels among drivers considered at fault (e.g., Waller, 1972; Borkenstein et al., 1974; Farris et al., 1976). Second, there is a direct association between BAC and both the risk and the severity of collisions when drinking frequency is controlled (Hurst, 1974; O.E.C.D., 1978). This dose–response relationship and the fact that the increased magnitude of risk as a function of BAC is very marked also provide strong support for a causal hypothesis. Third, a variety of case–control studies conducted at roadside show excessively high rates of legal impairment among cases compared to matched controls (e.g., Farris et al., 1976). Fourth, a study of fatal crashes with and without alcohol involvement (O'Day, 1970) found that the distribution of crashes by age was bimodal. When separate distributions were plotted for those crashes that were alcohol-related and those that were not, it was found that the peak for crashes involving nondrinking drivers occurred at age 18, whereas that for drinking drivers occurred at age 21, which was the legal drinking age at the time. These data suggest that alcohol is an independent risk factor for collisions. Finally, evidence from this and other legal impact studies suggests that changes in the legal drinking age affect the rate of alcohol-involved collisions among young drivers (Whitehead, 1977; Wagenaar, 1981).

The evidence we have examined strongly supports the hypothesis that alcohol impairs driving performance and thereby raises the risk of crashes to a considerable extent. Individual and environmental factors also play a role, but alcohol may be the most important single factor in elevating risk. Jones and Joscelyn (1977) conclude that the relationships between personality and stress variables and crashes involving alcohol are not sufficiently developed to be useful predictors of the risk of a crash. That alcohol is not involved in most crashes, particularly those that are less severe, is due in large part to the fact that most drivers have not been drinking or have consumed only small amounts of alcohol.

4.   SUMMARY AND CONCLUSIONS

Current knowledge on the subject of impaired driving derives from many sources that differ in fundamental ways. We have described these sources, assessed the validity and comparability of their data, and presented a variety of recent findings.

We now have considerable descriptive information about the characteristics and circumstances of nighttime driving, and about drivers involved in various types of collisions and drivers charged with legal impairment and the extent of alcohol involvement associated with each. Laboratory results suggest that certain complex behaviors that may represent those involved in real life driving are impaired at certain BACs. Case–control studies have provided evidence that alcohol is a strong risk factor for collisions. Although many methodologic issues limit the applicability and specificity of these findings, they can usefully be applied to related problems for, as Jones and Joscelyn (1978) stated, "The main value of research in the field lies not in providing irrefutable proof of hypotheses about drinking and driving, but in providing data on which to base informed decision-making."

Nevertheless, there are a number of specific gaps in our knowledge. The exact processes whereby alcohol causes impairment have not been clearly delineated. Laboratory evidence suffers from problems of validity, among others, and there is some doubt that findings can be applied to actual driving behavior. The epidemiologic literature is deficient in current, comprehensive studies that compare large samples of cases involved in collisions with matched controls (Jones and Joscelyn, 1978). We do not have recent North American data on rates of legal impairment among the general driving population on a 24-hr, 7-days-a-week basis. We do not know the precise reasons why samples of DWIs differ from legally impaired drivers selected at roadside. We have an idea of the many variables that are involved in the etiology of collisions generally and alcohol-related collisions specifically, but multivariate techniques are required to determine the pattern and importance of the various contributing factors operating simultaneously as well as their interaction.

We need behavioral data on the differential effects of alcohol on various age groups to confirm or refute the "inexperience hypothesis," which is used to explain higher rates of crashes at lower BACs among young drivers (Jones and Joscelyn, 1977). The role of physiological, psychological, and social factors that would explain sex differences in risk factors for crashes should be explored. The effect of personality and stress variables should be tested within the context of controlled studies that use measures of alcohol and psychological traits that are applicable at the time of the event (Jones and Joscelyn, 1977).

Knowledge of the subject of impaired driving has progressed enormously during the past two decades. Recent years have seen a growing sophistication of study design and an increasing integration of findings. The massive body of data that has been collected is used more often as a basis for developing theory.

Because of the ethical and logistical problems inherent in conducting true roadside experiments, future research will require imaginative designs that involve unobtrusive measures of impaired driving behavior.

Changes in the structure of the driving population and in drinking and driving patterns may well shift attention from young men to older drivers and women who are also at excessive risk of crashes when legally impaired. The detailed examination of these groups requires samples that are larger overall or that oversample in these categories.

The successful pursuit of knowledge on driver impairment and alcohol-related collisions will not only enhance the prospect of developing effective measures for prevention, but will also provide a model for quasiexperimental investigations that may be useful in other lines of research.

ACKNOWLEDGMENTS

The authors express their appreciation to Gerald J.S. Wilde and Evelyn R. Vingilis for many valuable comments and suggestions on an earlier version of this chapter.

REFERENCES

Allsop, R. E., 1966, Alcohol and Road Accidents, Road Research Laboratory Report No. 6, Ministry of Transport, Harmondsworth.

Attwood, D. A., Williams, R. D., and Madill, H. D., 1980, Effects of moderate blood alcohol concentrations on closed-course driving performance, *J. Stud. Alcohol* **41**(7):623.

Beitel, G. A., Sharp, M. C., and Glauz, W. D., 1975, Probability of arrest while driving under the influence of alcohol, *J. Stud. Alcohol* **36**:109.

Bonnichsen, R., and Goldberg, L., 1981, Large-scale breath-blood comparisons under field conditions: Methods, evaluation techniques and results, in: *Alcohol, Drugs and Traffic Safety*, Volume 1 (L. Goldberg, ed.) pp. 796–810, Almqvist & Wiksell International, Stockholm, Sweden.

Borkenstein, R. F., 1975, Problems of enforcement, adjudication and sanctioning, in: *Alcohol, Drugs and Traffic Safety* (S. Israelstam and S. Lambert, eds.), pp. 655–662, Addiction Research Foundation, Toronto.

Borkenstein, R. F., Crowther, R. F., Shumate, R. P., Ziel, W. B., and Zylman, R., 1964, The role of the drinking driver in traffic accidents, *Blutalkohol* **11**(Suppl. 1):1.

Borkenstein, R. F., Perrine, M. W., Van Berkom, L. C., and Crowther, R. F., 1974, Scientific Roadside Surveys of Alcohol in the Driving Population, Committee on Alcohol and Drugs, National Safety Council, Chicago.

Boshier, R., and Johnson, D., 1974, Does conviction affect employment opportunities? *Br. J. Criminol.* **14**(3):264.

Bradburn, N. M., and Sudman, S., 1979, *Improving Interview Method and Questionnaire Design*, Jossey-Bass, San Francisco.

Bragg, B. W. E., and Cousins, L. S., 1979, Changing the subjective probability of arrest for impaired driving, in: *Proceedings of the 7th International Conference on Alcohol, Drugs and Traffic Safety* (I. R. Johnston, ed.), pp. 649–654, Australian Government Publishing Service, Canberra.

Browning, J. J., and Wilde, G. J. S., 1975, Research in Drinking and Driving, Studies of Safety in Transport, Queen's University, Kingston, Canada.

Campbell, D. T., and Stanley, J. C., 1963, *Experimental and Quasiexperimental Designs for Research*, Rand McNally and Company, Chicago.

Carlson, W. L., 1972, Alcohol usage of the nighttime driver, *J. Safety Res.* **4:**12.

Carr, B., Borkenstein, R. F., Perrine, M. W., Van Berkom, L. C., Voas, R. B., et al., 1974, International Conference on Research Methodology for Roadside Surveys of Drinking-Driving, Alcohol Countermeasures Workshop Report No. DOT HS-801 220, National Highway Traffic Safety Administration, Washington, D.C.

Chapanis, A., 1967, The relevance of laboratory studies to practical situations, *Ergonomics* **10:**557.

Chi, L., Ferrence, R. G., and Whitehead, P. C., 1973, Characteristics of Impaired Drivers in London, Ontario, Substudy No. 563, Addiction Research Foundation, Toronto.

Clark, A. W., and Prolisko, A., 1979, Social role correlates of driving accidents, *Hum. Factors* **21**(6):655.

Damkot, D. K., 1981, Alcohol, task demands, and personality affect driving: Beware the interactions, in *Alcohol Drugs and Traffic Safety*, Volume 1 (L. Goldberg, ed.), pp. 923–937, Almqvist & Wiksell International, Stockholm, Sweden.

Damkot, D. K., Toussie, S. R., Akley, N. R., Geller, H. A., and Whitmore, D. G., 1977, On-the-Road Driving Behavior and Breath Alcohol Concentration, U.S. Department of Transportation, National Highway Traffic Safety Administration, Washington, D.C., Report No. DOT HS-802 264.

Douglass, R. L., Filkins, L. D., and Clark, F. A., 1974, The Effect of Lower Legal Drinking Age on Youth Crash Involvement, U.S. Department of Transportation, National Highway Traffic Safety Administration, Washington, D.C., Report No. DOT HS-801 213.

Dubowski, K. M., 1975, Recent developments in breath-alcohol analysis, in: *Alcohol, Drugs and Traffic Safety* (S. Israelstam and S. Lambert, eds.), pp. 483–494, Addiction Research Foundation, Toronto.

Duncan, D., and Vogel-Sprott, M., 1978, Drinking habits of impaired drivers, *Blutalkohol* **15:**252.

Duncan, J. A., 1979, Report on a survey of the effect of breath testing in Canberra, A.C.T., in: *Proceedings of the 7th International Conference on Alcohol, Drugs and Traffic Safety* (I. R. Johnston, ed.), pp. 481–493, Australian Government Publishing Service, Canberra.

Edelman, S., and Walker, D., 1977, *Impaired Driving: A Critical Review of the Research Literature*, Ministry of the Attorney-General, Vancouver, Canada.

Edwards, D. S., Hahn, C. P., and Fleishman, E. A., 1969, Evaluation of Laboratory Methods for the Study of Driver Behavior: The Relation between Simulator and Street Performance, American Institute for Research, Silver Springs, Maryland, Final Report.

Ennis, P., 1978, Drinking and Driving among Ontario Students in 1977, Substudy No. 938, Addiction Research Foundation, Toronto.

Farris, R., Malone, T. B., and Lilliefors, H., 1976, A Comparison of Alcohol Involvement in Exposed and Injured Drivers, Phases I and II, National Highway Traffic Safety Administration, Washington, D.C., Technical Report DOT HS-801 826.

Foley, J. P., Glauz, W. D., and Sharp, M., 1976, Profile analysis of persons arrested for drunk driving, *Hum. Factors* **18**(5):455.

Freedman, K., Henderson, M., and Wood, R., 1973, Drinking and Driving in Sydney: A Community Survey of Behaviour and Attitudes: Report 1: An Overview of Sex and Age Differences, Traffic Accident Research Unit 1/73, Department of Motor Transport, New South Wales, Australia.

Goldberg, L., 1981, Random road tests in non-accident-involved drivers: Epidemiological data, differential characteristics and role of alcoholism, in: *Alcohol Drugs and Traffic Safety*, Volume 1 (L. Goldberg, ed.), pp. 846–859, Almqvist & Wiksell International, Stockholm, Sweden.

Harger, R. N., 1975, The Hamburg, W. Germany, 1973 report questioning the reliability of breath alcohol instruments for medicolegal use in traffic cases, in: *Alcohol, Drugs and Traffic Safety* (S. Israelstam and S. Lambert, eds.), pp. 583–596, Addiction Research Foundation, Toronto.

Hickey, M., Hayden, P. M., Layden, M. T., and Hearne, R., 1975, Alcohol and driving, *J. Irish Med. Assoc.* **68**(23):583.

Hurst, P. M., 1974, Epidemiological aspects of alcohol in driver crashes and citations, in: *Alcohol, Drugs and Driving* (M. W. Perrine, ed.), National Highway Traffic Safety Administration, Technical Report, DOT HS-801 096.

Hurst, P. M., and Darwin, J. H., 1977, Estimation of alcohol for nonrespondents in roadside breath surveys, *Accident Anal. Prev.* **9**(2):119.

Hyman, M. M., 1968, Accident vulnerability and blood alcohol concentrations of drivers by demographic characteristics, *Q. J. Stud. Alcohol* (Suppl. 4):34.

Hyman, M. M., Helrich, A. R., and Besson, G., 1972, Ascertaining police bias in arrests for drunken driving, *Q. J. Stud. Alcohol* **33**:148.

Institute for Road Safety Research (SWOV), 1977, Drinking by Motorists, SWOV, Voorburg, The Netherlands.

Interministerial Committee on Drinking–Driving, 1979, The 1979 Ontario Roadside BAC Survey: Summary Report, Ministry of the Solicitor-General, Toronto.

Jones, A. W., 1976, Precision, accuracy and relevance of breath alcohol measurements, in: *Alcohol, Drugs and Driving* (M. Mattila, ed.), Volume II of *Modern Problems of Pharmacopsychiatry*, S. Karger, Basel.

Jones, R K., and Joscelyn, K. B., 1977, 1977 Report on Alcohol and Highway Safety, Technical Volume, The University of Michigan, Highway Safety Research Institute, Ann Arbor.

Jones, R. K., and Joscelyn, K. B., 1978, Alcohol and Highway Safety 1978: A Review of the State of Knowledge, Summary Volume, The University of Michigan, Highway Safety Research Institute, Ann Arbor.

Kannemann, K., and Warren, R. A., 1980, A method for estimating the BAC distribution among fatally injured drivers, *Accident Anal. Prev.* **12**(4):247–257.

Lacey, J. H., Stewart, J. R., and Council, F. M., 1979, Development of predictive models to identify persons at high risk of alcohol related crash involvement, in: *Proceedings of the 7th International Conference on Alcohol, Drugs and Traffic Safety* (I. R. Johnston, ed.), pp. 107–119, Australian Government Publishing Service, Canberra.

Lance, J.-M. R., and Romeder, J.-M., 1979, La fraction de la mortalité routière attribuable à l'alcool, *Can. J. Public Health* **70**:310.

Laurell, H., 1979, Effects of small doses of alcohol on driver performance in emergency traffic situations, in: *Proceedings of the 7th International Conference on Alcohol, Drugs and Traffic Safety* (I. R. Johnston, ed.) pp. 157–167, Australian Government Publishing Service, Canberra.

Lehman, R. J., Wolfe, A. C., and Kay, R. D., 1975, A Computer Archive of ASAP Roadside Breathtesting Surveys, 1970–1974, Highway Safety Research Institute, Ann Arbor.

Lemert, R., 1966, Strategies of research design in the legal impact study, *Law Soc. Rev.* (November), **1**(1):111–132.

Levett, J., and Karras, L., 1975, Errors in current alcohol breath analysis, in: *Alcohol, Drugs and Traffic Safety* (S. Israelstam and S. Lambert, eds.) pp. 527–532, Addiction Research Foundation, Toronto.

Lovell, W. S., 1972, Breath tests for determining alcohol in the blood, *Science* **178**:264.

Lovibond, S. H., 1979, The effects of alcohol on skilled performance, in: *Proceedings of the 7th International Conference on Alcohol, Drugs and Traffic Safety* (I. R. Johnston, ed.) pp. 168–173, Australian Government Publishing Service, Canberra.

Maisto, S. A., Sobell, L. C., Zelhart, P. F., Connors, G. J., and Cooper, T., 1979, Driving records of persons convicted of driving under the influence of alcohol, *J. Stud. Alcohol* **40**(1):70.

Moskowitz, H., 1974, Alcohol influences upon sensory motor function, visual perception, and attention, in: *Alcohol, Drugs and Driving* (M. W. Perrine, ed.), National Highway Traffic Safety Administration, Washington, D.C.

Noordzij, P. C., 1975, Comparison of blood and breath testing under field conditions, in: *Alcohol, Drugs and Traffic Safety* (S. Israelstam and S. Lambert, eds.), pp. 553–560, Addiction Research Foundation, Toronto.

Oates Jr., J. F., 1974, Factors Influencing Arrests for Alcohol-Related Traffic Violations, National Highway Traffic Safety Administration, Report No. DOT HS-801 230, Washington, D.C.

O'Day, J., 1970, Drinking involvement and age of young drivers in fatal accidents, *HIT LAB Rep.* **October:**13–14.

Organisation for Economic Co-operation and Development, 1978, Road Research: New Research on the Role of Alcohol and Drugs in Road Accidents, O.E.C.D., Paris (September).

Pelz, D. C., and Schuman, S. H., 1974, Drinking, hostility and alienation in driving of young men, in: *Proceedings of the 3rd Annual Alcoholism Conference of the National Institute on Alcohol Abuse and Alcoholism*, June 1973 (M. E. Chafetz, ed.), National Institute of Alcohol Abuse and Alcoholism, Rockville, Maryland, DHEW Publication No. (ADM) 75-137.

Penttila, A., Aas, K., Nevala, P., Piipponen, S., and Pikkarainen, J., 1981, Drunken driving in Finland. II. The profile on frequency based on road-side surveys, in: *Alcohol, Drugs and Traffic Safety*, Volume 1 (L. Goldberg, ed.), pp. 234–250, Almqvist & Wiksell International, Stockholm, Sweden.

Penttila, A., Maki, M., and Pikkarainen, J., 1979, Drunken driving in Finland. Late-night, weekend incidence in five communities in 1978, *J. Traffic Med.* **7**(2):30.

Perrine, M. W., 1970, Identification of personality, attitudinal and biographical characteristics of drinking drivers, *Behav. Res. Highway Safety* **2**(4):207.

Perrine, M. W., 1971, Methodological Considerations in Conducting and Evaluating Roadside Research Surveys, National Highway Traffic Safety Administration, Report No. DOT HS-800 471, Washington, D.C.

Perrine, M. W., 1974, Alcohol influences upon driving-related behavior: A critical review of laboratory studies of neurophysiological, neuromuscular and sensory activity, in: *Alcohol, Drugs and Driving* (M. W. Perrine, ed.), National Highway Traffic Safety Administration Report No. DOT HS-801 096, Washington, D.C.

Pliner, P., and Cappell, H., 1977, Drinking, driving, and the attribution of responsibility, *J. Stud. Alcohol* **38**(3):593.

Rockerbie, R. A., Martin, G. R., and Parkin, H. E., 1981, Blood alcohol in hospitalized traffic crash victims, in: *Alcohol, Drugs and Traffic Safety*, Volume 1 (L. Goldberg, ed.), pp. 165–173, Almqvist & Wiksell International, Stockholm, Sweden.

Rosenberg, N., Laessig, R. H., and Rawlings, R. R., 1974, Alcohol, age and fatal traffic accidents, *Q. J. Stud. Alcohol* **35**:473.

Ryan, G. A., and Salter, W. E., 1979, Drinking habits, social characteristics and blood alcohol concentrations of road crash casualties, in: *Proceedings of the 7th International Conference on Alcohol, Drugs and Traffic Safety* (I. R. Johnston, ed.), Australian Government Publishing Service, Canberra.

Schliewen, R. E., 1979, 1974 National Roadside Survey, Data Analysis and Report, Transport Canada, Road and Motor Vehicle Traffic Safety Branch, Ottawa.

Selzer, M. L., and Vinokur, A., 1974, Life events, subjective stress, and traffic accidents, *Am. J. Psychiatry* **131**(8):903.

Shaffer, J. W., Towns, W., Schmidt Jr., C. W., Fisher, R. S., and Zlotowitz, H. I., 1974, Social adjustment profiles of fatally injured drivers: A replication and extension, *Arch. Gen. Psychiatry* **30**(4):508.

Shults, S. D., and Layne Jr., N. R., 1979, Age and BAC when arrested for drunken driving and public drunkenness, *J. Stud. Alcohol* **40**(5):492.

Simpson, H. M., Warren, R. A., Pagé-Valin, L., and Collard, D., 1978, Analysis of Fatal Traffic Crashes in Canada, 1976. Focus: The Impaired Driver, Traffic Injury Research Foundation of Canada, Ottawa.

Smart, R. G., 1969, Are alcoholics' accidents due solely to heavy drinking? *J. Safety Res.* **1**(4):170.

Smart, R. G., and Schmidt, W., 1969, Physiological impairment and personality factors in traffic accidents of alcoholics, *Q. J. Stud. Alcohol* **30**:440.

Smith, G. A., Wolynetz, M. S., and Wiggins, T. R. I., 1976, "Drinking Drivers in Canada: A National Roadside Survey of the Blood Alcohol Concentrations in Nighttime Canadian Drivers," Transport Canada, Road and Motor Vehicle Traffic Safety Branch, Ottawa.

Sobell, M. B., Sobell, L. C., and Samuels, F. H., 1974, Validity of self-reports of alcohol-related arrests by alcoholics, *Q. J. Stud. Alcohol* **35**:276.

Stroh, C. M., 1974, Alcohol and Highway Safety; Roadside Surveys of Drinking-Driving Behavior: A Review of the Literature and a Recommended Methodology, Transport Canada, Road and Motor Vehicle Traffic Safety Branch, Ottawa.

Traffic Injury Research Foundation, 1975, Analysis of Fatal Traffic Crashes in Canada, 1973. Focus: The Impaired Driver, Traffic Injury Research Foundation of Canada, Ottawa.

Van Ooijen, D., 1979, The effects of a new DWI law, in: *Proceedings of the 7th International Conference on Alcohol, Drugs and Traffic Safety* (I. R. Johnston, ed.), pp. 471–480, Australian Government Publishing Service, Canberra.

Vingilis, E. R., 1982, Sex and age characteristics of suspected impaired drivers, *Accident Anal. Prev.* **14**:425.

Wagenaar, A. C., 1981, Effects of an increase in the legal minimum drinking age, *J. Public Health Policy* **2**(3):206.

Waller, J. A., 1972, Factors associated with alcohol and responsibility for fatal highway crashes, *Q. J. Stud. Alcohol* **33**:160.

Waller, J. A., 1975, Epidemiologic issues about alcohol, other drugs and highway safety, in: *Alcohol, Drugs and Traffic Safety* (S. Israelstam and S. Lambert, eds.), pp. 3–12, Addiction Research Foundation, Toronto.

Waller, J. A., 1979, Analytic issues in studying the interaction of alcohol and other drugs in highway crashes, in: *Proceedings of the 7th International Conference on Alcohol, Drugs and Traffic Safety* (I. R. Johnston, ed.), pp. 15–23, Australian Government Publishing Service, Canberra.

Ward, R. M., and Spronz, W., 1976, A geographical analysis of drunken drivers, *J. Stud. Alcohol* **37**(7):997.

Warren, R. A., 1980, Treatment of Missing Data in Road Safety Investigations: Methodological Innovations. Paper presented at the 17th Annual Meeting of the Traffic Injury Research Foundation of Canada, Toronto.

Warren, R. A., and Simpson, H. M., 1976, The Young Driver Paradox, Traffic Injury Research Foundation, Ottawa.

Warren, R. A., and Simpson, H. M., 1978. Impaired Driving, Technical Report Series No. 8, Health and Welfare Canada, Ottawa.

Warren, R. A., and Simpson, H. M., 1981, Effects of measurement assumptions on estimations of risk of collisions, Traffic Injury Research Foundation, Ottawa, Canada.

Webster's, 1962, *Webster's New World Dictionary*, Nelson, Foster and Scott Ltd., Toronto.

Wechsler, H., and McFadden, M., 1979, Drinking among college students in New England: Extent, social correlates and consequences of alcohol use, *J. Stud. Alcohol* **40**(11):969.

Whitehead, P. C., 1975, DWI programs: Doing what's in or dodging what's indicated? *J. Safety Res.* **7**(3):127.

Whitehead, P. C., 1977, *Alcohol and Young Drivers: Impact and Implications of Lowering the Drinking Age*, Monograph Series, No. 1, Health Protection Branch, Health and Welfare Canada, Ottawa.

Wilde, G. J. S., 1974, Alcohol and Traffic Accidents: A Review in Quest of Remedies, Alcohol and Highway Safety, Ministry of Transport, Ottawa, Canada.

Willett, T. C., 1964, *Criminal on the Road*, No. 12, International Library of Criminology, Delinquency and Deviant Behaviour, Tavistock Publications, London.

Wolfe, A. C., 1974, 1973 U.S. National Roadside Breathtesting Survey: Procedures and Results, Highway Safety Research Institute, Ann Arbor, Michigan.

Wolfe, A. C., 1975, Characteristics of late-night, weekend drivers: Results of the U.S. national roadside breath testing survey and several local surveys, in: *Alcohol, Drugs and Traffic Safety* (S. Israelstam and S. Lambert, eds.), pp. 41–50, Addiction Research Foundation, Toronto.

Wolfe, A. C., and Chapman, M. M., 1973, An analysis of drinking and driving survey data, *HIT LAB Rep.* 4(4):1.

Yoder, R. D., and Moore, R. A., 1973, Characteristics of convicted drunken drivers, *Q. J. Stud. Alcohol* 34:927.

Zelhart, P. F., 1972, Types of alcoholics and their relationship to traffic violations, *Q. J. Stud. Alcohol* 33:811.

Zelhart, P. F., and Schurr, B. C., 1977, People who drive while impaired, a review of research on the drinking-driving problem, in *Alcoholism: Physiological and Psychological Bases* (N. Estes and E. Heinemann, eds.), Mosby, St. Louis.

Zusman, M. E., and Huber, J. D., 1979, Multiple measures and the validity of response in research on drinking drivers, *J. Safety Res.* 11(3):132.

Zylman, R., 1972, Race and social status discrimination and police action in alcohol-affected collisions, *J. Safety Res.* 4(2):75.

9

The Contribution of Prospective Studies to the Understanding of Etiologic Factors in Alcoholism

GEORGE E. VAILLANT

1. INTRODUCTION

The dimension of time is crucial to the comprehension of unfolding natural phenomena. Thanks to Newton and Einstein this fact will not surprise workers in the natural sciences, but the effects of time continue to bewilder social scientists. Through their interest in static and statistical relationships between individuals and between groups, social scientists sometimes lose track of dynamisms within the single individual. Such dynamisms can only be elucidated by combining prospective study with focus upon case histories. Such study can be best accomplished by personal interviews at multiple points in time. As water refracts the passage of light and so produces visual illusions, so do lives passing through time bend memory and understanding and thus produce cognitive illusions. In examining how prospective studies dispel such illusions, the intent of this review will be to be representative rather than exhaustive.

In masterful reviews, Rutter (1981) and Robins (1979) have pointed out what we generally may expect to learn from prospective studies. This essay will elaborate upon these generalizations as they pertain to alcoholism in particular. I shall address five areas that Robins and Rutter have singled out as being peculiarly amenable to elucidation by prospective study. First, prospective study can dispel preferential recollection, especially as memory distorts the association between alcoholism and alleged childhood and familial antecedents. Second,

---

GEORGE E. VAILLANT ● Dartmouth Medical School, Hanover, New Hampshire .

prospective study of community samples can dispel enduring mythologies that are maintained by nonresponse bias. Such bias is inherent in the cross-sectional study of that subsample of alcoholics who come to clinics. Third, in the case of highly correlated symptoms, prospective study can elucidate "cart and horse" relationships; this can reveal which symptom occurred first. Fourth, prospective studies not only validate diagnosis but lend coherence to an evolving symptom picture. Finally, only prospective design clarifies etiologic factors which affect prognosis.

The studies upon which I shall focus include six prospective studies of the premorbid characteristics of alcoholics (McCord and McCord, 1960; Robins et al., 1962; Jones, 1968, 1971; Loper et al., 1973; Vaillant, 1980a; and Vaillant and Milofsky, 1982a). These studies are summarized in Table 1. I shall also focus upon six outcome studies of 6 or more years' duration which were not always prospective in design but which illuminate clinical course (Voegtlin and Broz, 1949; Sundby, 1967; Goodwin et al., 1971; Ojesjo, 1981; Pendery et al., 1982; Vaillant, 1983). By including the dimension of time, each of these 12 studies helps to dispel common illusions about alcoholism. They are summarized in Table 2.

The selection of the 12 studies is admittedly arbitrary. The first criterion, of course, is prospective design. In a prospective design, the subjects are initially studied as research subjects and the initial measures are recorded; they are then followed forward in time. In an outcome or follow-back study the subject is studied in time present and the past is reconstructed from old records that may or may not have been gathered with the researcher's question in mind. A second criterion for selection was length of study (at least 8 years). A third criterion was that the study focus on alcohol abuse rather than use; and a fourth, that the data gathered be relevant to issues of etiology. Among the more important studies excluded on the basis of the criteria were those by Myerson and Mayer (1966), Lundquist (1973), Bratfos (1974), Fillmore (1975), and Hyman (1976).

2. PREFERENTIAL RECOLLECTION

The fallibility of human memory is a stumbling block to understanding the etiology of many disorders, but in no case is this observation more striking than in alcoholism. In 1941 Paul Schilder wrote, "The chronic alcoholic person is one who from his earliest childhood has lived in a state of insecurity" (p. 290). There is little in the retrospective literature, that is, the literature based upon the memory of alcoholics, that would contradict Schilder's thesis. But the illness of alcoholism profoundly distorts the individual's own recollection of relevant childhood variables.

Each of the studies in Table 1 contradicts common suppositions about the contribution of childhood to alcoholism. However, the studies by the McCords

Table 1. Prospective Studies of the Development of Alcoholism

Study	Age at follow-up	Original sample	Lost to follow-up	Alcohol abusers	Not also clearly antisocial	Proportion of alcohol abusers also dependent[a]	Recent interviews	Multiple contacts over time
McCord and McCord (1960)	30–35	325	70	29	11	2/3	No	No
Robins et al. (1962)	c. 43	367	81	78[b]	29	1/3	Yes	No
Jones (1968)	36–38	106	40	6	6	?	Yes	Yes
Jones (1971)	36–38	106	61	3[c]	3	?	Yes	Yes
Loper et al. (1973)	30–35	[d]	[d]	38	38	All	No	No
Vaillant (1980)	55	204	2	26	25	1/3	Yes	Yes
Vaillant and Milofsky (1982a)	47	456	56	110	90	2/3	Yes	Yes

[a] These proportions are our rough estimates of the severity of the alcohol abuse in the reviewed studies based on the authors' criteria for inclusion.
[b] This figure includes the men who, because they met Robins' diagnostic criteria for schizophrenia and sociopathy, were excluded from her later report (Robins, 1966). The figure excludes women and controls who abused alcohol.
[c] In a more recent report (Jones, 1981), this has been increased to 8.
[d] This was a follow-back study.

Table 2. Six Long-Term Follow-Up Studies of Alcohol Abuse

Study and nature of sample	Treatment	Type and length of follow-up	Size of original sample	Attrition (percent)		Number of survivors followed	Outcome for survivors (percent)		
				Lost or refused	Dead		Abstinent	Asymptomatic drinkers	Still alcoholic
Voetglin and Broz (1949): Private inpatient; good prognosis; ages 30–50	Emetine aversion and follow-up	Mail, 10 years	?	?	?	104		22[a]	78
Sundby (1967): Clinic; poor prognosis; all classes; ages 30–55	Nonspecific	Record search 20–35 years	1722	2	62	632		64[a]	36
Goodwin et al. (1971): Prison; alcoholic only by history; ages 20–35	Nonspecific	Interviews, 8 years	c.111	13	5	93	8	33	59
Ojesjo (1981): Community; good prognosis; age 46 ± 20	Nonspecific	Interviews, 15 years	96	0	26	71		32[a]	68
Pendery et al. (1982):	Behavior therapy in social drinking	Interviews and record search, 10 years	20	0	30	14	43	7	50
Vaillant (1983): Inpatient; poor prognosis; public clinic; ages 30–50	Detoxification; AA-oriented; follow-up	Interviews, 8 years	106	6	27	71	39	6	55

[a] The authors did not distinguish between remission achieved via abstinence or via return to asymptomatic drinking.

and by Robins focused selectively upon premorbidly antisocial children and were not designed to study the childhoods of alcoholics in general. The studies by Jones (1971, 1981) contained a very small number of subjects, and those by Vaillant (1980) and Loper et al. (1973) were confined to college graduates. Thus, perhaps the study in Table 1 best suited to test Schilder's generalization was the study by Vaillant and Milofsky (1982a). They undertook a reexamination of the junior high school boys whom the Gluecks (1950) had used as a control group in their monograph *Unraveling Juvenile Delinquency*. Originally, this sample of inner city men had been selected in early adolescence for nondelinquency. The men were personally reinterviewed at times—ages 25, 31, and 47—well past the peak age for onset for alcoholism. Although chosen at age 14 ± 2 for the absence of obvious delinquency, eventually 19% of the sample spent time in jail; this datum suggests that the sample was only modestly biased towards good behavior. Because the sample had been matched with a delinquent sample for IQ, ethnicity, and residence in a high-crime neighborhood, the average IQ of the sample was 95. Only 48% graduated from high school and 61% of the men had at least one parent who had been born in a foreign country.

At the time of the original study the boys, their parents and their teachers were individually interviewed. Public records were searched for evidence of alcoholism, criminal behavior, and mental illness in relatives. Over the three generations, multiple social agency reports were available on most families. Ninety percent of surviving subjects were reinterviewed at ages 25 and 31 (Glueck and Glueck, 1968) and 83% of survivors were reinterviewed at age 47. Alcoholism was defined by both the criteria of Cahalan (1970) and the *Diagnostic and Statistical Manual III (DSM III)*.

Two major scales were used to assess the quality of the subjects' childhoods by raters blind to adult outcome. The first was a scale of childhood environmental strengths (Vaillant, 1974). This scale gave points for what went well in childhood rather than what went wrong. The second scale was of childhood environmental weaknesses (Vaillant and Vaillant, 1981). This scale was based on 25 objective criteria reflecting the Gluecks' more impressionistically defined delinquency prediction scale (Glueck and Glueck, 1950). This scale included five items reflecting gross lack of familial cohesion (e.g., nine or more social agency contacts, raised for more than 6 months apart from both parents); 10 items reflecting gross lack of affection and supervision by the mother, and 10 items on gross lack of supervision and affection from the fathers. Other scales measured familial alcoholism, social class, and boyhood employment.

In this study by Vaillant and Milofsky, few significant differences in the childhoods of alcoholics and nonalcoholics were observed. For example, although as adults many more alcoholics than nonalcoholics were in social class V, there had been no such differences observed in the social class of their parents. Nor were there differences in IQ between alcoholics and nonalcoholics. Although as adults many more alcoholics than nonalcoholics were diagnosed as mentally ill, as children the alcoholics had not been rated more emotionally disturbed nor had they come more often from multiproblem families. As adults, alcoholics were much less likely to have had stable employment than nonalcoholics, but in childhood there had been no differences in their capacity for schoolwork or parttime employment. There were no significant between-group differences in the subjects' relationships with their mothers in terms of either maternal affection or maternal supervision. Given that retrospective studies often report difficulties along these lines, the implication is that alcoholics, after the fact, may exaggerate childhood difficulties to explain alcoholism-engineered difficulties.

There were statistically significant differences between childhoods of alcoholics and nonalcoholics in that the former experienced fewer childhood environmental strengths and less-stable familial cohesion and relationship with the father. However, when parental alcoholism was controlled even these differences disappeared. Put differently, of 51 men who had *few* childhood environmental weaknesses but who did have an alcoholic parent, 27% became alcohol-dependent as adults; of the 56 men with many environmental weaknesses but no alcoholic parent only 5% became alcohol-dependent.

Besides familial alcoholism, the other major childhood variable that predicted adult alcoholism was ethnicity (Vaillant and Milofsky, 1982a). The Gluecks' sample was well-suited for the contrast of familial differences in cultural attitudes toward alcohol use. The boys themselves shared the same urban environment, the same community patterns of alcohol price and availability of alcohol, and roughly the same peer groups. In contrast, their parents, most of whom were foreign-born came from very diverse cultural backgrounds. Admittedly, the relationship between alcohol use and culture is extremely complex (Heath, 1975; Marlatt and Rohsenow, 1980; and Stivers, 1976), but in the study by Vaillant and Milofsky (1982a) this relationship was grossly simplified. Namely, ethnicity was crudely scaled to contrast cultures that forbid drinking in children but condone drunkenness in adults (e.g., Irish and Anglo-American) with cultures that teach children how to drink responsibly but condemn adult drunkenness (e.g., Jewish and Italian) (Jellinek, 1960; Pittman and Snyder, 1962). The 75 men of Irish extraction were seven times more likely to manifest alcohol dependence than the 130 men of Italian, Syrian, Jewish, Greek, or Portugese extraction.

The lack of correlation between bleak childhood and adult alcoholism was confirmed in a second study (Vaillant, 1980) of a very different socioeconomic sample. The second study was a 40-year prospective study of 268 college graduates who were extensively studied in college and reinterviewed at ages 30 and 47 (Vaillant, 1977). Attrition was negligible. As in the Gluecks' inner city study (Vaillant and Milosky, 1982a), bleak childhood in the college sample significantly predicted poor adult mental health but not alcoholism. In these college men the absence of environmental strengths predicted, and presumably played a causal role in, the development of "oral traits" (pessimism, passivity, self-doubt, and heightened dependency) and also predicted increased adult use of prescription drugs, even when psychoactive drugs were excluded (Vaillant, 1983). If oral dependent traits did not predict alcohol abuse, once men began to abuse alcohol then both oral dependent traits and use of mood-altering drugs became more common.

A second retrospective perception that could not be confirmed by prospective study was the assertion that alcoholics often abuse alcohol from their initial contact with the drug. A retrospective study by Ullman (1953) as well as the testimonials of Alcoholics Anonymous speakers suggest that alcohol abuse can begin after an alcoholic's first drink. In both of the prospective studies by Vaillant included in Table 1 alcohol dependence appeared very much to resemble tobacco dependence; nobody "has to smoke" after a few weeks of cigarette use. Progression from asymptomatic asocial drinking to frank alcohol abuse to frank alcohol dependence occurred gradually over a span of 3–30 years. Before real difficulties in control could be discerned, many of the alcohol abusers in the college sample drank socially for as long as 20 years. Admittedly, there is enormous individual variation in the evolution of alcoholism, both in the rapidity of onset of abuse and in the "progression" of symptoms and the eventual severity of alcohol dependence. Compared to alcoholism in the well-socialized, upper middle class, college sample, alcoholism in the most sociopathic members of the Gluecks' inner city sample was a far more extreme disorder and one with a much more rapid onset (Vaillant, 1983). However, rather than postulate, as does Goodwin (1979), a genetic difference between sociopathic alcoholics and those with a less rapid onset, an environmental hypothesis may be offered. Among antisocial adolescent personalities, alcohol intake, like many other mood-altering behaviors, is used from the beginning to alter consciousness, to obliterate conscience, and to defy social canons. Such use of a potentially addictive substance can only hasten the onset of dependence. For environmental reasons, the children of alcoholic parents are often antisocial (Haberman, 1966).

The third illusion that can be dispelled by prospective study is that alcohol is a good tranquilizer. If the illusion were true, alcohol should achieve what the alcoholic retrospectively recollects that alcohol achieves. Heavy alcohol ingestion should raise self-esteem, alleviate depression, reduce social isolation, and abolish anxiety. However, if alcoholics are observed during periods of active drinking and their mood is recorded, then the alcoholics' retrospective accounts of what happened are found to be at variance from what was actually observed. In actual fact, such prospective empirical studies reveal that alcohol fairly consistently increases depressive affect and suicidal ideation, interferes with normal sleep patterns, decreases self-esteem, and, (less consistently) produces increased anxiety and social isolation (McNamee et al., 1968; Tamerin et al., 1970; Nathan et al., 1970; Allman et al., 1972; Logue et al., 1978). The "anxiety" that alcohol is most effective in relieving is the tremulousness, fearfulness, and dysphoria produced by brief abstinence in alcohol-dependent individuals. To assert that alcohol is a tranquilizer in this latter instance is to confuse cause and effect.

3. SELECTION BIAS AND ENDURING MYTHS

The second broad set of illusions dispelled by prospective study are those that result from failures of deduction rather than distorted memory. Such failures in reasoning arise from many sources. One arises from regarding the characteristics of individuals who attend clinics as representative of all individuals afflicted with a given disease. First, individuals who frequently seek help from clinics—whether for heart disease, tuberculosis, alcoholism or any other chronic illness—tend to be more dependent, more physically ill, and more psychologically vulnerable. Such individuals are oversampled. Second, clinical samples tend to exclude those who die, spontaneously recover, or have alternative social supports. In contrast, prospective follow-up of an entire cohort (i.e., beginning with a normal population and then following it until those at risk have developed the disease) avoids such pitfalls.

As a result of such sampling bias, three premorbid personality types have been repeatedly postulated to play an etiologic role in alcoholism: the emotionally insecure, anxious, and dependent (Simmel, 1948; Blane, 1968); the chronically depressed, (Winokur et al., 1969); and the sociopathic (Robins, 1966). Although early studies of "the prealcoholic personality" have been seriously challenged (Syme, 1957) reviews of more recent retrospective studies (Tahka, 1966; Blum, 1966) agree that premorbidly alcoholics are passive, dependent, egocentric, sociopathic, intolerant of anxiety, lacking in self-esteem, and frightened of intimacy. However, all of these characteristics are common in individuals who are frequent clinic and emergency room attenders for any reason; all have been called into question by the prospective studies in Table 1.

In 1956, the McCords began a very imaginative follow-up of the 325 treated members of the Cambridge Somerville Study (Powers and Witmer, 1959). This study had initially been conceived to examine in a controlled fashion the effect of a 5-year counseling program on antisocial grammar school boys. The original design had been to counsel for 5 years both a predelinquent group and a control group of fifth-grade boys. Since each boy was seen at least weekly by a counselor who kept extensive progress notes, these notes and the original family investigation undertaken to assess delinquency risks allowed the McCords to make extensive judgments about the premorbid characteristics of these boys and their families (McCord and McCord, 1960). Fifteen years later, when the "boys" were then aged between 30 and 35, raters (blind to earlier judgments) obtained evidence for alcoholism from public records, clinics, probation officers and social agencies. An alcoholic was "one whose drinking has become a source of community or family difficulties or one who has recognized that excessive drinking is a primary problem to him" (p. 98).

In their landmark study, the McCords refuted several hypotheses regarding the etiology of alcoholism, refutations that have been upheld in subsequent prospective studies. They observed that, contrary to prior impressions, children

with nutritional disorders, glandular disorders, "strong inferiority feelings," pho-
bias, and "more feminine feelings" were *not* more likely to develop alcoholism.
More important, boys with "strong encouragement of dependency" from their
mothers and manifest oral tendencies, (thumbsucking, playing with their mouths,
early heavy smoking, and compulsive eating) were actually *less* likely to develop
alcoholism. Contrary to popular belief, prealcoholics were outwardly more self-
confident, less disturbed by normal fears, more aggressive and active, and more
heterosexual. In one stroke, the McCords' prospective study brought into question many of the salient traits of the hypothetical prealcoholic personality.

At the Institute of Human Development at Berkeley, Jones (1968, 1971)
obtained similar results. She studied a very small but repeatedly interviewed
sample of middle-class young people. She compared heavy problem drinkers
with moderate and light drinkers. All her subjects had been extensively studied
in junior high school and high school and then again in early and middle adulthood
using personality traits defined by the "Q-sort" (Block, 1961). Jones compared
the high school evaluations of the 6 men who became heavy drinkers with those
of the 17 who remained moderate drinkers. In high school the heavy drinkers
were characterized as "out of control, rebellious, pushing limits, self indulgent
and assertive." Among the girls, the heavy drinkers were more likely in high
school to be "expressive, attractive, poised, and buoyant" (Jones, 1971). In no
way could Jones' alcoholics be described as particularly dependent or anxious.

Another etiologic illusion results from enthusiastic clinic treatment, for well-
entrenched belief systems can lead to misleadingly favorable outcomes, even in
quite well controlled studies. These results, in turn, can lead to erroneous conclusions regarding etiology. Hope, selection bias, and observer bias all play a
role. Such studies often obscure the multifactorial etiology of alcoholism. Long-
term follow-up studies, of the sort described in Table 2, have been required to
dispel the illusions.

Let me cite three examples that suggested that simple learning and conditioning paradigms could explain the complex phenomenon of alcoholism. In
1944, Shadel reported that emetine aversion therapy was an unusually effective
treatment for alcoholism. However, when Voegtlin and Broz (1949) followed
up the same patients 7–10 years later, only 22% of Shadel's patients were still
regarded as being in remission. Again, in 1956 in a well-controlled study at the
Menninger Clinic, Wallerstein (1956) suggested that disulfiram (Antabuse®)
seemed valuable in the treatment of alcoholism. But more recent reviews of
prospective studies by Viamontes (1972), Mottin (1973), and Costello (1975)
all suggest that disulfiram *per se* adds little to favorable long-term outcome.

Still more recently, the Sobells' comparatively well controlled study suggested that a major etiologic factor in alcoholism was a failure to self-monitor
blood alcohol levels. The Sobells' effort to teach alcoholics to drink socially
provided the illusion that increasing the alcoholics' awareness of such internal
cues would facilitate asymptomatic drinking. Their findings at two years were

impressive (Sobell and Sobell, 1976). However, a 10-year re-follow-up study by Pendery and colleagues (1982) suggests that the Sobells' faith in the value of training alcoholics to return to controlled drinking was too optimistic; Table 2 indicates their long-term results were no better than the natural history of the disorder. As had been the case at the Shadel and Menninger Clinics, the results reported by the Sobells appeared to have more to do with nonspecific effects of enthusiasm.

It is equally true, however, that the apparent futility of treating alcoholics, especially those who come to public clinics and emergency rooms, is also a myth resulting from selection bias and from cross-sectional designs which tend to count relapses rather than remissions. For example, an unpublished review of admissions to the Springfield, Massachusetts, freestanding detoxification center (Carlson, 1979) illustrates this point. During a 78-month interval, roughly 5000 clients underwent more than 19,000 detoxifications. One-eighth of these 19,000 admissions encompassed the easily forgotten 2500 clients who never returned. One-eighth of these admissions included the 25 indelibly remembered subjects who returned 60 times or more. Thus, the 25 most intractable alcoholics were admitted as often as, and became far more deeply etched in clinician's consciousness than, the 2500 alcoholics who never came back and yet who may have included the best outcomes.

Lastly, prospective studies of alcoholism suggest that the *lifetime* prevalence of alcoholism is roughly twice the figure arrived at by cross-sectional studies. For example the proportion of alcohol abusers among the Gluecks' inner city men (Vaillant, 1983) seemed much higher than the national pattern revealed by the more representative American drinking practices (ADP) study which was derived from a national probability sample of the entire country (Cahalan et al., 1969). While 47% of the Gluecks' inner city men met Cahalan's criteria for heavy drinking (three or more drinks a day), this was true for only 12% of American adults reported by the ADP: an apparent fourfold difference. However, when alcohol use by a comparable ADP subsample—white men in their fifth decade of life—was examined, 30% met the criterion for heavy drinking. In similar fashion, use of alcohol by the inner city men at age 47 revealed that *currently* only 32% were heavy drinkers. By age 47, many of the heavy drinkers among the inner city men had returned to a pattern of more moderate drinking, or, in order to control their alcohol abuse, 10% of the entire sample had become abstainers.

4. ELUCIDATING "CART AND HORSE" RELATIONSHIPS

A third value of prospective study is that it clarifies the association between primary and secondary etiologic variables. For example, even the McCords, with their prospective study, allowed deductive error to come between them and their prospective data. They suggested that heightened dependency was a cause of alcoholism, and wrote, "We believe that the confirmed alcoholic increases his intake of alcohol because intoxification satisfies his dependency urges and obliterates reminders of his own inadequacies. We assume that his character is organized around a quest for dependency" (p. 156). However, their study provided no direct evidence to support this hypothesis. It took several later prospective studies (Kammeier et al., 1973; and Vaillant, 1980) to confirm that dependency, like heightened anxiety, in alcoholics was secondary to the disorder, alcoholism, and that alcoholism was not merely a symptom of the dependent personality.

Similarly, reactive depression is so frequently associated with alcoholism that psychiatric texts have consistently suggested that depression was a major cause of alcoholism. However, the prospective studies by Kammeier and co-workers (1973) and Pettinatti et al. (1982) both suggest that the reverse was true. The former authors contrasted the scores on the Minnesota Multiphasic Personality Inventory (MMPIs) of 38 men in college with MMPIs obtained many years later when the college men were admitted to an alcohol clinic. The alcoholics' mean depression scale on the MMPI was elevated only in the latter instance. Pettinatti made the same observation in reverse. At the time of alcoholism treatment the MMPI depression scale of alcoholics who were to achieve stable abstinence was pathologically elevated. Four years later, their mean MMPI depression scale values had returned to normal. Alcoholism causes depression.

Kammeier and co-workers also compared the MMPIs of their 38 college students who later developed alcoholism with MMPIs of 148 matched male classmates (Loper et al., 1973). They observed that the prealcoholics were more compulsive, nonconforming, and gregarious and were more likely to answer "true" to items such as "In school I sometimes was sent to the principal for cutting up" and "I like dramatics." Such investigations are in keeping with findings by Jones (1968) and Vaillant (1983). However, in the same prospective study of MMPIs the authors noted that, whereas in college the original composite MMPIs of the 38 prealcoholic men had been within normal limits, when the men were hospitalized for alcoholism, their MMPIs were significantly elevated not only on depression, but also to pathologic levels on the psychopathic deviancy and paranoia scales. Once alcoholic these men's composite profile revealed "the neurotic patterns consistent with self-centered, immature, dependent, resentful, irresponsible people who were unable to face reality." Once they had developed alcoholism, the men were far more likely ($p < 0.01$) to answer "false" to "I am happy almost all the time" and "true" to "I shrink from facing a crisis," "I am high-strung," and "I am certainly lacking in self confidence." In other words, the men developed alcoholism first and then conformed to the hypothetical alcoholic personality.

Prospective design has provided a further check on the above findings. For years it had been noted that alcoholics, even when they achieved sobriety, remained psychologically very unstable (Gerard et al., 1962; Pattison et al.,

1968). If after 4 years Pettinatti's abstinent alcoholics appeared normal, were they not, perhaps, very different from most abstinent alcoholics? Studying the Gluecks' inner city men, Vaillant and Milofsky (1982b) observed that the 21 former alcoholics who had achieved the most stable abstinence (mean duration of 10 years) did not significantly differ in psychopathology from men who had never abused alcohol. Equally important, however, the currently abstinent alcoholics did not differ in terms of childhood risk factors for adult psychopathology from the men who continued actively to abuse alcohol and continued to manifest an unusual amount of psychopathology.

However, still another question demands answer. If unhappy childhood and adolescence does not lead to adult alcoholism, cannot *adult* depression and dependency still lead to alcoholism? Vaillant's (1980) prospective study of college men followed until age 60 suggests that even in this case alcoholism may be more horse than cart. Estimates of psychopathology, chronic depression, and dependence were derived from data on men between the ages of 20 and 47. However, since 58% of the men classified as alcohol abusers were not so classified until after age 45 it was possible to study the effect of adult psychopathology on alcoholism. Bleak childhoods and psychological instability in college did predict and presumably played a causal role in the development of "oral" traits (pessimism, passivity, self doubt, and heightened dependency) in young adult life. As young adults men who displayed many such "oral" traits also showed evidence of personality disorder (i.e., a lifelong difficulty with loving, perseverance, and postponement of gratification). They were also more anxious and inhibited about expressing aggression. However, between the ages of 20 and 47, the men who eventually abused alcohol did not exhibit significantly more premorbid evidence of dependence or personality disorder than men who continued to drink asymptomatically until age 60. Once college men began to abuse alcohol, however, oral dependent traits were very common.

Prospective study of the same adult sample allowed another personality difference that supposedly distinguishes adult alcoholics from nonalcoholics to be called into question. Using the thematic aperception test (TAT), McClelland and colleagues (1972) had suggested that latent aggressive needs were increased among heavy alcohol abusers. As with so many personality factors, however, this observation could not be confirmed prospectively. The TAT was administered to Vaillant's college men at age 30 and they were scored for need aggression by a McClelland-trained rater (Vaillant, 1980). The scores did not predict subsequent alcohol abuse.

Error in conceptualizing the etiology of alcoholism has also resulted from exaggerating the causal association of a relatively rare condition with a more common condition. For example, because of the observed association of alcohol abuse with bipolar affective disorder, some investigators suggested that alcoholism might be a variant of major depressive disorder (Winokur et al., 1969). However, in an impressive 30-year follow-up study of the lives of more than

1700 Scandinavian alcoholics, Sundby (1967) determined that the occurrence of psychotic depression (0.35%) was, if anything, less than that of the lifetime prevalence of the disorder in the general population. The finding suggests that if alcoholism is often observed in affectively disordered patients, it is because such patients are particularly likely to elicit psychiatric attention rather than because alcoholism is frequently caused by affective disorders. Sundby's finding, which suggests that major depressive disorder and alcoholism are quite independent disorders, has received recent confirmation from a variety of sources (Morrison, 1974; Schlesser et al., 1980).

Again, alcoholism is a far more common condition than sociopathy, but sociopathy has often been assigned a major etiologic role in alcoholism. Schuckit (1973) has succinctly summarized the etiologic possibilities linking alcoholism with sociopathy: (1) sociopaths abuse alcohol as one symptom of their antisocial personality; (2) alcoholics are sociopathic as a consequence of primary alcohol dependence; and (3) a common factor may underlie both conditions.

How one selects one's sample will determine which of Schuckit's etiologic possibilities is correct. In Table 1, Robins' (1966) sample was selected from a child guidance clinic where a majority of the male clients were referred for antisocial behavior. The McCords' study was also biased in favor of including predelinquent youth. In both studies, preexisting antisocial behavior appeared to be a major etiologic factor in adult alcoholism. However, among the Gluecks' relatively nondelinquent inner city sample, premorbid antisocial behavior appeared to be an etiologic variable in only one-eighth of alcohol abusers (Vaillant and Milofsky, 1982a). In prospective studies of middle-class samples antisocial behavior appeared to be almost always cart relative to the horse of alcohol abuse (Kammeier et al., 1973; Vaillant, 1980a).

Short-term follow-up studies (Gibbs and Flanagan, 1977) have uniformly suggested that antisocial behavior is a negative prognostic factor in alcoholism or, as Goodwin (1979) has suggested, may identify a more severe and genetically determined form of alcoholism. However, in their 8-year follow-up study of felons, Goodwin and co-workers (1971) observed a long-term outcome as favorable as any reported in Table 2. They also observed that neither the number of symptoms nor the severity of alcohol abuse correlated with a felon's eventual recovery from alcoholism. Similarly, the contrast of Vaillant's 35-year study of 26 upper-middle class alcoholics with the 25 most sociopathic alcoholics among the Gluecks' inner city sample revealed that at last contact a greater proportion of the alcoholic sociopaths had achieved stable abstinence than the alcoholic college men (Vaillant, 1983).

Finally, prospective study, combined with multivariate analysis, has allowed etiologically primary variables to be separated from statistically associated but causally unimportant secondary variables. The McCords (1960) concluded that immigrant parents, Catholicism, and low social class protected the Cambridge and Somerville youths in their study against the development of alcoholism. In so doing, the McCords failed to note that in the cities of Cambridge and Somerville in 1940 such individuals were predominantly first- and second-generation Italians. The possibility existed that Italian-American drinking practices rather than the other characteristics to which the McCords attributed etiologic importance might have protected their subjects from alcohol abuse.

Vaillant and Milofsky (1982a) took advantage of the fact that, through public records, the Gluecks had followed their junior high school sample back for three generations both in terms of ethnicity and familial alcoholism. When 113 other psychosocial variables originally coded by the Gluecks (1950) were assessed for their value in predicting adult alcohol abuse, six of the Gluecks' items—father's alcoholism, marital conflict, lack of maternal supervision, many moves, no attachment to father, and no family cohesiveness—were very significantly ($p < 0.001$) correlated with subsequent development of alcohol abuse. These six dichotomously rated Glueck items were almost identical to the predictor variables identified by the McCords as having primary etiologic impact on alcoholism. Thus, they were used to create a composite "McCord" predictor variable. When entered first into multiple regression analysis, this composite "McCord" variable explained 7% of the observed variance in adult alcoholism—a rather large percentage in view of the time that had elapsed. However, once the "McCord" variable was entered, both the number of alcoholic relatives and Mediterranean (largely Italian) ethnicity could still explain an *additional* 8% of variance. In contrast, if the theoretically antecedent variables—family history of alcoholism and ethnicity—were entered first into the regression equation, then the secondary "McCord" variables explained little further variance. The tentative conclusion of this observation would be that cultural and hereditary variables that are specifically related to alcohol abuse are of greater etiologic importance than nonspecific familial instability.

Prospective studies have also been used to great advantage by the Jessors (1975), by Kandel and her associates (Margulies et al., 1977), and Robins (1972) to study the onset of alcohol use and abuse in young adults. Each investigator has been able to separate out statistically primary etiologic variables, especially peer group variables, from other variables which, although strongly associated with deviance, seemed of less etiologic importance to alcoholism *per se*.

5. THE PROBLEM OF EVOLVING SYMPTOMS

One reason why the etiology of alcoholism is so confused is that investigators who scrutinize the behavior of alcoholics for short periods of time come to doubt the existence of alcoholism as a stable entity. Indeed, the diagnosis of alcoholism depends so much upon definition that some workers believe that there are as many alcoholisms as there are problem drinkers. Certainly, the elderly retired stockbroker who first comes to medical attention due to cirrhosis and polyneuritis appears to suffer a different disorder from the young motorcyclist who comes to legal attention due to drunken episodes of belligerence and multiple fractures.

To address the problem of different alcoholisms, Jellinek (1960) devised his typology of alcohol abuse: alpha (symptoms and psychological, but not physical, dependence), beta (medical symptoms but no physical dependence), gamma (symptoms *and* physical dependence), delta (physical dependence but few or no symptoms), and epsilon (binge drinking). While serving as a useful cross-sectional classification, Jellinek's scheme breaks down if studied longitudinally. Over time, patterns of alcohol abuse do not remain constant. The beta alcoholic may return to social drinking, the alpha alcoholic may progress to become a gamma alcoholic, and the epsilon alcoholic may become abstinent and vanish from clinical sight. What I am suggesting, then, is that to examine the course of an alcoholic's life under too high magnification is to perpetuate the illusion that clinical course cannot be categorized.

If viewed over the short-term, alcoholism appears extremely unstable. For example, in their month-by-month study of 100 alcoholics, Orford and Edwards (1977) called into question the validity of any black and white definition of the disorder. Every month they interviewed their alcoholic subjects and their wives. On the average, wives believed that their husbands engaged in drinking in only 31 of the 52 weeks and for only 23 of those weeks did they regard their husband's drinking as unacceptable. Thus, during the year under study, the average alcoholic was abstinent for 21 weeks and engaged in perfectly acceptable "social" drinking for 8 weeks more. The standard deviations were small. In other words, on a day-by-day basis many alcoholics do not have trouble with their drinking. What is equally important, however, is that in contrast to nonalcoholics, alcoholics *do* have difficulty in the long run. At the year's end, only 20 of Orford and Edwards' alcoholics appeared in clear clinical remission and 55 of their men could be unambiguously classified as problem drinkers. In the study of alcoholism, time is a critical dimension.

In an 8-year follow-up of 100 clinic patients (Vaillant, 1983), 95 relapsed to alcohol dependence: a criterion often used to indicate clear failure of treatment. Within the same 8-year period, however, 59 of the same sample achieved at least 6 months of abstinence: a criterion often used to indicate stable recovery. On the one hand, after the index admission 6 of the 29 men who eventually achieved 3 or more years of stable abstinence had required 10 or more detoxifications. On the other hand, 15 of the 27 most chronic alcoholics achieved at one time 4 or more consecutive months of community abstinence. In other words, outcomes that after 8 years appeared clearly differentiated could not have been distinguished if studied in cross-section.

Jellinek's (1952) concept of a natural progression of the symptoms underlying the "disease" of gamma alcoholism endured for many years. In the past decade, however, there has appeared a compelling and coherent body of empirical work that contradicts belief in the orderly evolution of alcoholism (Knupfer,

1972; Cahalan and Room, 1974; Fillmore, 1975; Clark, 1976; Clark and Cahalan, 1976; Room, 1977; Roizen et al., 1978).

> Summarizing this work, Clark and Cahalan (1976) write: The common conception of alcoholism as a disease fails to cover a large part of the domain of alcohol problems and a more useful model would be to place further emphasis on the development and correlates of particular *problems* related to drinking, rather than assuming that alcoholism as an underlying and unitary progressive disease is a source of most alcohol problems (1976, p. 258).

Prospective study helps to clarify the relationship between alcoholism and alcohol-related problems. Aamark (1951) and Keller (1975) both elaborate on the distinction between the unstable *problem drinker* studied by Clark and Cahalan and the *chronic alcoholic* described by Jellinek and Ojesjo. The modal "problem" drinker is aged 25–35, is married, is working, and has never been treated for alcoholism. His or her use of alcohol is markedly responsive to environmental factors and can, over time, become either more or less symptomatic. In such individuals, symptoms of alcohol abuse are likely to be "disjunctive," by which I mean that the presence of a given symptom of alcohol abuse will not significantly predict the presence of other symptoms that (according to Jellinek's sequence of progression) might theoretically precede it.

In contrast, the modal "chronic alcoholic" is seen less commonly, but his or her alcohol abuse has evolved and, in so doing, has become far less plastic. He or she is 10 years older, aged 35–45, and has unstable marital and employment status. Such individuals may have sought treatment for alcoholism and exhibit a pattern of alcohol abuse that is relatively insensitive to environmental variables. They often become stably abstinent but rarely evolve into patterns of asymptomatic drinking. In such individuals, the appearance of a given symptom will be statistically associated with the preexistence of earlier symptoms in the theoretic chain of progression. Obviously, to elucidate such a relationship a prospective view is imperative.

In illuminating the reversibility of the 25- to 35-year-old "problem drinker" the 8-year follow-up by Goodwin and co-workers (1971) is noteworthy. It reflects the lowest death rate and the highest rate of return to asymptomatic drinking observed in the long-term follow-ups summarized in Table 2. The felons who made up the sample were very young (average age, 27) and, unlike the men in most other samples, they had not sought treatment for alcohol dependence. Rather, when interviewed in prison, they had merely reported a past history of alcohol-related problems. Goodwin's study is heuristically important because it underscores a fundamental principle involved in the reversibility of alcohol abuse. By inadvertently selecting alcoholics who had abused alcohol for only a short time and with little physical dependence, Goodwin and his co-workers were able to identify problem drinkers, a large proportion of whom were to return to asymptomatic drinking. The most common reason given by the felons for returning to asymptomatic drinking was environmental changes, usually marriage and/or increasing family responsibility.

In illuminating the relative irreversibility of the less common, older gamma alcoholic into which *some* problem drinkers evolve, the problem-by-problem analysis of alcohol abuse by 47-year-old men is instructive (Vaillant et al., 1982). The degree to which alcohol-associated social deviance among the Gluecks' inner city men correlated with medical evidence for alcoholism was examined. The correlation of *symptoms ever* was substituted for the correlation of symptoms currently present. When such a longitudinal view of alcoholism was substituted for a cross-sectional view, there did not appear to be many different alcoholisms. The clinician's "disease" and the sociologist's "continuum of drinking behaviors" appeared quite congruent. When the lifetime prevalence of five items that define alcohol dependence from the vantage point of the medical model (morning drinking, alcohol-related medical problems, going on the wagon, clinical diagnosis of alcoholism, and self-admission of loss of control) were correlated with five items that identify problem drinking from the social deviance model (alcohol-related arrests, occupational problems, social problems, marital problems, and fights) the correlations between individual symptoms ranged from 0.4 to 0.6. When the *DSM III*'s definition of physiological dependence was correlated with the 11-point Cahalan scale of socially deviant drinking behaviors, the r was 0.87. Admittedly, only empirical longitudinal study can confirm whether such high correlations will be observed in countries such as Portugal and France where many individuals are alleged to be physiologically dependent on alcohol but are asymptomatic in terms of social deviance.

Longitudinal studies also help to avoid the distorting effects of the law of initial values; the tendency of extreme physiological or psychological values to regress toward the mean over time. Alcoholics tend to present themselves for medical attention at their clinical nadir and at the time of their most extreme exacerbation of symptoms. Thus, if follow-up consists of a brief and rigid time frame, alcoholic patients will appear to improve regardless of the treatment employed. Eighteen months after admission 67% of the Rand Report patients were reported to have fewer alcohol-related problems than they had during the 1 month before admission (Armor et al., 1978). However, when the Rand investigators extended their follow-up to 4 years and substituted a 6-month period instead of a 1-month period, much of the improvement that they had originally noted appeared evanescent (Polich et al., 1980).

But even a 4-year period of observation is too short. In their heuristically very important study of alcohol use by a community sample of randomly selected white males, Clark and Cahalan (1976) noted that respondents who reported loss of control often had alcohol-related problems. However, a majority of their subjects who reported alcohol-related problems did not report loss of control. Indeed, "loss of control" reported at the beginning of the observation period correlated with "loss of control" 4 years later with an r of only 0.13. How can such findings be reconciled with Jellinek's concept of alcoholism as a progressive disease?

Clearly, methodologic considerations are crucial. On the one hand, the course of a chronic relapsing disease may appear very unstable if many mild cases are included; if data are gathered by questionnaires; if deaths are excluded; if periods of observation are short; if syndromes are broken down into individual symptoms; and if integrated individual case histories are ignored. On the other hand, the course of a chronic disease may appear stable and progressive if only severe cases are included; if data are gathered by skilled clinical interviews; if all deaths are reported; if symptoms—however individually unstable—are treated as clusters; if long periods of observation are used; and if individual lives rather than statistical analyses are scrutinized. The first method was employed by Clark and Cahalan (1976); it is valuable to epidemiology and to understanding the behavior of heterogeneous populations. The second method, employed by Ojesjo (1981), is valuable to clinical medicine and in understanding population subgroups.

One difficulty with the first method is that, by depending upon statistical analyses of data derived from self-administered questionnaires, Cahalan and co-workers lost the power of the clinical case history. For example, Clark (1976) reports that 18 out of 29 self-acknowledged binge drinkers reported no loss of control. Such an observation may seem reasonable enough in a computer analysis of results of self-administered questionnaires, but the same observation during an interview might provoke incredulity in a clinician and, at the very least, further questions. From a clinical vantage point, binge drinking and loss of control are usually synonymous. The self-administered questionnaire, in turn, has the advantage of being free of the clinician's biases. Alcoholism needs to be studied from many perspectives.

The study by Ojesjo (1981) is the longest prospective study of alcoholics that has been based upon a totally unselected community sample. It provides a much more stable view of alcoholism than that provided by Clark and Cahalan. Ojesjo began with a representative community cohort drawn from the district of Lundby in Sweden. Twenty-five years before, Essen-Möller (1956) had selected the sample for an epidemiologic study of mental illness. After 10 years, Hagnell and Tunving (1972) again followed-up Essen-Möller's cohort with only 2% attrition. Within this sample they identified 96 alcohol abusers. The average age of these alcohol abusers was 47 and half of the 96 would have met the *DSM III* criteria for alcohol dependence.

After the elapse of another 15 years, Hagnell's sample was followed up again by Ojesjo. At each time period, Ojesjo categorized the subject's problem drinking as alcohol abuse, alcohol dependence, or chronic alcoholism (dependence with serious medical sequelae). At first glance, these categories seemed most unstable over time. Of 49 men categorized as alcohol abusers in 1957, only 4 were still categorized as alcohol abusers in 1972. Of 29 men classified as alcohol-dependent in 1957, only 8 were still classified as alcohol-dependent in 1972. Concealed in this apparent instability was clear support for Jellinek's concept of progression. Among the 49 men who were counted as alcohol abusers in 1957, 17 men were no longer classified as alcohol abusers in 1972 because their alcoholism had progressed either to death or to chronic alcoholism. Another 25 had achieved stable remission, usually through abstinence. Of the 29 alcohol-dependent men 13 had progressed to death or chronic alcoholism, 4 had returned to stable abstinence, and, as noted above, 8 remained alcohol-dependent. Thus, after 15 years, only 4 alcohol-dependent men contradicted Jellinek's concept of progression by being reclassified in 1972 in the less severe category of alcohol abuse.

Finally, because both individuals and alcohol abuse develop over time, erroneous etiologic considerations arise from the cross-sectional comparison of individuals who are at different stages of development. For example, Jessor and Jessor (1977) report that adolescent girls who mature early are more likely to abuse alcohol than those who mature late. However, when Magnusson and Stattin (1983) prospectively followed adolescent women with early and late menarches, they found that marked between-group differences in alcohol abuse observed at age 15 had vanished by age 25. Such findings suggest that early maturity *per se* may not be an etiologic factor in alcoholism.

6. ETIOLOGY AND PROGNOSIS

Prospective longitudinal study is as necessary to understand why alcoholics stop drinking as it is to understand why they begin. Many premorbid factors associated with social stability, especially with occupational and marital stability (Bromet et al., 1977; Costello, 1975, 1980), predict compliance and favorable short-term response to treatment. Exhaustive reviews by Gibbs and Flanagan (1977) and Baekeland et al. (1975) reveal that early age of onset, low social class, social alienation, broken marriage, many arrests, and sociopathy militate against a favorable response to treatment. Such data might suggest that these negative factors play an etiologic role in the more severe or malignant form of alcoholism.

However, placing undue emphasis upon factors that predict the short-term prognosis of alcoholism but do not affect its long-term course has tended to mislead us about its etiology. Elucidating the etiology of favorable prognosis is confounded by the fact that clinical populations contain a disporportionate number of skid row residents. Such individuals often have severe social deficits predating their alcoholism: schizophrenia, mental retardation, or childhood foster care. The fact that such socially deprived patients make repeated visits exaggerates the relationship between social incompetence and intractable alcoholism. When general population studies of alcoholism are undertaken, premorbid social adjustment fails to predict long-term abstinence and, when skid row alcoholics are excluded, premorbid psychological soundness and social competence may not be necessary for ultimate recovery in alcoholism (Vaillant and Milofsky, 1982b). Among the Gluecks' sample of inner city youth, the 21 men who achieved abstinence of at least 3 years could not be distinguished premorbidly from the 35 men whose alcoholism had become relentlessly more symptomatic until the present. The blindly assessed childhoods of these two very different alcoholic outcome groups seem roughly comparable. Certainly, in terms of those childhood variables most important in predicting mental health (Vaillant and Vaillant, 1981), there was little difference between the securely abstinent and the progressive alcoholics. Maternal supervision, boyhood competence, childhood weaknesses, and IQ did not differentiate the 21 securely abstinent alcoholics from even the 21 men among the progressive alcoholics who showed the greatest social incapacitation secondary to their drinking. Ethnicity, early termination of education, and a family history of alcoholism predicted who would develop alcoholism, but even these risk variables did not predict who would then recover. Among the Gluecks' inner city men, 25 were classified as both sociopaths and alcoholics; 48% were currently abstinent. In contrast, 40 inner city alcohol abusers manifested no antisocial symptoms except for their heavy drinking, but only 28% were currently abstinent.

Clearly, such findings require confirmation. Nevertheless, they suggest that alcoholism may be different from psychiatric conditions such as sociopathy, schizophrenia, and reactive depression. In these latter conditions, long-term clinical course is powerfully affected by premorbid adjustment. In alcoholism, as long as the nervous system is spared, the severity of symptoms does not alter the likelihood of eventual stable abstinence.

Again, Costello (1980) observed that although multimodality inpatient treatment and subsequent attendance at alcohol outpatient clinics was associated with good outcome, the apparent causal association may be illusory. The two premorbid variables, marriage and stable employment, predicted both conscientious clinic attendance and remission from alcoholism. When premorbid social stability was controlled, identified components of inpatient treatment explained little further independent variance in outcome.

Is the apparent effectiveness of Alcoholics Anonymous (AA) based upon a similar illusion? A prospective 8-year study of clinic alcoholics reported by Vaillant (1983) examined the relationship between AA attendance, premorbid variables, and outcome. At the time of first inpatient admission (1972) 100 alcoholics were studied in terms of premorbid variables known to affect prognosis. These patients were then evaluated every 18 months for 8 years. At the end of 8 years (1980), half of the 29 patients with stable remissions (3 years or more) had made 300 or more visits to Alcoholics Anonymous. This was true of only one of the 37 active alcoholics. Whereas social stability in 1972 predicted stable remission in 1980, social *instability* in 1972 predicted heavy use of AA over the intervening 8 years. Thus, 32 patients attended AA meetings 100 or more times (mean 600 visits) between 1972 and 1980; the number of those individuals with stable psychosocial adjustment went from 2 in 1972 to 15 in 1980. Put differently, premorbidly socially stable alcoholics tended to become abstinent without Alcoholics Anonymous; but if socially unstable alcoholics are to recover, AA attendance may be a causally important intervening variable. Clearly, without a control group denied access to AA such a study does not prove a causal relationship, but the findings are far more convincing than if a prospective design had not been employed.

7. SUMMARY

In conclusion, several caveats seem in order. In adult life variables other than those discussed in this paper possess great etiologic importance, for the etiology of alcoholism is nothing if not multifactorial. Nor do I wish to imply that prospective studies are without flaws. They are always unwieldy, out of date, subject to the caprice of historical monument, and difficult to manipulate experimentally. Most of the prospective studies in this essay have used subject contacts that were separated by several years; thus, many of the data reported in longitudinal studies are to some degree retrospective.

I do not wish to suggest that cross-sectional methods are not of equal value. Let me cite just a few seminal etiologic studies that have been cross-sectional or of a relatively brief duration and yet have contributed in a major way to our understanding of alcoholism. The demographic variables elucidated by Cahalan (1970); the importance of occupation as elucidated by Plant (1979); the clinical significance of social peer groups as developed by Jessor and Jessor (1975); the importance of disposable income, urbanization, and availability of alcohol as demonstrated by Smart (1977); the effect of social instability (documented by Pitman and Snyder, 1962) and the effects of attribution and expectancy summarized by Marlatt and Rohsenow (1980) are as important to understanding the etiology of alcoholism as any of the variables discussed in this chapter.

If I point out the limits of the brief follow-up, I must at the same time acknowledge that the studies by the Rand group and by Cahalan and Clark have been most valuable in underscoring that a black-and-white model of progressive alcoholism is untenable. One of the difficulties of studying case histories over time is that clinicians may become unduly entranced by small numbers of patients selected from biased samples. Cross-sectional, cross-cultural studies of large cohorts have done much to correct such prejudices.

If we are to understand the etiology of alcoholism, a balanced view is necessary. Cross-sectional and longitudinal studies each have unique contributions to make to our understanding. This chapter has, by design, focused upon the contribution of longitudinal studies. Clearly, however, creative syntheses of cross-sectional and longitudinal studies will illuminate many problems in alcohol research more brightly than could either method alone.

ACKNOWLEDGMENT

This work was supported by The Grant Foundation, Inc., the Spencer Foundation, and research grant AA-01372 from the National Institute on Alcoholism and Alcohol Abuse.

REFERENCES

Aamark, C., 1951, A study in alcoholism, *Acta Psychiatr. Scand.* (Suppl. 70).

Allman, L. R., Taylor, H. A., and Nathan, P. E., 1972, Group drinking during stress: Effects on drinking behavior, affect, and psychopathology, *Am. J. Psychiatry* **129**:669–678.

Armor, D. J., Polich, J. M., and Stanbul, H. B., 1978, *Alcoholism and Treatment*, Wiley, New York.

Baekeland, F., Lundwall, L., and Kissin, B., 1975, Methods for the treatment of chronic alcoholism: a critical appraisal, in: *Research Advances in Alcohol and Drug Problems*, Volume 2, (R. J. Gibbons, Y. Israel, H. Kalant, R. E. Popham, W. Schmidt and R. G. Smart, eds.), Wiley, New York.

Blane, H. T., 1968, *The Personality of the Alcoholic: Guises of Dependency*, Harper and Row, New York.

Block, J., 1961, *The Q-sort Method in Personality Assessment and Psychiatric Research*, Charles C. Thomas, Springfield, Illinois.

Blum, E. M., 1966, Psychoanalytic views of alcoholism, *Q. J. Stud. Alcohol* **27**:259–299.

Bratfos, O., 1974, *The Course of Alcoholism: Drinking, Social Adjustment and Health*, Universitet Forlaget, Oslo, Norway.

Bromet, E., Moos, R., Bliss, F., and Wothmann, C., 1977, Posttreatment functioning of alcoholic patients: Its relation to program participation, *J. Consult. Clin. Psychol.* **45**:829–842.

Cahalan, D., 1970, *Problem Drinkers: A National Survey*, Jossey Bass, San Francisco.

Cahalan, D., Cisin, I. H., and Crossley, H. M., 1969, *American Drinking Practices: A National Survey of Behavior and Attitudes*, Monograph 6, Rutgers Center for Alcohol Studies, New Brunswick, New Jersey.

Cahalan, D., and Room, R., 1974, *Problem Drinkers Among American Men*, Rutgers Center for Alcohol Studies, New Brunswick, New Jersey.

Carlson, M., 1979, Personal communication, Alcoholism Services of Greater Springfield, Massachusetts.

Clark, W. B., 1976, Loss of control, heavy drinking, and drinking problems in a longitudinal setting, *J. Stud. Alcohol* **37**:1256–1290.

Clark, W. B., and Cahalan, D., 1976, Changes in problem drinking over a four-year span, *Addict. Behav.* **1**:251–259.

Costello, R. M., 1975, Alcoholism treatment and evaluation II: Collation of two year follow-up studies, *Int. J. Addict.* **10**:857–867.

Costello, R. M., 1980, Alcoholism treatment effectiveness: Slicing the outcome variance pie, in: *Alcoholism Treatment in Transition*, (S. G. Edwards and M. Grant, eds.), Croom Helm, London.

Essen-Möller, E., 1956, Individual traits and morbidity in a Swedish rural population, *Acta Psychiatr. Scand.* (Suppl. 100).

Fillmore, K. M., 1975, Relationships between specific drinking problems in early adulthood and middle age: An exploratory 20 year follow-up study, *Q. J. Stud. Alcohol* **36**:882–907.

Gerard, D. L., Saenger, G., Wile, R., 1962, The abstinent alcoholic, *Arch. Gen. Psychiatry* **6**:83–95.

Gibbs, L., and Flanagan, J., 1977, Prognostic indicators of alcoholism treatment outcome, *Int. J. Addict.* **12**:1097–1141.

Glueck, S., and Glueck, E., 1950, *Unraveling Juvenile Delinquency*, Commonwealth Fund, New York.

Glueck, S., and Glueck, E., 1968, *Delinquents and Nondelinquents in Perspective*, Harvard University Press, Cambridge, Massachusetts.

Goodwin, D. W., 1979, Alcoholism and heredity, *Arch. Gen. Psychiatry* **36**:57–61.

Goodwin, D. W., Crane, J. B., and Guze, S. B., 1971, Felons who drink: An 8-year follow-up, *Q. J. Stud. Alcohol* **32**:135–147.

Haberman, P. W., 1966, Childhood symptoms in children of alcoholics and comparison group parents, *J. Marriage Family* **28**:152–154.

Hagnell, O., and Tunving, K., 1972, Prevalence and nature of alcoholism in a total population, *Soc. Psychiatry* **7**:190–201.

Heath, D. B., 1975, A critical review of ethnographic studies of alcohol use, in, *Research Advances in Alcohol and Drug Problems* (R. J. Gibbins, Y. Israel, H. Kalant, R. E. Popham, W. Schmidt, and R. C. Smart, eds.), Wiley, New York.

Hyman, M. M., 1976, Alcoholics 15 years later, *Ann. N.Y. Acad. Sci.* **273**:613–623.

Jellinek, E. M., 1952, Phases of alcohol addiction, *Q. J. Stud. Alcohol* **13**:673–684.

Jellinek, E. M., 1960, *The Disease Concept of Alcoholism*, Hillhouse Press, New Haven, Connecticut.

Jessor, R., and Jessor, S. L., 1975, Adolescent development and the onset of drinking, *J. Stud. Alcohol* **36**:27–51.

Jones, M. C., 1968, Personality correlates and antecedents of drinking patterns in adult males, *J. Consult. Clin. Psychol.* **36**:2–12.

Jones, M. C., 1971, Personality antecedents and correlates of drinking patterns in women, *J. Consult. Clin. Psychol.* **36**:61–69.

Jones, M. C., 1981, Midlife patterns: Correlates and antecedent, in: Present and Past in Middle Life, (D. Eichorn, J. Clausen, N. Haan, M. Honzik, and P. Mussen, eds.), Academic Press, New York.

Kammeier, M. L., Hoffmann, H., and Loper, R. G., 1973, Personality characteristics of alcoholics as college freshman and at time of treatment, *Q. J. Stud. Alcohol* **34**:390–399.

Keller, M., 1975, Problems of epidemiology in alcohol problems, *J. Stud. Alcohol* **36**:1442–1451.

Knupfer, G., 1972, Ex-problem drinkers, in: *Life History Research in Psychopathology*, vol. 2, (M. Roff, L. N. Robins, and H. Pollack, eds.), University of Minnesota Press, Minneapolis.

Logue, P. E., Gentry, W. D., Linnoila, M., and Erwin, C. W., 1978, Effects of alcohol consumption on state anxiety changes in male and female nonalcoholics, *Am. J. Psychiatry* **135**:1079–1081.

Loper, R. G., Kammeier, M. L., and Hoffmann, H., 1973, M.M.P.I. characteristics of college freshman males who later became alcoholics, *J. Abnorm. Psychol.* **82**:159–162.

Lundquist, G. A. R., 1973, Alcohol dependence, *Acta Psychiatr. Scand.* **49**:332–340.

Magnusson, D., Stattin, H., 1982, Short-term and long-term consistency: a methodological problem, in: *Personality Development as a Person–Environment Interaction* (D. Magnusson and V. L. Allen, eds.), Academic Press, New York.

Margulies, R. Z., Kessler, R. C., and Kandel, D. B., 1977, A longitudinal study of onset of drinking among high school students, *J. Stud. Alcohol* **38**:897–912.

Marlatt, G. A., and Rohsenow, D. J., 1980, Cognitive processes in alcohol use: Expectancy and the balanced placebo design, in: *Advances in Substance Abuse, Behavioral and Biological Research* (N. K. Marlow, ed.), JAI Press, Greenwich, Connecticut.

McClelland, D. C., Davis, W. N., and Kahn, R., 1972, *The Drinking Man*, Free Press, New York.

McCord, W., and McCord, J., 1960, *Origins of Alcoholism*, Stanford University Press, Stanford, California.

McNamee, H. B., Mendelson, J. H., and Mello, N. K., 1968, Experimental analysis of drinking patterns of alcoholics, concurrent psychiatric observations, *Am. J. Psychiatry* **124**:1063–1069.

Morrison, J. R., 1974, Bipolar affective disorder and alcoholism, *Am. J. Psychiatry* **131**:1130–1133.

Mottin, J. L., 1973, Drug induced attenuation of alcohol consumption, *Q. J. Stud. Alcohol* **34**:444–472.

Myerson, D. J., and Mayer, J., 1966, Origins, treatment and destiny of skid row alcoholic men, *N. Engl. J. Med.* **275**:419–424.

Nathan, P. E., Titler, N. A., Lowenstein, L. M., Solomon, P., and Rossi, A. M., 1970, Behavioral analysis of chronic alcoholism, *Arch. Gen. Psychiatry* **22**:419–430.

Ojesjo, L., 1981, Long-term outcome in alcohol abuse and alcoholism among males in the Lundby general population, Sweden, *Br. J. Addict.* **76**:391–400.

Orford, J., and Edwards, G., 1977, *Alcoholism*, Oxford University Press, New York.

Pattison, E. M., Headley, E. B., Gleser, G. C., and Gottschalk, L. A., 1968, Abstinence and normal drinking: An assessment of changes in drinking patterns in alcoholics after treatment, *Q. J. Stud. Alcohol* **29**:610–633.

Pendery, M. L., Maltzman, I. M., and West, L. J., 1982, Controlled drinking by alcoholics? New findings and a reevaluation of a major affirmative study, *Science* **217**:169–175.

Pettinatti, H. M., Sugerman, H., Maurer, H. S., 1982, Four year MMPI changes in abstinent and drinking alcoholics, *Alcohol. Clin. Exp. Res.* **6**:487–494.

Pittman, D. J., and Snyder, C. R., 1962, *Society, Culture and Drinking Patterns*, Wiley, New York.

Plant, M. L., 1979, *Drinking Careers*, Tavistock, London.

Polich, J. M., Armor, D. J., and Braiker, H. B., 1981, *The Course of Alcoholism*, Wiley, New York.

Powers, E., and Witmer, H., 1959, *An Experiment in the Prevention of Delinquency*, Columbia University Press, New York.

Robins, L. N., Bates, W. N., and O'Neal, P., 1962, Adult drinking patterns of former problem children, in: *Society, Culture, and Drinking Patterns* (D.J.P. Pittman and C.R. Snyder, eds.), Wiley, New York.

Robins, L. N., 1966, *Deviant Children Grown Up: A Sociological and Psychiatric Study of Sociopathic Personality*, Williams and Wilkins, Baltimore.

Robins, L. N., 1979, Longitudinal methods in the study of normal and pathological development, in: *Grundlagen und Methoden der Psychiatrie* (G. Assad et al., eds.), Springer-Verlag, Heidelberg.

Robins, L. N., 1972, An actuarial evaluation of the causes and consequences of deviant behavior in young black men, in: *Life History Research in Psychopathology*, volume 2 (M. Roff, L. N. Robins, and M. Pollack, eds.), University of Minneapolis Press, Minneapolis.

Roizen, R., Cahalan, D., and Shanks, P., 1978, "Spontaneous remission" among untreated problem drinkers, in: *Longitudinal Research on Drug Use* (D. B. Kandel, ed.), Wiley, New York.

Room, R., 1977, Measurement and distribution of drinking patterns and problems in general populations, in: *Alcohol Related Disabilities* (G. Edwards, M. M. Gross, and M. Keller, eds.), WHO, Geneva, Offset Publication #32.

Rutter, M., 1981, Longitudinal studies: A psychiatric perspective, in: *Prospective Longitudinal Research* (S.A. Mednick and A.E. Baert, eds.), Oxford University Press, Oxford.

Schilder, P., 1941, The psychogenesis of alcoholism, *Q. J. Stud. Alcohol* **2**:277–292.

Schlesser, M. A. G., Winokur, G., and Sherman, B. M., 1980, Hypothalamic–pituitary–adrenal axis activity in depressive illness. *Arch. Gen. Psychiatry* **37**:737–743.

Schuckit, M., 1973, Alcoholism and sociopathy—diagnostic confusion, *Q. J. Stud. Alcohol* **34**:157–164.

Shadel, C. A., 1944, Aversion treatment of alcohol addiction, *Q. J. Stud. Alcohol* **5**:216–228.

Simmel, E., 1948, Alcoholism and addiction, *Psychoanal. Q.* **17**:6–31.

Smart, R. G., 1977, The relationship of availability of alcoholic beverages to per capita consumption and alcoholism rates, *J. Stud. Alcohol* **38**:891–896.

Sobell, M. B., and Sobell, L. C., 1976, Second year treatment outcome of alcoholics treated by individualized behavior therapy: Results, *Behav. Res. Ther.* **14**:195–215.

Stivers, R., 1976, *A Hair of the Dog*, Pennsylvania State University Press, University Park, Pennsylvania.

Sundby, P., 1967, *Alcoholism and Mortality*, Universitets Forlaget, Oslo, Norway.

Syme, L., 1957, Personality characteristics and the alcoholic, *Q. J. Stud. Alcohol* **18**:288–301.

Tahka, V., 1966, *The Alcoholic Personality*, Foundation for Alcohol Studies, Helsinki, Finnish.

Tamerin, J. S., Weiner, S., and Mendelson, J. H., 1970, Alcoholics' expectancies and recall of experiences during intoxication, *Am. J. Psychiatry* **126**:1697–1704.

Ullman, A. D., 1953, The first drinking experience of addictive and of normal drinkers, *Q. J. Stud. Alcohol* **14**:181–191.

Vaillant, G. E., 1974, Natural history of male psychological health, II: Some antecedents to healthy adult adjustment, *Arch. Gen. Psychiatry* **31**:15–22.

Vaillant, G. E., 1977, *Adaptation to Life*, Little Brown, Boston.

Vaillant, G. E., 1980, Natural history of male psychological health, VIII, antecedents to alcoholism and "orality," *Am. J. Psychiatry* **17**:181–186.

Vaillant, G. E., 1983, *Natural History of Alcoholism*, Harvard University Press, Cambridge, Massachusetts.

Vaillant, G. E., and Vaillant, C. O., 1981, Natural history of male psychological health, X: Work as a predictor of positive mental health, *Am. J. Psychiatry* **138**:1433–1440.

Vaillant, G. E., and Milofsky, E. S., 1982a, The etiology of alcoholism: a prospective viewpoint, *Am. Psychol.* **37**:494–503.

Vaillant, G. E., and Milofsky, E. S., 1982b, Natural history of male alcoholism, IV: Paths to recovery, *Arch. Gen. Psychiatry* **39**:127–133.

Vaillant, G. E., Gale, L., and Milofsky, E., 1982, Natural history of male alcoholism, II: Relationship between different diagnostic dimensions, *J. Stud. Alcohol* **43**:216–232.

Viamontes, J. A., 1972, Review of drug effectiveness in the treatment of alcoholism, *Am. J. Psychiatry* **128**:1570–1571.

Voegtlin, W. L., and Broz, W. R., 1949, The conditioned reflex treatment of chronic alcoholism, X: An analysis of 3125 admissions over a period of ten and a half years, *Ann. Intern. Med.* **30**:580–597.

Wallerstein, R. W., 1956, Comparison study of treatment methods for chronic alcoholism, *Am. J. Psychiatry* **113**:228–233.

Winokur, G., Clayton, D. J., and Reich, T., 1969, *Manic Depressive Illness*, C. V. Mosby, St. Louis.

10

Treatment of Nonopiate Dependency

Issues and Outcomes

BARRY S. BROWN

It has been widely recognized that the accelerated efforts to deal with drug abuse that marked the early 1970s were weighted heavily toward a concern with heroin use. Jaffe (1979) has reported that this focus on heroin in the development of a United States treatment program was associated with the public concern about the relationship between heroin use and crime, and with the belief that "new 'technologies' for treating heroin users" could help to reduce that criminal activity. Nonopiate drug abuse, by comparison, was seen as being of a different order of magnitude, at least in terms of its capacity to constitute a threat to public safety, and did not initially attract the resources available to heroin addiction treatment. By 1973, however, a relatively short time after the initial expansion of the American treatment effort, there came to be a focus on polydrug abuse: the abuse of drugs other than the opiates or alcohol (Benvenuto and Bourne, 1975; Wesson and Smith, 1979; DuPont, 1978).

While the initial effort with regard to heroin involved greatly accelerated activity in both treatment and research, the later thrust in polydrug—or nonopiate—drug abuse was more largely a research effort. It first involved the exploration of treatment needs, and the response to treatment, of nonopiate drug users in 12 settings (Wesson et al., 1978). Without denigrating the importance of that national research effort, it seems apparent that the core treatment response to the problem of drug abuse had already been put in place and that these later initiatives involved efforts to make relatively minor modifications of that system. In that context it becomes important to explore the potential for that programming developed for the opiate-using client to meet the needs of the non-opiate-using client.

BARRY S. BROWN ● National Institute on Drug Abuse, Rockville, Maryland.

The goals of drug abuse treatment programming are stated most frequently in behavioral terms. They are the diminution of drug use, reduced criminality, and increased productivity: employment, schooling, homemaking (Sells et al., 1977). As Sells (1979) has noted, these goals of treatment have won easy acceptance from the various drug abuse treatment programs. The easy acceptance of those outcome criteria has probably been the key to the development of an easily understood, widely regarded, and highly consistent treatment evaluation effort for and by the drug abuse treatment community. However, the behavioral outcome criteria selected would seem to reflect a programmatic emphasis on clients' prosocial community functioning—on social deficit—and a corresponding lack of emphasis on clients' psychological functioning,—psychological deficit.

This emphasis in the treatment of the drug abuse client appears well-suited to an earlier view of the needs and functioning of addict-clients. In the early 1970s, many believed that heroin availability was a fact of inner city life. Becoming caught up in a cycle of heroin addiction and crime was not viewed as signaling psychopathology, but as signaling limited or antisocial opportunity. The addict was portrayed as requiring a job and/or skills training, confrontations to clarify his or her maladaptive functioning, pull-ups and/or directive counseling to correct inappropriate behavior and encourage the adoption of prosocial, or acceptable, behaviors. Throughout, it seemed reasonable to assume a certain homogeneity among addict-clients. Addict–clients could be safely characterized as predominantly male, young, urban, of relatively low education, generally lacking in work experience/skills and involved in significant criminal activity (Robins, 1979). Peer pressure, not psychopathology, explained their investment in heroin (Chein et al., 1964; Brown et al., 1971; Nurco, 1979).

There appears no such easy claim to uniformity among nonopiate clients. Moreover, it has been argued that there appears little reason to believe that the treatment programs structured to respond to the needs of the opiate user are equally capable of meeting the needs of the nonopiate client (Avery et al., 1978). While characteristics of the opiate user in treatment have been changing (Graham et al., 1976), nonopiate drug abusers have generally been seen as representing a far wider spectrum of the general population than have opiate users (Wesson and Smith, 1979; Robins, 1979). Theoretical and empirical categories of non-opiate substance abuse have been developed in terms of age, sex, income, combinations of psychosocial variables, referral source, drug type, and etiology. These differing categories of nonopiate abuse have been seen as calling for different treatment responses. In addition, several investigators have reported a high degree of psychopathology among drug abuse clients (Benvenuto and Bourne, 1975; Wesson et al., 1975; McKenna, 1978, 1979; Stauss et al., 1977; Smart and Jones, 1970; Brook et al., 1974). Indeed, Grant et al. (1976, 1978) and Adams et al. (1975) have suggested possibly irreversible organic impairment associated with heavy use of barbiturates and nonbarbiturate hypnotics. Other investigators have reported significant investments in nonopiate drug abuse among persons admitted to psychiatric facilities (Blumberg et al., 1971; Hall et al., 1977; Fischer et al., 1975; McLellan and Druley, 1977; Alterman et al., 1982) frequently with an emphasis on amphetamines (Rockwell and Ostwald, 1968; Razani et al., 1975).

Moreover, there is evidence that the percentage of nonopiate drug use clients relative to the percentage of opiate drug use clients has been increasing. From the first quarter of 1977 through the third quarter of 1980, there was a near-steady increase in number of clients admitted to drug abuse treatment from 47,000 to more than 60,000. During that time, opiate use went from 64.6% of all admissions to 54.9% of all admissions (NIDA, 1981). Except for persons using barbiturates and inhalants, admissions for all nonopiate drugs rose steadily during that same time. Admission of patients using marijuana went from 10.2% in the first quarter of 1977 to 16.2% for the third quarter of 1980. Thus, the total number of opiate-using clients remained relatively stable, the total number of nonopiate clients rose sharply.

Recent U.S. national surveys (Fishburne et al., 1980; Johnston et al., 1981) suggest that selected types of nonopiate drug use are increasing dramatically. The study by Fishburne et al. (1980) is the last reported survey in a series and involves a large number (7224) of face-to-face interviews with a national probability sample. The authors report significant increases in lifetime use of cocaine and of hallucinogens between 1977 and 1979 for all persons aged 18 and older. Moreover, lifetime use of hallucinogens is also reported as significantly greater in 1979 than in 1977 for persons aged 12–17.

Johnston et al. (1981) report findings for a large random sample of high school seniors (17,500), that also constitutes the authors' latest report in a series of such surveys. They found significant increases in the nonmedical use of stimulant drugs only. Indeed, they report some diminution in measures of recent use of marijuana. Like Fishburne et al., they find lifetime prevalence for all other drug categories remaining relatively stable over time.

1.   EFFECTIVENESS OF TRADITIONAL MODALITIES

Four treatment modalities have been viewed as central to drug abuse treatment. Again, given the overriding concern with heroin use, it is perhaps not surprising that two of these are oriented exclusively to the opiate client: methadone maintenance and methadone detoxification. A third, the therapeutic community, grew out of the experiences of opiate users seeking a strategy for achieving long-term abstinence. Only the fourth modality, outpatient drug-free, was developed without specific reference to the opiate user. Perhaps partially as a consequence of the nonspecificity of its target population, that modality remains the most heterogeneous of the four treatment forms. Given the importance of therapeutic community and outpatient drug-free programming as treatment forms, initial focus will be placed on the evaluative studies that have explored those modalities.

Again, as befits the emphasis on heroin addiction, the most comprehensive follow-up of drug abuse treatment paid relatively little attention to nonopiate drug abuse clients. The data on drug abuse clients admitted to 50 programs between 1969 and 1973 were limited largely to opiate users, reflecting the client population of the time (Simpson et al., 1976). In later analysis, Simpson et al. (1980) identified a subsample of "nonaddicts" among clients admitted to their national sample of treatment programs. Nonaddicts were described as persons who had never used opiates daily prior to admission to treatment. Thus, the nonaddict sample was largely—but not exclusively—a nonopiate sample. In assessing outcome 1 year posttreatment, Simpson et al. (1980) reported that 47% of outpatient drug-free nonaddict clients and 42% of therapeutic community nonaddict clients could be described as abstinent for both nonopiate and opiate drugs and as showing "little or no criminality" 1 year posttreatment. It is noteworthy that the nonaddict group contained higher percentages of success for both modalities considered than did either of two addict groups constructed (current addicts and former addicts).

In a study assessing treatment impact at three therapeutic communities and a single methadone maintenance program, Sheffet et al. (1980) found that program graduates did not differ from "all other admissions" in terms of type of drug use. Types of drugs explored were heroin, barbiturates, cocaine, and alcohol.

Neither the studies produced by Sells, Simpson, and their colleagues nor the more limited investigation by Sheffet and his colleagues examined findings for nonopiate users as a separate group. However, both studies reported overall positive findings in terms of posttreatment outcome for the combined populations of opiate and nonopiate clients.

The Treatment Outcome Prospective Study, or TOPS (Research Triangle Institute, 1981), a large scale follow-up effort now in process, gives promise of yielding data with regard to the community functioning of nonopiate as well as opiate clients. In a preliminary analysis, Ginzburg et al. (1982) reported that 93.6% of marijuana users retained in therapeutic community programs for 3 months showed a "large reduction" in marijuana use compared to 23.5% of marijuana users seen for 3 months in outpatient settings. It is important to note that the samples were not made up of subjects using marijuana only or of subjects claiming marijuana as their primary drug, but consisted of subjects using marijuana at least twice weekly.

Keil et al. (1979) examined intreatment outcomes for nonopiate drug abusers admitted to three program types: therapeutic community programs, halfway house programs, and "counseling/psychotherapy," i.e., outpatient, programs. The investigators report lower rates of intreatment productive activity and intreatment arrests in therapeutic community as compared to halfway house and outpatient programs almost certainly as a consequence of the differences in opportunity available to clients in therapeutic community programs compared to other treatment forms. In general, the authors view the demographic and background characteristics of nonopiate clients as more significant to treatment outcome than the treatment forms themselves. Thus, pretreatment productive activity is associated with intreatment productive activity and lessened likelihood of arrest; pretreatment criminal activity is negatively associated with intreatment productive activity and, somewhat less strongly, is associated with intreatment arrests. The authors do note that time in treatment is associated with intreatment productive activity.

In their assessments of the response of nonopiate drug abuse clients to outpatient drug free treatment, investigators are generally pessimistic. Anderson et al. (1972) reported that only 9.6% of their clients reported for two outpatient visits, and only 1 of 208 for regularly scheduled therapy. Tennant (1979) reported difficulty in retaining nonopiate clients in treatment: 65.2% left treatment within 3 months and 47.8% left within the first treatment month. Tennant reports that only 28.3% of clients remained in treatment and were abstinent after 90 days. Avery et al. (1978), reporting on a follow-up of multiple-drug-abusing clients, found that 90% continued to use drugs illicitly 5 months into treatment although the investigators also found a marked decline in the use of sedative/hypnotics and amphetamines, and a smaller decline in the use of marijuana.

Hubbard et al. (1978), Wesson et al. (1974), and Wesson and Smith (1979) raise the possibility that some number of nonopiate (and other) drug abusers are invested in efforts to self-medicate and that treatment programs should explore the use of phenothiazines and lithium carbonate and/or tricyclic antidepressants as appropriate. Similarly, Wesson et al. (1974) recommend psychotherapy in addition to supportive social services and Wesson and Smith (1979) have highlighted biofeedback as a potential aid.

Several authors have drawn distinction between two general types of nonopiate and/or multiple drug abusing individuals: the sophisticated or street-wise drug abuser and the naive or straight drug abuser (Wesson and Smith, 1972; Wesson et al., 1975; Carlin and Stauss, 1978; Podell et al., 1978). Carlin and Stauss (1978) report the "streetwise social–recreational drug user" as the most frequently encountered client in the clinic they studied, with the older "straight self-medicating" client the next most frequently encountered. They report further that the groups demonstrate different psychopathologies and, similarly to Lachar et al. (1978), Kornblith (1981), and Wesson et al. (1975), they call for flexibility in treatment according to the different client types. In this regard, note can be taken of the concern expressed by Chambers and Brill (1971) and Podell et al.

(1978) that nonopiate drug abusers be treated separately from opiate users to prevent their deepened involvement in the drug abuse culture.

In general, studies focusing on the effectiveness of therapeutic community programs also tend to combine drug use groups and pay far more attention to the contributions to outcome of client variables other than type of drug use at admission. Several of the studies that do break out the nonopiate drug abuse client focus on retention as a prime issue. Wesson et al. (1974) report a particularly low rate of retention (13%) for 1 month or more for nonopiate clients admitted to one therapeutic community. However, Wexler and DeLeon (1977) found retention unrelated to drug of abuse. Harris et al. (1980), in a study comparing opiate and nonopiate dropouts and dismissed clients, found nonopiate dropouts to be more likely to report a previous suicide attempt than the other groups although opiate clients dismissed for disciplinary reasons were a close second. In a related study Hunter et al. (1978) reported nonopiate treatment completers to show an initially greater incongruity between self-concept and idealized self than did a sample of opiate treatment completers.

Finally, in any discussion of dropouts from therapeutic communities, one must note the DeLeon and Rosenthal (1979) comment that retention in treatment is a means to an end and that many clients who are not retained to graduation nonetheless are, in fact, retained in treatment for prolonged periods and are found to be functioning effectively in the community. Indeed, several studies have supported a view of therapeutic community dropouts as functioning effectively in the community where those clients were retained for an extended period (Simpson, 1981; DeLeon and Andrews, 1978; Collier and Hijazi, 1974; Wexler and DeLeon, 1977; DeLeon, 1982).

Stephenson et al. (1977) found behavioral success unrelated to primary drug pretreatment in a 1-year follow-up of therapeutic community clients ($n = 77$). However, at least 55% of the sample used opiates, as well as nonopiates, before treatment.

DeLeon (1982) had opportunity to compare the response to therapeutic community programming of nonopiate and opiate drug abusers considered separately. DeLeon investigated the community functioning of two groups of former therapeutic community residents. One group, resident during 1970–1971, was followed up 5 years later while the second group, in residence in 1974, was followed up 2 years later. For the 1970–1971 cohort, DeLeon found that 11.8% of opiate residents were retained to graduation compared to 13.9% of nonopiate residents. For the 1974 cohort, 12.1% of opiate users were retained to graduation compared to 17.9% of nonopiate nonalcoholic residents. For clients with marijuana as their primary drug at admission, 20.3% were retained to graduation.

At 5-year follow-up (3 years for dropouts) the 1970–1971 nonopiate sample ($n = 16$) showed a diminution in daily marijuana use from 43.8% pretreatment to 18.8% at follow-up. Abstinence from marijuana increased from 43.8 to 62.5%.

At the same time abstinence from any use of opiates by the nonopiate sample increased from 20.0 to 75.0%.

At 2-year follow-up the 1974 nonopiate nonalcoholic sample ($n = 59$) showed a diminution in daily marijuana use from 33.8% pretreatment to 13.6% at follow-up. Abstinence from marijuana increased from 23.7 to 37.3%. Abstinence from opiates increased from 69.5 to 89.7%. Alcohol use was also assessed in the 1974 samples. In the 2-year posttreatment follow-up, daily use of alcohol was found to decline among nonopiate residents from 20.3 to 11.2% while abstinence from alcohol use remained constant at 25.0%.

DeLeon was also able to consider therapeutic community residents who presented marijuana as their primary drug of abuse ($n = 30$) independently of other nonopiate subjects. DeLeon found that daily use of marijuana declined from 56.7% pretreatment to 16.7% at 2-year follow-up. Abstinence from marijuana use increased from 0.0 to 26.7%. For this same sample abstinence from use of any opiates increased from 79.3 to 89.7%; abstinence from alcohol increased from 20.7 to 34.5% while daily use remained relatively stable, decreasing from 20.7 to 17.2%.

DeLeon reports further that about 40% of other nonopiate nonalcoholic residents ($n = 38$) were abstinent for their primary drug of abuse at follow-up while daily use decreased from about 72% to approximately 18%. Although alcohol use remained constant for this sample at 75%, daily alcohol use declined from 32.0 to 5.6%.

While modification in use of the drug of choice is impressive for nonopiate subjects, reduction in opiate use by the opiate samples is even more impressive. For the 1970–1971 opiate sample, abstinence increased from 0.5% at admission to 81.6% at follow-up. For the 1974 opiate sample, abstinence from opiates increased from 0.0% at admission to 79.6% at follow-up.

Indeed, using composite measures of program success, DeLeon reports "the 1974 non-opioid abusers yielded the lowest success rates in either cohort." Thus, he reports an overall success rate of 54.4% for opiate residents, 37.0% for nonopiate users other than those admitted with a primary problem of either marijuana or alcohol, and less than 10% for marijuana residents.

In a related study by Vaglum and Fossheim (1980), a therapeutic community program was modified to offer three treatment forms varying in degrees of family therapy, individual therapy and confrontation. Evaluating clients ($n = 160$) up to 4 years postadmission, they found that users of psychedelic drugs, unlike opiate users, were most responsive to a program of family and individual therapy provided in a supportive environment.

In sum, there appears reason to question the effectiveness of traditional treatment forms with nonopiate clients, although a definitive judgement is not possible at this time. Nonetheless, it appears that other treatment initiatives, more largely targeted to nonopiate drug abuse, are worthy of study. Consequently, it is important to examine findings from the several studies designed to explore the efficacy of alternative treatment settings and of additional service components developed or adapted for use with nonopiate clients.

2. ALTERNATIVE TREATMENT SETTINGS

Two settings providing services to nonopiate drug abuse clients have received particular attention. The psychiatric service/community mental health center has been seen as providing treatment to a significant number of nonopiate clients (Safer and Sands, 1979). In addition, school settings have sometimes modified their procedures to provide services to drug abusers as well. Finally, consideration will be given to those initiatives that involved a dramatic restructuring of existing, more traditional, drug treatment programs.

In spite of the likelihood that significant numbers of nonopiate drug abusers are receiving treatment in mental health settings, virtually no study has been made of the treatment process or of treatment outcome for clients being seen in those settings. In one such study, Klinge et al. (1977) reported evidence of diminished psychological disturbance and improved school attendance in a sample of adolescent drug abusers. Unfortunately, the small sample size ($n = 12$) precluded meaningful statistical analysis.

Studies involving school-based interventions (Gottheil et al., 1977; NIDA, 1980) suggest the utility of further work in that area. Both studies examined the impact of richly augmented academic programs. Both studies reported marked positive change in adolescents' functioning, one in terms of academic performance as well as drug use (NIDA, 1980) and the other (Gottheil et al., 1977) in terms of school attendance only. However, only Gottheil and his colleagues made use of a control group.

Waal (1980) described the development of novel treatment structures sometimes in conjunction with existing programs and sometimes as alternatives to those programs. Those efforts, initiated in Scandinavian countries, made use of a variety of community agents to support or replace traditional institutional programming. The different projects involved foster families, clients' work in the community on behalf of community members, and use of drug-free collectives. Preliminary findings were reported for three projects involving systematic interaction of youthful drug abusers with community members. The three studies involved minimum follow-up periods of $1^1/_2$ years after admission to treatment.

Findings from the three studies suggested that between 67 and 80% of clients were drug-free and functioning effectively (in a prosocial manner) at follow-up. Again, the studies suffer from the lack of a comparison group invested in traditional treatment programming that might help clarify the impact of this innovative treatment form. Nonetheless, the findings are suggestive. Those findings, and the logic supporting a need for aftercare efforts to help clients test and refine new coping strategies in the community, would both seem persuasive arguments for further work in this area.

Just as aftercare initiatives merit further study to determine what they can add to traditional programming, so too comparative study of the efficacy of nontraditional and traditional treatment settings is needed. It would seem particularly important to investigate the impact of community mental health centers. An undetermined number of nonopiate clients are already selecting those programs for themselves; moreover, those programs place a greater emphasis on dealing with psychological deficit than do many traditional programs. Likewise, the role that the school can play in the rehabilitation of youthful drug abuse clients demands further study.

3. INNOVATIVE TREATMENT COMPONENTS

A variety of studies have explored treatment techniques designed to meet the special needs of nonopiate drug abuse clients, with a particular emphasis on adolescents. Those initiatives have concerned the use of individual and group psychotherapy, family therapy, and behavior modification strategies.

Given that reality therapy was developed initially for delinquent youth (Glasser, 1965), it would seem particularly appropriate to assess the impact of that strategy with youthful nonopiate drug abusers. Bratter (1972, 1973, 1974) has reported the effectiveness of reality therapy/confrontation techniques with adolescent drug abusers although Schuster (1978) found treatment staff unable to implement a reality therapy system in a residential treatment program. Bratter (1973) found that 70.3% of drug abusing adolescents ($n = 64$), including some using opiates, were found to be drug-free for follow-up periods of up to 4 years.

There is a dearth of study on individual and group counseling/therapy strategies with nonopiate users and opiate users alike. In spite of the emphasis on group counseling strategies with drug abuse clients, virtually no data exist regarding the impact of group techniques. While confrontation strategies are considered by many significant, if not essential, to drug abuse treatment others argue that confrontation must be modified to embrace more largely supportive features (Rachman and Heller, 1976). Study is needed to clarify the comparative contributions of supportive group intervention and confrontational techniques. Similarly, additional study is needed contrasting reality therapy techniques with more traditional psychotherapy efforts. Given the preponderance of paraprofessional counseling staff in drug abuse programs, it is particularly important to understand the efficacy of reality therapy and other techniques as might be practiced by those counseling staffs.

The role of family therapy in drug abuse treatment has been well-reviewed in several surveys of that literature (Baither, 1978; Sowder et al., 1980; Stanton, 1979a,b). However, as Stanton (1979b) has discussed, that literature reveals relatively few quantitative efforts and only a handful of studies involving comparison or control groups. Nonetheless, the potential of family therapy is suggested by the youthfulness of a significant number of the nonopiate clients as well as findings regarding the importance of familial support in overcoming drug abuse (Levy, 1972; Macro Systems, 1975; Eldred and Washington, 1976). Understanding of the role of family therapy is further complicated by the paucity of studies isolating nonopiate from opiate users as well as the need to consider the several schools of family therapy separately.

Four follow-up studies of family therapy concentrated on a nonopiate—and youthful—drug-abusing population and made use of some type of comparison group. Three (Wunderlich et al., 1974; Stecker, 1979; Szapocnik, 1982) describe positive results while one conducted by Winer (1974) found no differences between family therapy and comparison groups. In the study by Wunderlich and his colleagues (1974), court-referred drug abusers whose parents were provided group therapy ($n = 100$) were compared to nondrug cases whose parents were not involved in treatment ($n = 100$). The drug-using youngsters showed a lower rate of arrest, were more likely to remain in school, and showed a lower rate of disciplinary problems. Stecker (reported in Stanton, 1979b) conducted a follow-up study of adolescent and young adult clients whose families were involved in structural family therapy ($n = 16$). Follow-up occurred between 1 and 8 months after the last family therapy session. Stecker reported that 14 of the identified patients "showed improvement" with 10 (63%) drug-free. Stecker reports these results to be far superior to those obtained with clients showing the same characteristics but not receiving family therapy.

Szapocnik (1982) reports preliminary findings examining brief strategic family therapy and comparing that treatment form to "one-patient family therapy," i.e., family-centered treatment involving a single family member. Both treatment forms were found to be associated with dramatic diminution in adolescent clients' drug use and with improvement in family functioning. Later comparison to a control group of adolescents receiving group therapy is planned.

By comparison, Winer et al. (1974) conducted follow-up studies of 109 youngsters at periods 4–24 months after discharge. Winer examined four groups involving different types of family therapy and program involvement and two groups without family involvement. While the study suffered from various methodologic problems, it is noteworthy that all groups improved dramatically in terms of drug use and that one of the nonfamily conditions showed greatest change.

One is forced to conclude with Kaufman (1980) that the case for family therapy with substance abusers is not yet firmly established; the case for family therapy with regard to nonopiate drug abusers is in particular need of further study.

Several reviews have been undertaken examining the efficacy of behavioral treatment approaches with drug abuse clients (Callner, 1975; Gotestam et al., 1976; Krasnegor, 1980). Of the 14 studies identified that examined the use of behavioral treatment strategies with nonopiate drug abusers, 11 describe the impact of treatment on single individuals and 1 described treatment with two nonopiate clients. None of the studies involved comparisons with other treatment forms. Almost all (11 of 14) involved follow-up for at least 6 months.

Aversion therapy was employed in six studies. One (Spevack et al., 1973) used faradic aversion: electric shock. One (Blanchard et al., 1973) used chemical aversion therapy: Anectine® was used with paint taken by a sniffer to induce brief paralysis. Four (Kolvin, 1967; Anant, 1968; Duehn, 1978; Maletzky, 1974) used covert sensitization: verbal aversion therapy in which, in a step-by-step process, aversive imagery is linked with the drug use. Maletzky (1974) also employed a noxious odor in conjunction with more traditional covert sensitization techniques.

Covert extinction (imagining no response to the reinforcing stimulus) was used in one study (Gotestam and Melin, 1974).

Efforts to encourage alternative or competing behaviors in place of nonopiate drug use were explored in eight studies. Five (Kraft, 1969; Boer and Sipprelle, 1969; Suinn and Brittain, 1970; Matefy, 1973; Spevack et al., 1973) made use of systematic desensitization: a conditioning of relaxation responses to anxiety producing stimuli. However, all studies, except that by Kraft (1969), were concerned with reducing the anxiety seen as leading to LSD flashbacks or panic reactions. One study (Flannery, 1972) explored the use of covert conditioning: various strategies designed to permit an alternative response to the stimulus. Two studies (Boudin, 1972; Polakow and Doctor, 1973) made use of contingency contracting: making a contract with the client(s) to develop a relationship between performance of the problem behavior and a selected positive or negative consequence.

One study (Eriksson et al., 1975) made use of a token economy structure: an elaborately structured reward system for observed behaviors.

The studies using behavioral techniques are overwhelmingly positive in their reports of changed behavior. Omitting that research dealing with LSD sequelae, 10 of 11 studies report clients to be drug-free at time of follow-up although 1 of the 10 studies (Gotestam and Melin, 1974) reports one of five clients treated to have relapsed.

In spite of the positive findings and generally careful reporting of results, it is difficult to assess the real utility of behavioral approaches given the absence of control or comparison groups, the fact that data collection and analysis are routinely conducted by the therapists themselves, and the potential for the selective reporting of successes in the literature. In short, there is a need for systematic investigation in this potentially important area.

4. DRUG-SPECIFIC TREATMENT STRATEGIES

Some efforts have involved the development and testing of treatment strategies designed to meet the needs of clients categorized by the type of drug used.

Mason (1979) explored the efficacy of drug abuse treatment for inhalant abusers. She examined client data at six programs oriented to greater and lesser degrees to work with inhalant abuse. She also looked at national data and conducted client and staff interviews. Mason reports a depressing picture of staff ($n = 50$) pessimism supported by, and perhaps supporting, particularly early dropout from treatment by inhalant clients ($n = 117$).

Winn (1982) reports early positive results with inhalant abusers treated in a camp (therapeutic community) setting ($n = 65$) compared with other inhalant abusers randomly assigned to traditional drug abuse programming ($n = 54$). However, differences between groups were found to be sharply attenuated in community follow-up in spite of an aggressive aftercare program for the experimental group.

DeAngelis and Goldstein (1978) report results of an analysis of the functioning of a sample of PCP ("angel dust") users ($n = 45$) who were being seen in a therapeutic community described as offering a variety of supportive treatments in an environment characterized as dealing effectively with "hard-to-treat" adolescents. They report that PCP users were retained in treatment significantly longer than non-PCP users; however, it is unclear to what extent other client characteristics could account for differences in retention.

In a study of amphetamine users ($n = 438$) treated on an inpatient basis (length and type of care is unspecified), Frykholm (1980) reported that 12–18% of all amphetamine users admitted between 1970 and 1974, were drug-free 6 months postdischarge although 27–44% of first admissions for amphetamine abuse were drug-free 6 months after discharge. These findings were significantly better than those obtained for opiate users although again the role of demographic and background variables is not assessed.

Generally, reports of treatment specific to use of barbituric hypnotics, nonbarbituric hypnotics, and anxiolytics restrict themselves to discussions of detoxification strategies (Smith et al., 1979; MacKinnon and Parker, 1982). Wikler (1968) notes that psychotherapy appears in order after detoxification. However, as is the case with treatment of nonopiate drug abuse generally, there is a need for outcome study of this strategy.

5.   SUMMARY AND CONCLUSIONS

There is a compelling need to understand the effectiveness of traditional drug abuse programming—therapeutic communities and outpatient drug-free programs—with nonopiate drug abuse clients. Available findings, while not uniform, suggest that drug abuse programs are only moderately successful in working with those who abuse nonopiate drugs. As noted above, the core drug abuse modalities now available were developed for the urban young adult heroin user. Moreover, there appears reason to believe that those programs, while often mindful of psychological issues, placed a particular emphasis on social deficit.

The nonopiate client may well pose a markedly different set of problems. In part, some of the answers with regard to treatment efficacy as well as the nature of the treatment process with nonopiate drug abuse may be addressed through the Treatment Outcome Prospective Study.

In addition, however, it will be important to understand the process and efficacy of treatment provided through community mental health center programming: that programming specifically concerned with psychological deficit, as compared to traditional drug abuse programming. As a part of that effort, it will be important to explore the nature and extent of substance abuse shown by clients being seen in mental health settings.

Other settings, too, should be explored in terms of their contribution to the effective treatment of nonopiate drug abuse clients. School settings have an obvious potential for service delivery, given their capacity for early identification of drug abuse clients. Juvenile and adult correctional settings also provide obvious potential for the identification and treatment of nonopiate clients although their track record with opiate users at least is less than inspiring (Nash et al., 1981). Both of these settings require the development and testing of nontraditional treatment forms appropriate to those unique environments.

It will also be important to clarify more finely the behavioral and psychological issues that attend nonopiate drug use, and the differences in functioning of nonopiate and opiate clients. Again, those data should contribute to an understanding of the types of programming appropriate to the needs of nonopiate clients and the ways in which that programming may have to differ from the services provided for opiate users.

Nonetheless, our current knowledge regarding both nonopiate clients' characteristics and the effectiveness of traditional programming with nonopiate clients would seem to justify an exploration of the efficacy of selected treatment innovations. Individual, group, and family counseling and psychotherapy oriented toward the nonopiate client all merit further study. Aftercare strategies designed to allow nonopiate clients to be maintained drug-free and productive in the community similarly merit investigation.

In brief, the drug abuse client population has been changing from an opiate to an increasingly nonopiate one. However, neither the development of treatment alternatives nor the conduct of treatment research has kept pace with that change. It would seem urgent that effort now be made to improve our capacity to respond to the problem we have come to confront almost too late.

REFERENCES

Adams, K. M., Rennick, P. M., Schooff, K. G., and Keegan, J. F., 1975, Neuropsychological measurement of drug effects: Polydrug research, *J. Psychedel. Drugs* 7:151.

Alterman, A. I., Erdlen, D. L., Laports, D. J., and Erdlen, F. R., 1982, Effects of illicit drug use in an inpatient psychiatric population, *Addict. Behav.* 7:231.

Anant, S. S., 1968, Treatment of alcoholics and drug addicts by verbal aversion techniques, *Int. J. Addict.* **3**:381.

Anderson, W. H., O'Malley, J. E., and Lazare, A., 1972, Failure of outpatient treatment of drug abuse II. Amphetamines, barbiturates, hallucinogens, *Am. J. Psychiatry* **128**:1572.

Avery, R. F., Judd, L. L., Riney, W., and Takahashi, K., 1978, Long term follow-up on a multiple drug abusing population using a multi-level assessment approach, in: *Drug Abuse: Modern Trends, Issues, and Perspectives* (A. Schecter, H Alksne, and E. Kaufman, eds), pp. 686–695, Marcel Dekker, New York.

Baither, R. C., 1978, Family therapy with adolescent drug abusers: A review, *J. Drug Ed.* **8**:337.

Benvenuto, J., and Bourne, P. G., 1975, The federal polydrug abuse project: Initial report, *J. Psychedel Drugs* **7**:115.

Blanchard, E. B., Libet, J. M., and Young, L. D., 1973, Apneic aversion and covert sensitization in the treatment of a hydrocarbon inhalation addiction: A case study, *J. Behav. Ther. Exp. Psychiatry* **4**:383.

Blumberg, A. G., Cohen, M., Heaton, A. M., and Klein, D. F., 1971, Covert drug abuse among voluntary hospitalized psychiatric patients, *JAMA* **217**:1659.

Boer, A. P., and Sipprelle, C. N., 1969, Induced anxiety in the treatment for LSD effects, *Psychother. Psychosom.* **17**:108.

Boudin, H. M., 1972, Contingency contracting as a therapeutic tool in the decleration of amphetamine use, *Behav. Ther.* **3**:604.

Bratter, T. E., 1972, Group therapy with affluent, alienated, adolescent drug abusers: A reality therapy and confrontation approach, *Psychother. Theory Res. Pract.* **9**:308.

Bratter, T. E., 1973, Treating alienated, unmotivated, drug abusing adolescents, *Am. J. Psychother.* **27**:585.

Bratter, T. E., 1974, Reality therapy: A group psychotherapeutic approach with adolescent alcoholics, *Ann. N.Y. Acad. Sci.* **233**:104.

Brook, R., Kaplun, J., and Whitehead, P. C., 1974, Personality characteristics of adolescent amphetamine users as measured by the MMPI, *Br. J. Addict.* **69**:61.

Brown, B. S., Gauvey, S. K., Meyers, M. B., and Stark, S. D., 1971, In their own words: Addicts' reasons for initiating and withdrawing from heroin, *Int. J. Addict.* **6**:635.

Callner, D. A., 1975, Behavioral treatment approaches to drug abuse: A critical review of the research, *Psychol. Bull.* **82**:143.

Carlin, A. S., and Stauss, F. F., 1978, Two typologies of polydrug abusers, in: *Polydrug Abuse* (D. R. Wesson, A. S. Carlin, K. M. Adams, and G. Beschner, eds.), pp. 97–127, Academic Press, New York.

Chambers, C. D., and Brill, L., 1971, Some considerations for the treatment of nonnarcotic drug abusers, *Ind. Med.* **40**:29.

Chein, I., Gerard, D. L., Lee, R. S., and Rosenfeld, E., 1964, *The Road to H: Narcotics, Delinquency, and Social Policy,* Basic Books, New York.

Collier, W. V., Hijazi, Y. A., 1974, A follow-up study of former residents of a therapeutic community, *Int. J. Addict.* **9**:805.

DeAngelis, G. G., and Goldstein, E., 1978, Treatment of adolescent phencyclidine (PCP) abusers, *Am. J. Drug Alcohol Abuse* **5**:399.

DeLeon G., 1982, *The Therapeutic Community: Study of Effectiveness,* NIDA Monograph (November, 1982).

DeLeon, G., and Andrews M., 1978, Therapeutic community dropouts 5 years later: Preliminary findings on self-reported status, in: *A Multicultural View of Drug Abuse* (D. Smith, ed.), pp. 369–377, Schenkman, Cambridge, Mass.

DeLeon, G., and Rosenthal, M. S., 1979, Therapeutic communities, in *Handbook on Drug Abuse* (R. L. DuPont, A. Goldstein, J. O'Donnell, and B. S. Brown, eds.), pp. 39–47, Government Printing Office, Washington, D.C.

Duehn, W. D., 1978, Covert sensitization in group treatment of adolescent drug abusers, *Int. J. Addict.* **13**:485.

DuPont, R. L., 1978 Foreword, in: *Polydrug Abuse* (D. R. Wesson, A. S. Carlin, K. M. Adams, and G. Beschner, eds.), pp. XVII–XX, Academic Press, New York.

Eldred, C. A., and Washington, M. N., 1976, Interpersonal relationships in heroin use by men and women and their role in treatment outcome, *Int. J. Addict.* **11**:117.

Eriksson, J. H., Gotestam, K. G., Melin, L., and Ost, L., 1975, A token economy treatment of drug addiction, *Behav. Res. Ther.* **13**:113.

Fischer, D. E., Halikas, J. A., Baker, J. W., and Smith, J. B., 1975 Frequency and patterns of drug abuse in psychiatric patients, *Dis. Nerv. Syst.* **36**:550.

Fishburne, P. M., Abelson, H. I., and Cisin, I., 1980, *National Survey on Drug Abuse: Main Findings: 1979*, NIDA Monograph (1980).

Flannery, R. B., 1972, Use of covert conditioning in the behavioral treatment of a drug-dependent college dropout, *J Counseling Psychol.* **19**:547.

Frykholm, B., 1980, Changes in short-term prognosis—A comparison between Swedish amphetamine and opiate abusers, *Drug Alcohol Depend.* **5**:123.

Ginzburg, H. M., Craddock, S. G., Hubbard, R., and Glass, J., 1982, Characteristics, behaviors, and outcomes for marijuana users seeking treatment in drug abuse treatment programs, Cannabinoids 82 Meeting (August, 1982), Louisville, Kentucky.

Glasser, W., 1965, *Reality Therapy: A New Approach to Psychiatry*, Harper and Row, New York.

Gotestam, K. G., and Melin, L., 1974, Covert extinction of amphetamine addiction, *Behav. Ther.* **5**:90.

Gotestam, K. G., and Melin, L., and Ost, L., 1976, Behavioral techniques in the treatment of drug abuse: An evaluative review, *Addict. Behav.* **1**:205.

Gottheil, E., Rieger, J. A., Farwell, B., and Lieberman, D. L., 1977, An out-patient drug program for adolescent students: Preliminary evaluation, *Am. J. Drug Alcohol Abuse* **4**:31.

Graham, T. G., Brown, B. S., and DuPont, R. L., 1976, Characteristics of new admissions to a narcotics treatment program: 1970–74, *Int. J. Addict.* **11**:967.

Grant, I., Mohns, L., and Miller, M., 1976, A neuropsychological study of polydrug users, *Arch. Gen. Psychiatry* **33**:973.

Grant, I., Adams, K. M., Carlin, A. S., Rennick, P. M., Judd, L. L., Schooff, K., and Reed, R., 1978, Organic impairment in polydrug uses: Risk factors, *Am. J. Psychiatry* **135**:178.

Hall, R. C. W., Popkin, M. K., DeVaul, R., and Stickney, S. K., 1977, The effect of unrecognized drug abuse on diagnosis and therapeutic outcome, *Am. J. Drug Alcohol Abuse* **4**:455.

Harris, R., Linn, M. W., and Pratt, T. C., 1980, A comparison of dropouts and disciplinary discharges from a therapeutic community, *Int. J. Addict.* **15**:749.

Hubbard, B., Judd, L. L., and Avery, R., 1978, Pharmacotherapy of drug abusers: Adjunctive psychopharmacologic management of nonopiate mixed substance abusers in an outpatient setting, *Int. J. Addict.* **13**:383.

Hunter, K. I., Linn, M. W., and Harris, R., 1978, Self-concept and completion of treatment for heroin and nonheroin drug abusers, *Am. J. Drug Alcohol Abuse* **5**:463.

Jaffe, J. H., 1979, The swinging pendulum. The treatment of drug abusers in America, in: *Handbook on Drug Abuse* (R. L. DuPont, A. Goldstein, J. O'Donnell, and B. S. Brown, eds.), pp. 3–16, Government Printing Office, Washington, D.C.

Johnston, L. G., Bachman, J. G., and O'Malley, P. M., 1981, Highlights From Student Drug Use in America 1975–1981, NIDA Monograph (1981).

Kaufmann, E., 1980, Myth and reality in the family patterns and treatment of substance abusers, *Am. J. Drug Alcohol Abuse* **7**:257.

Keil, T. J., Rush, T. V., and Dickmann, F. B., 1979, Pretreatment roles and therapeutic environment as correlates of in-treatment client "success": The case of the nonopiate user, *Int. J. Addict.* **14**:569.

Klinge, V., Lennox, K., and Vaziri, H., 1977, Follow-up of adolescent drug abusers and nonusers previously hospitalized in an inpatient psychiatric facility, *Drug Forum* 6:143.

Kolvin, T., 1967, Aversive imagery treatment in adolescents, *Behav. Res. Ther.* 5:245.

Kornblith, A. B., 1981, Multiple drug abuse involving nonopiate, nonalcoholic substances. II. Physical damage, long-term psychological effects and treatment approaches and success, *Int. J. Addict.* 16:527.

Kraft, T., 1969, Successful treatment of a case of chronic barbiturate addiction, *Br. J. Addict.* 64:115.

Krasnegor, N. A., 1980, Analysis and modification of substance abuse, *Behav. Modif.* 4:35.

Lachar, D., Schooff, K., Keegan, J., and Gdowsky, C., 1978, Dimensions of polydrug abuse: An MMPI study, in: *Polydrug Abuse* (D. R. Wesson, A. S. Carlin, K. M. Adams, and G. Beschner, eds.), pp. 149–180, Academic Press, New York.

Levy, B., 1972, Five years after: A follow-up of 50 narcotic addicts, *Am. J. Psychiatry* 7:102.

Macro Systems, Inc., 1975, *Three-year Follow-up Study of Clients Enrolled in Treatment Programs in New York City: Phase III-Final Report,* NIDA Monograph (June, 1975).

MacKinnon, G. L., and Parker, W. A., 1982, Benzodiazepine withdrawal syndrome: A literature review and evaluation, *Am. J. Drug Alcohol Abuse* 9:19.

Maletzky, B. M., 1974, Assisted covert sensitization for drug abuse, *Int. J. Addict.* 9:411.

Mason, T., 1979, *Inhalant Use and Treatment,* Government Printing Office, Washington, D.C.

Matefy, R. E., 1973, Behavior therapy to extinguish spontaneous recurrences of LSD effects: A case study, *J. Nerv. Ment. Dis.* 156:226.

McKenna, G. J., 1978, The drug/alcohol/psychiatry interface, in: *Critical Concerns in the Field of Drug Abuse* (A. Schecter, H. Alksne, and E. Kaufman, eds.), pp. 894–900, Marcel Dekker, 1978.

McKenna, G. J., 1979, Psychopathology in drug dependent individuals: A clinical review, *J. Drug Issues* 9:197.

McLellan, A. T., and Druley, K. A., 1977, Non-random relation between drugs of abuse and psychiatric diagnosis, *J. Psychiatr. Res.* 13:179.

Nash, G., Foster, K., and Lynn, R., 1981, The Impact of Drug Abuse Treatment upon Criminality: A Look at 30 Treatment Programs in New Jersey, NIDA Monograph (August, 1981).

National Institute on Drug Abuse, 1980, *The Learning Laboratory: The Door—A Center of Alternatives,* Government Printing Office, Washington, D.C.

National Institute on Drug Abuse, 1981, *Trend Report: January 1977–September 1980,* Government Printing Office, Washington, D.C.

Nurco, D. N., 1979, Etiological aspects of drug abuse, in: *Handbook on Drug Abuse* (R. L. DuPont, A. Goldstein, J. O'Donnell, and B. S. Brown, eds.), pp. 315–324, Government Printing Office, Washington, D.C.

Podell, M. P., Trilling, F. G., Angle, H. V., and Moore, E.K., 1978, Treatment referral of the polydrug user, in: *Drug Abuse: Modern Trends, Issues and Perspectives* (A. Schecter, H. Alksne, and E. Kaufman, eds.), pp. 142–148, Marcel Dekker, New York.

Polakow, R. L., and Doctor, R. M., 1973, Treatment of marijuana and barbiturate dependency by contingency contracting, *J. Behav. Ther. Exp. Psychiatry* 4:375.

Rachman, A. W., and Heller, M. E., 1976, Peer group psychotherapy with adolescent drug abusers, *Int. J. Group Psychotherapy* 26:373.

Razani, J., Farina, F. A., and Stern, R., 1975, Covert drug abuse among patients hospitalized in a psychiatric ward of a university hospital, *Int. J. Addict.* 10:693.

Research Triangle Institute, 1981, *Summary and Implications: Client Characteristics, Behaviors and Intreatment Outcomes—1979 TOPS Admission Cohort,* RTI Monograph (August, 1981).

Robins, L. N., 1979, Addict careers, in: *Handbook on Drug Abuse* (R. L. DuPont, A. Goldstein, J. O'Donnell, and B. S. Brown, eds.), pp. 325–336, Government Printing Office, Washington, D.C.

Rockwell, D. A., and Ostwald, P., 1968, Amphetamine use and abuse in psychiatric patients, *Arch. Gen. Psychiatry* **18:**612.

Safer, J. M., and Sands, H., 1979, A Comparison of Mental Health Treatment Center and Drug Abuse Treatment Center Approaches to Nonopiate Drug Abuse, Government Printing Office, Washington, D.C.

Schuster, R., 1978, Evaluation of a reality therapy stratification system in a residential drug rehabilitation center, *Drug Forum* **7:**59.

Sells, S. B., 1979, Treatment effectiveness, in: *Handbook on Drug Abuse* (R. L. DuPont, A. Goldstein, J. O'Donnell, and B. S. Brown, eds.), Government Printing Office, Washington, D.C.

Sells, S. B., Demaree, R. G., Simpson, D. D., Joe, G. W., and Gorsuch, R. L., 1977, Issues in the evaluation of drug abuse treatment, *Prof. Psychol.* **8:**609.

Sheffet, A. M., Quinones, M. A., Doyle, K. M., Lavenhar, M. A., Nakah, A. E., and Louria, D. B., 1980, Assessment of treatment outcomes in a drug abuse rehabilitation network: Newark, New Jersey, *Am. J. Drug Alcohol Abuse* **7:**141.

Simpson, D. D., Savage, L. J., Joe, G. W., Demaree, R. G., and Sells, S. B., (1976), DARP Data Book-Statistics on Characteristics of Drug Users in Treatment during 1969–1974, IBR Report 76-4 (April, 1976).

Simpson, D. D., Savage, L. J., and Sells, S. B., 1980, Evaluation of Outcomes in the First Year after Drug Abuse Treatment: A Replication Study Based on 1972–1973 DARP Admissions, IBR Report 80-8 (May, 1980).

Simpson, D. D., 1981, Treatment for drug abuse. Follow-up outcomes and length of time spent, *Arch. Gen. Psychiatry* **38:**875.

Smart, R. G., and Jones, D., 1970, Illicit LSD users: Their personality characteristics and psychopathology, *J. Abnorm. Psychol.* **75:**286.

Smith, D. E., Wesson, D. R., and Seymour, R. B., 1979, The abuse of barbiturates and other sedative-hypnotics, in: *Handbook on Drug Abuse* (R. L. DuPont, A. Goldstein, J. O'Donnell, and B. S. Brown, eds.), pp. 233–240, Government Printing Office, Washington, D.C.

Sowder, B., Dickey, S., and Glynn, T. J., 1980, *Family Therapy: A Summary of Selected Literature*, Government Printing Office, Washington, D.C.

Spevack, M., Peihl, R., and Rowan, T., 1973, Behavior therapies in the treatment of drug abuse: Some case studies, *Psychol. Rec.* **23:**179.

Stanton, M. D., 1979a, Family Treatment of drug problems: A review, in: *Handbook on Drug Abuse* (R. L. DuPont, A. Goldstein, J. O'Donnell, and B. S. Brown, eds.), pp. 133–150, Government Printing Office, Washington, D.C.

Stanton, M. D., 1979b, Family Treatment approaches to drug abuse problems: A review, *Fam. Proc.* **18:**251.

Stauss, F. F., Ousley, N. K., and Carlin, A. S., 1977, Psychopathology and drug abuse: An MMPI comparison of polydrug abuse patients with psychiatric inpatients and outpatients, *Addict. Behav.* **2:**75.

Stephenson, N. L., Boudewyns, P. A., and Lessing, R. A., 1977, Long-term effects of peer group confrontation therapy used with polydrug abusers, *J. Drug Issues* **7:**135.

Stecker, H., 1979, Family therapy in nonopiate drug rehabilitation, reported in Stanton, M. D., *Fam. Proc.* **18:**263.

Suinn, R. M., and Brittain, J., 1970, The termination of an LSD "freak out" through the use of relaxation, *J. Clin. Psychol.* **26:**127.

Szapocnik, J., 1982, *Brief Strategic Family Therapy: A Structural System*, NIDA Grant Submission, Miami, Florida.

Tennant, F. S., 1979, Outpatient treatment and outcome of prescription drug abuse, *Arch. Intern. Med.* **139:**154.

Vaglum, P., and Fossheim, I., 1980, Differential treatment of young abusers: A quasi-experimental study of a "therapeutic community" in a psychiatric hospital, *J. Drug Issues* **10**:505.

Waal, H., 1980, Unconventional treatment models for young drug abuses in Scandinavia, *J. Drug Issues* **10**:441.

Wesson, D. R., and Smith, D. E., 1979, Treatment of the polydrug abuser, in: *Handbook on Drug Abuse*, (R. L. DuPont, A. Goldstein, J. O'Donnell, and B. S. Brown, eds.), pp. 151–157, Government Printing Office, Washington, D.C.

Wesson, D. R., Smith, D. E., and Lerner, S. E., 1975, Streetwise and nonstreetwise polydrug typology: Myth or reality, *J. Psychedel. Drugs* **7**:121.

Wesson, D. R., Smith, D. E., Lerner, S. E., and Kettner, V. R., 1974, Treatment of polydrug users in San Francisco, *Am. J. Drug Alcohol Abuse* **1**:159.

Wesson, D. R., Grant, I., Carlin, A. S., Adams, K. M., and Harris, C., 1978, Neuropsychological impairment and psychopathology, in: *Polydrug Abuse* (D. R. Wesson, A. S. Carlin, K. M. Adams, and G. Beschner, eds.), pp. 263–272, Academic Press, New York.

Wexler, H. K., and De Leon, G., 1977, The therapeutic community: Multivariate prediction of retention, *Am. J. Drug Alcohol Abuse* **4**:145.

Wikler, A., 1968, Diagnosis and treatment of drug dependence of the barbiturate type, *Am. J. Psychiatry* **125**:758.

Winer, L. R., Lorio, J. P., and Scraffort, I., 1974, Effects of treatment on drug abuser and family, Special Action Office for Drug Abuse Prevention Report 4 RG003 (1974).

Winn, J., 1982, Kukulu Kumuhana—Final Evaluation Report, NIDA Grant Report H81 DA 02056 (March, 1982).

Wunderlich, R. A., Lozes, J., and Lewis, J., 1974, Recidivism rates of group therapy participants and other adolescents processed by a juvenile court, *Psychother. Theory Res. Practice* **11**:243.

11

Less-Hazardous Tobacco Use as a Treatment for the "Smoking and Health" Problem

LYNN T. KOZLOWSKI

Tobacco is a dirty weed. I like it.
It satisfies no normal need. I like it.
It makes you thin, it makes you lean,
It takes the hair right off your bean.
It's the worst darn stuff I've ever seen.
I like it.
G. L. HEMMINGER, in *Penn State Froth,* 1915

1. INTRODUCTION

The health care industry cares about tobacco use mainly because it causes death and disability in users and perhaps in their associates (e.g., U.S.D.H.E.W., 1979, U.S.D.H.H.S., 1982). The war against tobacco use is at root a war of messages and recommendations about conduct. Persuasion is important in this particular war on drugs because the product in question is neither illegal nor difficult to obtain.

If tobacco kills or injures, then, assuming no redeeming values, the obvious message is to stop or not start using tobacco. Unfortunately, tobacco appears to have some redeeming value, if only to the dependent user who suffers without it (e.g., Schachter et al., 1977; Silverstein, 1982). Whatever the reasons, history shows that tobacco, once introduced to a culture, is never eliminated, even in

LYNN T. KOZLOWSKI ● Addiction Research Foundation, Toronto, Ontario, Canada. The views expressed in this publication are those of the author and do not necessarily reflect those of the Addiction Research Foundation.

highly coercive societies (Brooks, 1952). However, change or evolution has taken place in the types of tobacco that are most popular (Kozlowski, 1982a). Once one is forced to assume that tobacco has redeeming values (benefits), then one needs to open negotiations with the enemy to determine if some deal can be struck with those who do not wish or who are unable to give up tobacco use entirely. The goal of the negotiation is to make the best of a health risk by minimizing it, knowing that, for some, it can not be eliminated practically. Partial victory is substituted, where possible, for total defeat.

The less-hazardous-tobacco-use message is directed toward those who are unwilling or unable to stop using tobacco completely. Ideally, it acts when the antitobacco message (stop or don't start) fails, and it complements the antitobacco message; in practice, these two messages are competitive and troubled allies. At present, both the less-hazardous and the antitobacco messages indicate goals that we are struggling to find out how to attain. We are still trying to discover how best to prevent, stop, or modify tobacco use. [Incidentally, the protobacco response to the above messages is that we really don't know yet if tobacco is dangerous, so continue to use tobacco if you care to (Friedman, 1975).]

This chapter will argue that the use of less-hazardous tobacco, if prohibitionistic impulses can be put aside, may have an important role in the treatment of the smoking and health problem. Just as research efforts are needed to try to improve prevention and cessation techniques, they are needed to try to improve the techniques of less-hazardous tobacco use. (For a review of issues related to the application of less-hazardous-tobacco-use treatments, see Kozlowski, 1984.) The phrase "less-hazardous tobacco use" is meant to be inclusive. Cigarettes, for example, are the most hazardous tobacco products overall; yet even cigarettes can become a less-hazardous use of tobacco, if only a little of a few cigarettes is smoked each day. On the other hand, some less-hazardous tobacco products are less-hazardous in certain respects no matter how they are used: chewing tobacco, for example, carries no risks of fire and essentially no risk of lung disease.

For some workers concerned with smoking and health, the mission of this chapter is outrageous. For these individuals (and institutions), no tobacco product can be part of the treatment of the smoking and health problem; complete prevention and absolute cessation of all tobacco use are the only acceptable goals. The exclusive goal of exterminating all smoking and tobacco use is so prominent a feature of the contemporary discussion of "Smoking or Health" that it will be necessary to (1) try to account for the predominance of this goal and (2) confront the possible pitfalls of pursuing only this means of reducing the health consequences of smoking. To try to avoid some needless arguements, I will define how I am using some key terms, before entering into the debate.

Less-hazardous means reduced in risk or not as dangerous; it does not automatically mean safe or without risk. Tobacco use can refer either to (1) the type of product or (2) the nature of its use. This chapter is not mainly about the so-called "less-hazardous cigarette." Low-yield cigarettes will not be referred to as less-hazardous or safer cigarettes. Though they may indeed be less hazardous than high-yield cigarettes, this point is still controversial (e.g., Kozlowski et al., 1982b; Gerstein and Levison, 1982). Though low-yield cigarettes are low-yield when placed in the ports of smoking machines, they are not necessarily low-yield when placed in the mouths of smokers. In fact, the lowest-yield cigarettes (1 mg tar, 0.1 mg nicotine) can turn into medium- or high-yield cigarettes when a smoking machine assay is adjusted to simulate better the smoking behavior seen in a human smoker apparently bent on compensating for the reduced yields (Kozlowski et al., 1982c).

Although treatment is often a medical term, it is employed here in its more fundamental meaning as a way of dealing with something. The smoking problem and the tobacco problem are more general than the smoking and health problem. Some individuals view any form of tobacco use as a serious waste of time and resources and as an activity to be discouraged; these views would hold even if tobacco use posed no risk of disease or disability. If one believes that tobacco use, *per se*, is a problem to be eliminated, then less-hazardous tobacco use presents at least one problem too many. If one believes that tobacco is a problem primarily because of serious effects on health, then the reduction of the toxic consequences of tobacco use is a worthwhile goal. The smoking and health problem focuses on the damage to health caused by cigarettes.

In this chapter, addiction or compulsive drug use, per se, is not considered a major health problem, unless the drug-taking behavior causes serious physical, social, psychological, or behavioral disturbance. According to the *Diagnostic and Statistical Manual of the American Psychiatric Association (DSM-III)* (American Psychiatric Association, 1980), smoking becomes an official disorder (305.1x Tobacco Dependence) if serious attempts to stop or reduce tobacco use have been unsuccessful, if tobacco withdrawal occurs during tobacco abstinence, or if the tobacco use continues despite a serious physical condition (e.g., respiratory or cardiovascular disease) that the user knows is exacerbated by tobacco use. No mention is made of "impairment in social or occupational functioning." Tobacco dependence is, in fact, alone among the several substance use disorders described in *DSM-III* (alcohol, barbiturates, cocaine, opioids, amphetamines, phencyclidine, hallucinogens, cannabis) in that impairment in social or occupational functioning is not judged to be an "immediate and direct" result of the use of the substance.

2. BACKGROUND

Compassion and Venom

According to an Arabian story (Bain, 1896), the Prophet, Mahomet, rescued a snake from freezing by warming the snake against his body. The thankless snake bit him, but Mahomet sucked the venom from his wound and spat it upon the ground. On that spot, it was said, grew the first tobacco plant, combining the compassion of the prophet with the venom of the serpent. The quest for less-hazardous tobacco products has been directed toward reducing or eliminating the "venom" of tobacco, while at the same time keeping its "compassion." From a hedonistic perspective, then, less-hazardous tobacco use strives to maintain pleasure and minimize pain. From the perspective of the marketplace, the ideal less-hazardous product sells well, but does not kill the customers.

Low-yield cigarettes lack one of the key requirements of a less-hazardous tobacco product, in that they remove the "compassion" along with the "venom." Low-yield cigarettes—as designed, not necessarily as smoked—are little more than placebo cigarettes. Ultra-low-tar cigarettes are ultra-low-smoke cigarettes. As much as 80% of the smoke in each puff of a cigarette yielding 1 mg tar can be diluting air (Kozlowski, 1981b). On the assumption that some of the pharmacologic actions of tobacco are responsible for the "compassion" (nicotine is most often thought to be the key ingredient), an across-the-board reduction in drug delivery hardly qualifies as a practical strategy for producing an acceptable less-hazardous tobacco product.

Tobacco use is often understood in a much too simplified way. That harm and benefit, venom and compassion, can reside in the same product is readily appreciated in many areas of applied research, yet researchers in the tobacco area have tended either to identify tobacco as a killer or to deny that claim. Though many individuals die from both the direct and indirect use of automobiles, I can not recall hearing the argument that, therefore, all automobile use should be prevented or stopped. Because of the widely appreciated benefits of the automobile, the less-hazardous automobile movement has been more prominent than the antiautomobile movement.

Jumping to Exclusions

In their classic book on logic, Cohen and Nagel (1962) note: "One of the most fruitful sources of intellectual confusion is the too facile assumption that any two propositions which are not equivalent are mutually exclusive" (p. 68). At least one of the ramifications of this confusion can be seen in the ready employment of false dichotomous questions. Such questions make a practice of opposing issues that are neither exhaustive nor mutually exclusive (Fischer, 1970). Notice that the Royal College of Physicians in the United Kingdom entitled their recent monograph on the health consequences of tobacco use "Smoking or Health" (Royal College of Physicians, 1977). To return to the analogy with the automobile, it is as if a book were entitled "Driving or Health." In neither instance is the dichotomy justified. Not all drivers suffer ill-health as a consequence of automobile use; some do. Not all smokers suffer ill-health as a consequence of tobacco use; some do. Neither are all nonsmokers and nondrivers certain to be healthy.

False dichotomies deny the crucial middle ground and emphasize the extremes. They polarize a question. They add to the memorability of the question. They might provide an image around which to rally contributions and interest in a problem. Yet it is perilous for scientists to treat them as any more than slogans or entertainments. To use the false dichotomy to guide research on the problem is, in fact, to base one's exploration on an unsupportable premise.

Importance of Beliefs

Beliefs and values are the first principles from which the creation and eradication of social problems flow (Lindblom and Cohen, 1979). Outside of the sometimes idealized world of public health education, complicated beliefs and circumstances contribute to the valuation of tobacco. Those who thought that announcing that "smoke kills" would lead to an exodus from the bondage of tobacco use might also have predicted that the high risk of earthquakes should have emptied California. If a patient in your care or a loved one dies or suffers from a tobacco-related disease, the costs of the activity may overwhelm the benefits. If you support your family through the sale of tobacco (or if you have gone to college because of a scholarship from a tobacco company), the benefits of tobacco may be salient. If you are a smoker who feels some pleasure in smoking, then the threat of a future death from smoking might be countered persuasively with the conviction that one must, after all, die of something and that, despite the most pampered life in the world, an accidental death from any number of causes could lurk around the next corner.

Cigarette smoking is argued to be the "largest preventable cause of death in America" (U.S.D.H.E.W., 1979). The term "preventable" is a problem. The wish to prevent should not be confused with the ability to prevent. In later sections of this chapter, it will be argued that there are limits to the preventability of tobacco use. The limits on the preventability of tobacco use become some of the strongest arguments for developing less-hazardous modes of tobacco use. Beliefs and values influence what one chooses to be preventable. Many activities are preventable, given enough effort to prevent; however, drug use has not shown itself to be an area of easy prevention, despite large investments in wars on this or that drug. It is doubtful whether drug use is, in practice, preventable in a free society. (Prevention and deterrence are quite different concepts.) I would revise the quotation that opens this paragraph by stating that cigarette smoking is the largest cause of death that authorities are trying to prevent.

Scientific Haggling

Scientific dispute does not take place at a level above the usual mire of human argument. The mantle of science is worn by people whose conflicts with a colleague share much with arguments with a spouse. Such maneuvers as intimidation, threat, insult, belittling, evasion, and flattery, to give a partial list, are as readily found in rational argument as they are in everyday argument (Lakoff and Johnson, 1980). Lakoff and Johnson assert that both rational argument and haggling are grounded in the metaphor "argument is war." Anyone familiar with the manufacture of the scientific bullets used in the war against (and for) tobacco use should be aware that all is fair in love and rational argument, especially when an emotionally and economically charged issue is involved.

Some of the most striking reactions to the topic of less-hazardous tobacco use have not appeared in print, but have occurred behind the scenes at scientific and medical meetings. In 1980, for example, a scientific meeting was convened by a major voluntary agency in the United States, in part to help set research priorities on smoking and health. The chairman of the smoking group had prepared a position paper that was to focus discussion on a list of research topics. In the opening plenary session, the distinguished chairman was asked, "Why isn't the issue of doing research on less-hazardous tobacco use on the agenda?" With no hesitation, he responded, "Better men than I have been ruined for proposing such a thing!" Be assured that although this comment influenced the proceedings, it did not appear in them, and neither did less-hazardous tobacco use appear in the list of topics in need of research.

Fortunately, this section will not have to rely on undocumented anecdotes to establish the needed background. In 1978, *Science* (Marx, 1978) reported on an event that was distressing the highest "smoking or health" officials in Washington, D.C. The Secretary of Health, Education, and Welfare (HEW) had been mounting a vigorous, high-profile campaign against cigarette smoking. Plans were being made for the 1979 Report on Smoking and Health of the Surgeon General. This report was to be released on the 15th anniversary of the landmark 1964 Surgeon General's Report. The 1979 Report was to be roughly three times the size of the 1964 Report. The promotion of healthier "lifestyles" was fast becoming a popular activity: everywhere running shoes were filled with jogging feet.

In this atmosphere, a government scientist, Dr. Gio B. Gori (no less than the Deputy Director, Division of Cancer Cause and Prevention, National Cancer Institute) published a paper (with Cornelius Lynch) indicating that low-yield cigarettes, especially modern low-tar brands, were less hazardous than high-yield cigarettes. The paper encouraged smokers to wean themselves progressively to less-hazardous cigarettes as "an alternative to smoking cessation that is perhaps more effective than the self-denial approaches of current anti-smoking messages" (Gori and Lynch, 1978). Although Gori and Lynch were careful to avoid calling low-yield cigarettes "safe," the publicity surrounding the publication in the prestigious *Journal of the American Medical Association* announced that "tolerable" cigarettes were at last available. As described by Marx (1978), ". . . the suggestion by a government scientist that smoking might be 'tolerable' was not well received by health officials who were afraid it would undermine their anti-smoking efforts." [In an equally notable example of understatement, Gori wrote in a summary essay: "Public policy in smoking and health has been dominated for years by idealistic approaches with moderate sympathy for less-hazardous cigarettes" (Gori, 1980).]

Those who were upset about the Gori paper and its impact included: the Secretary of HEW, the Surgeon General, and the Directors of the National Cancer Institute (NCI) and the National Heart, Lung, and Blood Institute. Dr. Gori is reported to have told the press that the Secretary of HEW was trying to have the NCI fire or at least discipline him. At the time of the Marx report, Dr. Gori had been removed recently from command of the NCI Smoking and Health program. Dr. Gori no longer works for the NCI.

The Gori and Lynch (1978) and the earlier Gori (1976) papers did offer encouragement that those smokers who would not stop smoking completely could benefit from a switch to lower-yield cigarettes. Although this may be true under some circumstances, the Gori research has been the object of a great deal of scientific criticism (e.g., Gart and Schneiderman, 1979; Warner, 1979), and I would be toward the front of the line in criticizing the Gori work. One of my key objections concerns the uncritical acceptance of the lower yield ratings of recent cigarettes. Such acceptance places unwarranted confidence in the adequacy of the simulation of human smoking behavior by standard smoking machines (Kozlowski et al., 1980, 1982c; Kozlowski, 1981a, c).

Why were officials afraid that a "less-hazardous cigarette" message would undermine antismoking efforts? The official antismoking efforts were directed at smoking prevention and cessation. In other words, the only message they wished to present was, "If you don't smoke, don't start; if you do smoke, stop." The addition of a further clause, "If you must smoke, at least smoke a lower-yield cigarette," was intolerable. Why should this additional message cause so much trouble? Does such a message actually spoil an antismoking campaign?

3. ARGUMENTS AND EVIDENCE, NOT FACTS AND PROOF

When one deals with social problem-solving, despite the fondest wishes of practitioners and politicians, truly objective facts are rarely found. And without such facts, no incontrovertible proofs will be forthcoming. At most, one can argue and give evidence in support of the arguments (Lindblom and Cohen, 1979). So, when policymakers ask what should be done about the smoking and health problem, they should expect arguments and evidence rather than facts and proof. Though the policymakers and their advisors may wish to act as if revealed facts can lead to a course of action, the process depends unavoidably on arguments.

4. ARGUMENTS AGAINST ADVOCATING LESS-HAZARDOUS TOBACCO USE

I will try to present arguments against less-hazardous tobacco use that contain more meat than straw. (My position is not as partisan as might be supposed: a diverse clientèle needs to be served, and, for some, antismoking messages are probably the most useful prescription.) I will, however, feel no obligation in this section to take on the pure role of devil's advocate; some arguments for advocating less-hazardous tobacco will be touched upon while presenting the arguments against less-hazardous tobacco use.

Damage to Cessation and Prevention Efforts

It is widely believed that to advocate less-hazardous tobacco use is to undermine antismoking efforts. The rationales behind this belief are no doubt complicated and several. The most prominent concerns the information-processing abilities and motivation of smokers. Although support could be cited from psychological research on human information-processing abilities (e.g., Nisbett and Ross, 1980), the prevailing belief probably rests more squarely on common truisms such as "People believe only what they want to believe," "People want to have their cake and eat it too," or "People will want to take the easy way out." The smoker, it is thought, will gather from the less-hazardous tobacco message that it is acceptable to continue to use tobacco and will tend to ignore the advice that less-hazardous use should be employed only by those who cannot or will not give up tobacco use entirely. A related argument is that complicated messages will not be as persuasive or as memorable as simple messages: the less-hazardous use message, then, complicates the overall message to the detriment of the antismoking message.

These first two arguments concern problems with the reception of the smoking and health message. Another line of argument holds that in a world of limited resources one cannot do all that one might like to do to reduce the smoking and health problem. In terms of priority rankings, prevention and cessation activities are seen then as more important than reduced-risk activities.

Will Recruitment to Tobacco Use be Encouraged by the Availability of Less-Hazardous Tobacco Products?   No one knows the extent to which concern about the health consequences of tobacco use acts to deter those who are otherwise tempted to take up tobacco. One line of research does indicate that women, in particular, find it easier to take up smoking, given the modern, "milder" low-yield cigarettes (Silverstein et al., 1980, 1982). If advocacy of less-hazardous tobacco use adds enough recruits to the ranks of tobacco users, then the reduced risks to the individual user could be outweighed by the greater number of individuals at risk (see the section on the Prevention Paradox below). Trends in recruitment to tobacco use should be monitored.

Will Tobacco Users Use Less-Hazardous Products Instead of Quitting? No one knows how many smokers would have given up tobacco use entirely if they had not known of the option of less hazardous use. Some smokers might switch to low-yield cigarettes to allay the pesterings of associates about the health consequences of smoking. However, it is doubtful that these individuals would be willing to give up tobacco entirely, unless greater social pressure were put on them. The group of smokers to be most concerned about is those who would have been able to abstain if they had not been offered the promise of reduced-risk tobacco use.

I know of no estimates of how many individuals have been lost to smoking cessation or prevention because of the availability of presumably safer ways to use tobacco. A high priority should be given to empirical research that would estimate the size of the problem. Also, a high priority should be given to determining how to present the risks of tobacco use to individuals in ways that will have the greatest impact on health care decisions and health care behavior (cf. Slovic et al., 1977). Even if many individuals are lost to tobacco abstinence because of the less-hazardous tobacco use message, it does not follow that, therefore, the costs of the treatment outweigh the benefits.

Being Faithful to One's Job Description

It can be as important to know what is not part of one's job as it is to know what is part of one's job. Although tobacco once was a product that was crucial to the practice of the healing arts (Stewart, 1967), modern physicians believe that it is not within their job description to, in effect, advocate the use of any tobacco product: if less-hazardous tobacco use is to develop, it is thought to be up to the tobacco companies to be the advocates and developers. For the medical profession in general, tobacco has become an evil substance that is totally unfit for human consumption, unlike certain other potentially hazardous products (e.g., eggs, whole milk, salt, and sugar) about which the medical profession is willing to make recommendations concerning less hazardous use. For some reason, the tobacco industry has been especially easy to identify as the enemy, perhaps because of deep-set Calvinist convictions about the sin of drug use. Though a physician might be comfortable advising a low-salt or low-sugar diet (knowing that a no-added-salt or no-added-sugar diet, though probably less-hazardous, would receive little compliance), this same physician could not recommend the use of a less-dangerous tobacco product (knowing that abstinence may also result in little compliance). I juxtapose the tobacco and the food industries to illustrate an ironic inconsistency in the practice of public health and medicine: all these products may be optional but some are much more optional than others.

The Less-Hazardous Message is Already Well-Known

The antismoking messages of prevention and cessation may inadvertently and unavoidably support the cause of less-hazardous tobacco use. Smokers on their own might tend to adopt less hazardous uses of tobacco in response to clear messages that they should stop using tobacco. In other words, the message of less-hazardous tobacco use might occur by default as the antismoking message is spread.

Similarly, it can be argued that the less-hazardous tobacco use message is already well-known, because of the publicity surrounding the tar and nicotine yields of cigarettes. Standard tar and nicotine ratings have been supplied by governments, in part to encourage the use of lower tar and nicotine cigarettes by those who do not stop smoking (Friedman, 1975). The modern "tar derby" emphasizes that lower yield is better. Even if one considers the low-yield cigarette as the paragon of less hazardous tobacco products, the less-hazardous-tobacco-use message has been spread mainly in a superficial and dangerous way. (See Kozlowski, 1984, for a discussion of applications of less-hazardous tobacco therapies.) It is not enough simply to point a tobacco user to different products: advice and assistance should be given to help the user reduce exposure to toxic tobacco products.

Less-Hazardous Tobacco Use as Boondoggle

One argument against advocating less-hazardous tobacco use is that the promise of reduced risks is more apparent than real. In other words, the recommendation to use less-hazardous products should not be made because there are no truly less-hazardous tobacco products.

Low-Yield Cigarettes.   Low-yield cigarettes are at the same time the most popular and the most questionable of the presumably less-hazardous tobacco products. Some evidence indicates significant, but small, reductions in risk to be gained from a switch to low-yield cigarettes (e.g., Lee and Garfinkel, 1981; Vutuc and Kunze, 1982); other evidence indicates no reductions in risk (e.g., Castelli et al., 1981; Kaufman et al., 1983; Robinson et al., 1982). The present evidence is far from conclusive (Russell et al., 1980*b*; Kozlowski et al., 1982*b*; U.S.D.H.H.S., 1981). It is possible that long-term use of low-yield cigarettes is required before a beneficial reduction in smoke exposure is seen and that those who are forced to switch brands are less likely to compensate for reduced yields than those who switch on their own (Russell et al., 1982).

All cigarettes, along with other smoking tobaccos, make it difficult for users to know exactly what they are ingesting from these products (Kozlowski, 1984). It is not possible simply to read a product label and thereby know what one is getting from a cigarette or pipe. Actual smoke intake depends more on the details of a smoker's behavior (number of puffs, volume of puffs, depth of inhalation):

more on the smoker than on the product (Kozlowski, 1983). Cigarette smoke is more often inhaled than any other kind of tobacco smoke (U.S.D.H.E.W., 1979), and therefore cigarette smoke presents special risks to the lungs. There is also the question of risks to those who associate with or are exposed to smokers when they are smoking (U.S.D.H.H.S., 1982). Also, the use of smoking tobaccos carries the risk of fires, and resultant death and suffering for both active and passive smokers (Berl and Halpin, 1978).

Given (1) the controversy over the epidemiologic effects (in both active and passive users of tobacco smoke), (2) the difficulty in monitoring dosage, (3) the problem of inhalation, and (4) the issue of fire hazards, it is not easy to be sanguine about low-yield cigarettes as a treatment for the health consequences of smoking. Especially in light of other, more promising options for less-hazardous tobacco use, I am inclined to be very pessimistic about the value of low-yield cigarettes.

Even if low-yield cigarettes are poor less-hazardous tobacco products, it does not follow that other less hazardous tobacco treatments are therefore ineffective. The failure of one pharmaceutical is no grounds for closing down the pharmacopoeia.

Pipes and Cigars.   The available evidence suggests that pipes and cigars are less hazardous than cigarettes (Doll and Peto, 1976; U.S.D.H.E.W., 1979). People who start (and stay) with pipes or cigars tend not to inhale and hence reduce the exposure of their lungs to toxic smoke products. It is unclear whether smokers who turn from cigarettes to pipes and cigars continue the habit of inhaling. Some researchers have found inhalation among so-called secondary pipe or cigar smokers (Castledon and Cole, 1973; Turner et al., 1977, 1981); others have not found evidence of substantial inhalation by secondary pipe or cigar smokers (McCusker et al., 1982; Wald et al., 1981).

Even if secondary pipe and cigar smokers do inhale, the epidemiologic evidence indicates that, while these smokers are at greater risk of dying than are primary pipe or cigar smokers, they are at a lower risk than those who continue to smoke cigarettes (Doll and Peto, 1976). If people were encouraged from the start to smoke pipes or cigars rather than cigarettes, this problem of inhalation among pipe or cigar smokers might not arise.

Smokeless Tobaccos.   There is really no dispute about whether smokeless tobaccos present fewer hazards to the user than do smoking tobaccos (Harrison, 1964; Russell et al., 1980a). Smokeless tobaccos expose the lungs to essentially no tobacco toxins. No carbon monoxide and no tar is produced. The oral cancers associated with oral smokeless tobaccos are substantially less lethal and are more easily diagnosed than lung cancers (U.S.D.H.E.W., 1979). In addition, smokeless tobaccos pose no problems of second-hand smoke and no risks of fire. Clearly risks are reduced, but the residual risks are substantial enough to cause some authorities to refuse to advocate their use (Christen et al., 1979). A subclass of the boondoggle objection then is that reductions in risk are too small to warrant support. (For a discussion of nicotine-containing chewing gum as possibly the least hazardous of the smokeless "tobaccos," see Kozlowski et al., 1982*a*, and Kozlowski, 1984.)

5. ARGUMENTS FOR ADVOCATING LESS-HAZARDOUS TOBACCO USE

Job Descriptions Reconsidered

Health Care Providers.   Health care providers are sometimes preoccupied with their roles as opinion leaders and "moral forces" within their communities. Their function as authority figures can even obscure the more central parts of their job descriptions. Is it not best to try to do what one can to reduce death and disability in those who continue to use tobacco? Is not the reduction of death and disability a fundamental part of the job description of a health care provider?

Tobacco Product Providers.   For the health care provider to consider the support of less-hazardous tobacco use as a job for the tobacco industry may be naïve. It could be risky to leave the development of less-hazardous products to an industry whose life's blood is the cigarette. Although the tobacco industry is best-suited technically to developing less-hazardous tobacco products, it should not be forgotten that it has a business to protect. Also, the tobacco industry has steadfastly denied that tobacco use causes any medical problems; this hardly puts them in a position to invest much in the development of products that reduce hazards that they assert are not there to begin with.

Community Health and the Prevention Paradox

Physicians who are not specially trained in public health and preventive medicine are apt to make a category mistake when considering the issue of less-hazardous tobacco use. This category mistake [i.e., allocating concepts to a category to which they do not belong (Ryle, 1949)] consists of mistaking the public health issue for a personal health issue written large. As a matter of personal health care, for a physician to recommend the less-hazardous use of tobacco can be seen (and felt) as a failure to use the positive powers of one's practice.

It does not follow that a small benefit to the health of the individual will constitute a small benefit to the health of the community. Dr. William Castelli (1981) made the mistake of removing the less-hazardous tobacco use argument from the public health domain and placing it in the physician's office. Taking the most generous estimate from the report of Lee and Garfinkel (1981), Castelli noted that a pack-a-day smoker who switched from unfiltered to filtered cigarettes reduced his or her risk of lung cancer from 20 times that of a nonsmoker to 15 times that of a nonsmoker. Castelli wrote: "I do not personally get much satisfaction encouraging someone to pursue a habit which increases the risk of lung cancer 15 times" (p. 642). This does describe the situation from the physician's perspective; however, from the perspective of one interested in community health, the satisfactions may be obvious: if 2000 pack-a-day smokers had been dying each year from lung cancer, now 1500 smokers would be dying. Five hundred people would still be alive; 25% fewer smokers would be dying from lung cancer.

Rose (1981) describes the "prevention paradox"—"a measure that brings large benefits to the community offers little to each participating individual." A treatment that is worthwhile and practical for the community may have trivial influence on the individual. Conversely, the treatment that may have the most benefit for the individual may be impractical and hence of little use for the community. Rose uses the treatment of hypertension as an example. Extremely high blood pressure can be controlled with drugs, but relatively few individuals have extremely high blood pressure. If the average diastolic blood pressure of the community were reduced by just 7–8 mm Hg (say, by altering the diet), then the number of disorders due to blood pressure would decline as much as if all those with pressures of 105 mm Hg or more were treated in a 100% effective way. The less dramatic therapy reaps appreciable net benefit because of the large number of people involved with the treatment.

The continuing discussion of the low-yield cigarette has often ignored the relevance of this paradox for tobacco use (e.g., Marks, 1982). Basically, the principle behind the paradox is that small effects on a large enough scale can produce more net benefits than can effects of heroic proportions on a small scale. This principle is also manifest in Russell's (Russell et al., 1979) advice on the benefits of physicians' advice on smoking cessation. Each physician will have relatively little success in persuading patients to give up cigarettes, but given the number of physicians available to spread the word, the net effects could be many times larger than the effect of more expensive alternative therapies.

Of course, one of the key assumptions involved with employing the prevention paradox as an argument for less-hazardous tobacco use is that the number of people enjoying the small benefit must be large enough to add up to a substantial net benefit. One might think that if the prevention and cessation efforts became highly successful, there would be few tobacco users left to enjoy the small benefits of less-hazardous tobacco use; however, it must be remembered that for those individuals who continue to use tobacco, cessation is, by definition, not an alternative treatment to less-hazardous use.

The Limits of Prevention and Cessation

None of the arguments for the advocacy of less-hazardous tobacco use should be used to argue against the deployment of prevention and cessation programs. The less-hazardous tobacco use message has no war with the antismoking message; in fact, the relationship between the two is symbiotic. As noted above, the less-hazardous use message is best viewed as an effort to deal with the failures of other efforts.

Prevention programs in the schools have received great attention in recent years (e.g., Evans et al., 1981). Though these programs have shown some success in reducing recruitment to smoking, at least during the school years, no one would argue that any program has discovered a certain technique for drastically reducing the number of smokers in high school, say, below the level of 10% of the students. Similarly, formal smoking cessation treatment programs find in general that 80% of their clients will relapse to smoking within 1 year (Raw, 1978). One estimate of how well smokers succeed at stopping smoking after repeated attempts on their own indicates that about 40% will fail to abstain in the long run (Schachter, 1981).

One of the best studies of the overall impact of the antismoking campaign comes from a cohort analysis of smokers in the United States population (Warner and Murt, 1982). Based on the percentages of smokers in different age groups before the antismoking campaign really got started (before the 1964 Report of the U.S. Surgeon General), Warner and Murt estimated how many smokers would have been expected in these same age groups had the antismoking campaign not taken place. In 1964, 67% of the 21- to 24-year-old men were smokers; in 1975, only 41% of the 21- to 24-year-olds were smokers. In 1964, 42% of the 21- to 24-year-old women were smokers; in 1975, 34% were smokers. They estimate that, if it were not for the antismoking campaign, 61% of men 18–27 years old would have been smokers in 1978; only 39% of this group were smokers in 1978, a difference of 22 percentage points. For women of the same age, they estimate that 49% would have been smoking; 37% actually were smokers in 1978, a difference of 12 percentage points. Though these figures indicate substantial success for antismoking efforts, they also clearly show that a potential market exists for less-hazardous tobacco.

The Promise of Diminishing Returns.   Tobacco users differ in how dependent they are on tobacco. A number of studies have shown that cessation interventions are more successful with less-dependent tobacco users (e.g., Fagerstrom, 1982; Kozlowski et al., 1981). One of the clearest implications of this finding is that the pool of continuing smokers is becoming more likely to contain more-dependent tobacco users. [The population of smokers is made up increasingly of fewer and heavier smokers, (U.S.D.H.H.S., 1981)]. In other words, the antismoking campaign has probably tended to remove those who are most easily removed from the ranks of smokers. Those who remain are likely to be a hard core of recalcitrant and perhaps "reactant" smokers. Reactance is a technical term that refers to an individual's assertion of freedom of action when faced with attempts to restrict that freedom (Brehm, 1966). Less-hazardous tobacco use may be one of the few treatments available for these smokers.

A Question of Class.   Recently, there has been a growing concern about the social inequalities of health care delivery systems (e.g., Morris, 1980). If one looks carefully at smoking statistics, it is apparent that, in general, those of lower socioeconomic status are more likely to use tobacco than are those of higher socioeconomic status. [An exception is that higher-class women are smoking more than lower-class women (U.S.D.H.E.W., 1979)]. If one looks at some especially disadvantaged groups, one finds, for example, that in Canada only 23% of teenagers (ages 15–19) who are still in school are daily smokers, whereas 48% of those who are no longer attending school (essentially high school dropouts) are daily smokers (Health and Welfare Canada, 1981). The same report finds that those with low levels of education, and those who are unemployed or in low-status jobs, are more likely to be current daily smokers. Moody (1980) finds that those from lower socioeconomic groups also take more puffs per cigarette and are exposed to more daily tar than are those from higher socioeconomic groups.

Has the antismoking campaign been less successful in reaching the lower classes? Has the antismoking campaign been less successful with those of low socioeconomic status that it has reached? Are those of low socioeconomic status more likely to be dependent on tobacco? Certainly it is fair to say that a school dropout may miss out on many of the antismoking efforts in the schools. When one has lost a job, is one also inclined to hold on to the compassion to be found in tobacco? Whatever the reasons, socioeconomic lines have indicated systematic limits to the power of current antismoking efforts.

6.   THE LIMITS OF LESS-HAZARDOUS TOBACCO USE

Prevention and cessation efforts are not alone in having limited power and success. Much of the speculation about less-hazardous tobacco use as a treatment has not, in fact, received empirical test. Some of these good ideas may not work in practice.

The epidemiology of tobacco-related diseases can only serve as a guide to possible treatments. Epidemiologic samples are self-selected as tobacco users: epidemiologic studies generally show no more than correlations between tobacco use and disease. No one knows, for example, how the health consequences of smoking might change in a group of cigarette smokers who were randomly assigned to take up pipe smoking. No one knows how many participants in such a study would be able to comply with their instructions. We do have evidence that secondary pipe and cigar smokers do have less risk of disease than those who continue to smoke cigarettes (Doll and Peto, 1976); but we do not know if the change to pipes or cigars is the cause of the reduced risk. Perhaps those cigarette smokers who do change to pipes or cigars are very different (e.g., constitutionally) than those who continue as cigarette smokers (Seltzer, 1972). Despite these reservations, the epidemiologic literature does form the basis for predictions about less-hazardous tobacco use.

Technical versus Behavioral Interventions

Technical interventions depend upon changes in the tobacco product. Behavioral interventions depend upon changes in conduct. If, for example, a less-toxic tobacco tar could be developed, then, one might find reductions in lung cancer incidence even if there were no changes in tar intake. It has been argued that modern tars are less toxic (milligram for milligram) than the tars of the 1950s and that this reduced toxicity might account for the reduced incidence (e.g., Gori, 1976). If modern tars are less toxic or can be made less toxic, one would have a less-hazardous tobacco use treatment for the smoking and health problem that would (assuming that the compassion remained) pose essentially no problems of patient compliance.

The ideal technical intervention involves the modification of a product that the tobacco user is already using. Being able to reduce the intrinsic risks of a product that tobacco users will not use provides little treatment for the tobacco and health problem. Some behavioral interventions might be directed to persuading the tobacco user to use a less-hazardous product. Other behavioral interventions will be directed to the less-hazardous use of the product currently being used. Each of these kinds of behavioral intervention is truly easier described than done. For a discussion of some of the challenges involved with behavioral interventions, see Kozlowski (1984).

Diet, Drugs, Occupation, and the Risks of Tobacco

There may be adjunctive ways to engage in less-hazardous tobacco use. The epidemiologic literature suggests that it would be advisable for smokers to change other behaviors to reduce the health consequences of tobacco use. This literature is, for the most part, suggestive rather than conclusive.

Those who work with asbestos and smoke cigarettes are at especially high risk of lung disease (see U.S.D.H.H.S., 1982, for a review). Similarly, cigarette smoking and birth control pills may act synergistically to increase the risk of cardiovascular disease in women (see U.S.D.H.H.S., 1980). Tobacco and alcohol appear to act synergistically to increase the risk of cancers of the mouth, pharynx, larynx, and esophagus (see U.S.D.H.H.S., 1982, for review). It is possible that a continuing tobacco user could reduce the health consequences of tobacco use by being careful to avoid alcohol, asbestos, and birth control pills: in terms of practicality, it should not be prejudged which of these activities is optional for given individuals. As a positive measure to reduce the risks of cancer in the tobacco user, there is growing evidence that a diet rich in pro-vitamin A has a protective effect against lung cancer (Doll and Peto, 1981; Shekelle et al., 1981). Less-hazardous tobacco use might, then, be established by modifying (1) the tobacco use, (2) a cofactor for risk, or (3) both.

7. LEGITIMIZING THE TOPIC AND THE NEED FOR RESEARCH

The arguments for and against advocating less-hazardous tobacco use have certainly not been exhaustive. This chapter has tried to legitimize the study of less-hazardous tobacco use as a beneficial treatment for the smoking and health problem. Despite the impression that some antismoking readers might have, I am uneasy about a blanket endorsement of the less-hazardous-tobacco-use therapy. Data may indeed emerge in the future that will show the less-hazardous movement to have been ill-advised. Current research should, however, not fear to show both the advantages and disadvantages of all aspects of the war against tobacco-related maladies.

As is the case in many areas of applied research, in this area it is not possible to wait until all of the data are in to decide what should be done about less-hazardous tobacco use. As a self-administered therapy, "less-hazardous tobacco use" exists already. Many tobacco users will not be persuaded to give up their use of tobacco, despite the best efforts of the antismoking campaign. If there are ways to reduce obvious errors in this self-administered therapy, then the consumers of these therapies should know about them (Kozlowski, 1982b). Without research on the would-be forms of less-hazardous tobacco use, we are not able to establish their actual, rather than supposed, net worth.

ACKNOWLEDGMENTS

The author thanks R. Frecker, C. P. Herman, S. Herling, L. Jelinek, M. Pope, and K. Wagner for their assistance.

REFERENCES

American Psychiatric Association, 1980, *Diagnostic and Statistical Manual of Mental Disorders*, 3rd. Ed. American Psychiatric Association, Washington, D.C.
Bain, J., 1896, *Tobacco in Song and Story*, Caldwell, New York.
Berl, W. G., and Halpin, B. M., 1978, Human fatalities from unwanted fires, The Johns Hopkins University—Applied Physics Laboratory, December, p. 8.
Brehm, J. W., 1966, *A Theory of Psychological Reactance*, Academic Press, New York.
Brooks, J. C., 1952, *The Mighty Leaf: Tobacco through the Centuries*, Little Brown, Boston.
Castelli, W. P., 1981, Filter cigarettes and heart disease, *Lancet* 2:642.
Castelli, W. P., Dawber, T. R., Feinleib, M., Garrison, R. J., McNamara, P. M., and Kannel, W. B., 1981, The filter cigarette and coronary heart disease: The Framingham study, *Lancet* 2:109.

Castledon, C. M., and Cole, P. V., 1973, Inhalation of tobacco smoke by pipe and cigar smokers, *Lancet* **2**:21–22.

Christen, A. G., Armstrong, W. R., and McDaniel, R. K., 1979, Intraoral leukoplakia, peridontal breakdown, and tooth loss in a snuff dipper, *J. Am. Dent. Assoc.* **98**:584–586.

Cohen, M. R., and Nagel, E., 1962, *An Introduction to Logic,* Harcourt, Brace and World, New York.

Doll, R., and Peto, R., 1976, Mortality in relation to smoking: 20 years observation on male British doctors, *Br. Med. J.* **2**:1525–1536.

Doll, R., and Peto, R., 1981, The causes of cancer: Quantitative estimates of avoidable risks of cancer in the United States today, *J. Natl. Cancer Inst.* **66**(6):1192–1308.

Evans, R., Hill, P. C., Raines, B. E., and Henderson, A. H., 1981, Current behavioral, social, and educational programs in control of smoking: A selective critical review, in: *Perspectives on Behavioral Medicine* (S. M. Weiss, J. A. Herd, and B. H. Fox, eds.), pp. 261–284, Academic Press, New York.

Fagerström, K.-O., 1982, A comparison of psychological and pharmacological treatment in smoking cessation, *J. Behav. Med.* **5**(3):343–351.

Fischer, D. H., 1970, *Historian's Fallacies: Toward a Logic of Historical Thought,* Harper and Row, New York.

Friedman, K. M., 1975, *Public Policy and the Smoking-Health Controversy,* D. C. Heath and Company, Lexington, Mass.

Gart, J. J., and Schneiderman, M. A., 1979, "Low-risk" cigarettes: The debate continues, *Science* **204**:690–691.

Gerstein, D. R., and Levison, P. K., eds., 1982, *Reduced Tar and Nicotine Cigarettes: Smoking Behavior and Health,* Committee on Substance Abuse and Habitual Behavior and Social Sciences and Education, National Research Council, National Academy Press, Washington, D. C.

Gori, G. B., 1976, Low-risk cigarettes: A prescription, *Science* **194**:1243–1245.

Gori, G. B., 1980, A summary appraisal, in: *Banbury Report 3. A Safe Cigarette?* (G. B. Gori, and F. G. Bock, eds.), Cold Spring Harbor Laboratory, Cold Spring Harbor, New York.

Gori, G. B., and Lynch, C. J., 1978, Toward less hazardous cigarettes, *JAMA* **240**:1255–1259.

Harrison, D. F. N., 1964, Snuff—its use and abuse, *Br. Med. J.* **2**:1649–1651.

Health and Welfare Canada, 1981, *The Health of Canadians. Report of the Canada Health Survey,* Minister of Supply and Services, Ottawa, Canada. (Catalogue 82-538E).

Kaufman, D. W., Helmrich, S. P., Rosenberg, L., and Miettinen, A. S., 1983, Nicotine and carbon monoxide content of cigarette smoke and the risk of myocardial infarction in young men, *N. Engl. J. Med.* **308**:409–413.

Kozlowski, L. T., 1981*a*, Tar and nicotine delivery of cigarettes: What a difference a puff makes, *JAMA* **245**(2):158–159.

Kozlowski, L. T., 1981*b*, Smokers, non-smokers, and low-tar smoke, *Lancet* **1**:508.

Kozlowski, L. T., 1981*c*, Application of some physical indicators of cigarette smoking, *Addict. Behav.* **6**:213–219.

Kozlowski, L. T., 1982*a*, The determinants of tobacco use: Cigarette smoking in the context of other forms of tobacco use, *Canad. J. Public Health* **73**:236–241.

Kozlowski, L. T., 1982*b*, *Tar and Nicotine Ratings May Be Hazardous to Your Health,* Alcoholism and Drug Addiction Research Foundation, Toronto.

Kozlowski, L. T., 1983, Perceiving the risks of low-yield ventilated-filter cigarettes: The problem of hole-blocking, in: *Proceedings of the International Workshop on the Analysis of Actual vs. Perceived Risks,* (V. Covello, W. G. Flamm, J. Rodricks, and R. Tardiff, eds.), Plenum Press, New York.

Kozlowski, L. T., 1984, Pharmacological approaches to smoking modification, in: *Behavioral Health: A Handbook of Health Enhancement and Disease Prevention* (J. R. Matarazzo, N. E. Miller, S. M. Weiss, J. A. Herd, and S. M. Weiss, eds.), John Wiley, New York.

Kozlowski, L. T., Rickert, W., Robinson, J., and Grunberg, N. E., 1980, Have tar and nicotine yields of cigarettes changed? *Science* **209**:1550–1551.

Kozlowski, L. T., Director, J., and Harford, M. A., 1981, Tobacco dependence, restraint, and time to the first cigarette of the day, *Addict. Behav.* **6**:307–312.

Kozlowski, L. T., Appel, C.-P., Frecker, R. C., and Khouw, V., 1982a, Nicotine, a prescribable drug available without prescription, *Lancet* **1**:334.

Kozlowski, L. T., Frecker, R. C., and Lei, H., 1982b, Nicotine yields of cigarettes, plasma nicotine in smokers and public health, *Prev. Med.* **11**:240–244.

Kozlowski, L. T., Rickert, W. S., Pope, M. A., Robinson, J. C., and Frecker, R. C., 1982c, Estimating the yield to smokers of tar, nicotine, and carbon monoxide from the "lowest-yield" ventilated-filter cigarettes, *Br. J. Addict.* **77**:159–165.

Lakoff, G., and Johnson, M., 1980, "Metaphors We Live By," University of Chicago Press, Chicago.

Lee, P. N., and Garfinkel, L., 1981, Morality and type of cigarette smoked, *J. Epidemiol. Commun. Health* **35**:16–22.

Lindblom, C. E., and Cohen, D. K., 1979, "Usable Knowledge," Yale University Press, New Haven, Connecticut.

Marks, L., 1982, Policies and postures in smoking control, *Br. Med. J.* **284**:391–395.

Marx, J. L., 1978, Health officials fired up over "tolerable" cigarettes, *Science* **201**:795–798.

McCusker, K., McNabb, E., and Bone, R., 1982, Plasma nicotine levels in pipe smokers, *JAMA* **248**(5):577–578.

Moody, P. M., 1980, The relationship of quantified human smoking behavior and demographic variables, *Soc. Sci. Med.* **14A**:49–54.

Morris, J. N., 1980, Are health services important to the people's health? *Br. Med. J.* **280**:167–168.

Nisbett, R., and Ross, L., 1980, "Human Inference: Strategies and Shortcomings of Social Judgement," Prentice Hall, Englewood Cliffs, New Jersey.

Raw, M., 1978, The treatment of cigarette dependence, in: *Research Advances in Alcohol and Drug Problems*, Vol. 4, (Y. Israel, F. B. Glaser, H. Kalant, R. E. Popham, W. Schmidt, and R. G. Smart, eds.), pp. 441–485, Plenum Press, New York.

Robinson, J. C., Young, J. C., and Rickert, W. S., 1982, A comparative study of the amount of smoke absorbed from low yield ('less hazardous') cigarettes. Part 1: Non-invasive measures, *Br. J. Addict.* **77**:383–397.

Rose, G., 1981, Strategy of prevention: Lessons from cardiovascular disease, *Br. Med. J.* **282**:1847–1851.

Royal College of Physicians, 1977, *Smoking or Health*, Pitman, London.

Russell, M. A. H., Wilson, C. Taylor, C., and Baker, C. D., 1979, Effect of general practitioners' advice against smoking, *Br. Med. J.* **2**:231–235.

Russell, M. A. H., Jarvis, M. J., and Feyerabend, C., 1980a, A new age for snuff? *Lancet* **1**:474–475.

Russell, M. A. H., Jarvis, M., Iyer, R., and Feyerabend, C., 1980b, Relation of nicotine yield of cigarettes to blood nicotine concentrations in smokers, *Br. Med. J.* **280**:972–976.

Russell, M. A. H., Sutton, S. R., Iyer, R., Feyerabend, C., and Vesey, C. J., 1982, Long-term switching to low-tar low-nicotine cigarettes, *Br. J. Addict.* **77**:145–158.

Ryle, G., 1949, *The Concept of Mind*, Barnes and Noble, New York.

Schachter, S., 1981, Self-treatment of smoking and obesity, *Canad. J. Public Health* **72**:401–406.

Schachter, S., Silverstein, B., Kozlowski, L. T., Perlick, D., Herman, C. P., and Liebling, B., 1977, Studies of the psychological and pharmacological determinants of smoking, *J. Exp. Psychol. [Gen.]* **106**:3–40.

Seltzer, C. C., 1972, Differences between cigar and pipe smokers in healthy white veterans, *Arch. Environ. Health* **25**:187–191.

Shekelle, R. B., Liu, S., Raynor, W. J., Jr., Lepper, M., Maliza, C., Rossof, A. H., Paul, O., Shyrock, A. M., and Stamler, J., 1981, Dietary vitamin A and risk of cancer in the Western Electric study, *Lancet* **2**(8267):1185–1190.

Silverstein, B., 1982, Cigarette smoking, nicotine addiction, and relaxation, *J. Pers. Soc. Psychol.* **42**(5):946–950.

Silverstein, B., Feld, S., and Kozlowski, L. T., 1980, The availability of low-nicotine cigarettes as a cause of cigarette smoking among teenage females, *J. Health Soc. Behav.* **21**:383–388.

Silverstein, B., Kelly, E., Swan, J., and Kozlowski, L. T., 1982, Physiological predisposition toward becoming a cigarette smoker: Experimental evidence for a sex difference, *Addict. Behav.* **7**:83–86.

Slovic, P., Fischoff, B., and Lichtenstein, S., 1977, Behavioral decision theory, *Annu. Rev. Psychol.* **28**:1–39.

Stewart, G. G., 1967, A history of the medicinal uses of tobacco, *Med. Hist.* **11**:228–268.

Turner, J. A. McM., Sillett, R. W., and McNicol, M. W., 1977, Effect of cigar smoking on carboxyhaemoglobin and plasma nicotine concentrations in primary pipe and cigar smokers and ex-cigarette smokers, *Br. Med. J.* **2**:1387–1389.

Turner, J. A. McM., Sillett, R. W., and McNicol, M. W., 1981, The inhaling habits of pipe smokers, *Br. J. Dis. Chest* **75**:71–76.

U. S. Department of Health, Education and Welfare, Public Health Service, 1979, *Smoking and Health: A Report of the Surgeon General*, (Publication No. (PHS) 79-50066), U. S. Government Printing Office, Washington, D. C.

U. S. Department of Health and Human Services, Public Health Service, 1980, *The Health Consequences of Smoking for Women*, U. S. Government Printing Office, 1980 0-326-003, Washington, D. C.

U. S. Department of Health and Human Services, Public Health Service, 1981, *The Health Consequences of Smoking: The Changing Cigarette*, U. S. Government Printing Office, Washington, D. C.

U. S. Department of Health and Human Services, Public Health Service, 1982, *The Health Consequences of Smoking: Cancer*, U. S. Government Printing Office, 1982 0-367-198/579, Washington, D. C.

Vutuc, C., and Kunze, M., 1982, Lung cancer in women in relation to tar yields of cigarettes, *Prev. Med.* **11**:713–716.

Wald, N. J., Idle, M., Boreham, J., Bailey, A., and Van Vunakis, H., 1981, Serum cotinine levels in pipe smokers: Evidence against nicotine as a cause of coronary heart disease, *Lancet* **2**:775–777.

Warner, K. E., 1979, Toward less hazardous cigarettes, *JAMA* **241**:2143.

Warner, K. E., and Murt, H. A., 1982, Impact of the anti-smoking campaign on smoking prevalence: A cohort analysis, *J. Public Health Policy* **3**:374–390.

Index